Methods in Molecular Biology

Series Editor
John M. Walker
School of Life and Medical Sciences
University of Hertfordshire
Hatfield, Hertfordshire, UK

For further volumes:
http://www.springer.com/series/7651

For over 35 years, biological scientists have come to rely on the research protocols and methodologies in the critically acclaimed *Methods in Molecular Biology* series. The series was the first to introduce the step-by-step protocols approach that has become the standard in all biomedical protocol publishing. Each protocol is provided in readily-reproducible step-by step fashion, opening with an introductory overview, a list of the materials and reagents needed to complete the experiment, and followed by a detailed procedure that is supported with a helpful notes section offering tips and tricks of the trade as well as troubleshooting advice. These hallmark features were introduced by series editor Dr. John Walker and constitute the key ingredient in each and every volume of the *Methods in Molecular Biology* series. Tested and trusted, comprehensive and reliable, all protocols from the series are indexed in PubMed.

The Blood-Brain Barrier

Methods and Protocols

Edited by

Nicole Stone

Discovery Pharmacology, COMPASS Pathways PLC, London, UK

Humana Press

Editor
Nicole Stone
Discovery Pharmacology
COMPASS Pathways PLC
London, UK

ISSN 1064-3745 ISSN 1940-6029 (electronic)
Methods in Molecular Biology
ISBN 978-1-0716-2288-9 ISBN 978-1-0716-2289-6 (eBook)
https://doi.org/10.1007/978-1-0716-2289-6

This Humana imprint is published by the registered company Springer Science+Business Media, LLC part of Springer Nature.
The registered company address is: 1 New York Plaza, New York, NY 10004, U.S.A.

Preface

The blood-brain barrier (BBB) is a complex interface that separates the systemic circulation from the central nervous system (CNS). Its structure is comprised of specialized cells, namely brain microvascular endothelial cells, astrocytes, and pericytes, along with neurons and microglial cells that partake in BBB maintenance. In addition, given that the brain is an immune privileged site, the BBB plays an important role in restricting peripheral immune cells which would otherwise gain access to the CNS. This book aims to support early-career researchers, as well as giving novel insights to more experienced BBB scientists. It collates a wide range of methodologies which will aid any researcher in the fascinating world of the blood-brain barrier. Every chapter included has an overall introduction into the specific area, a list of tools, and reagents, together with a detailed, step-by-step procedure in order to successfully reproduce each author's rigorous methodologies. Furthermore, where possible, each chapter will include diagrams to aid researchers on experimental setup and troubleshooting. I hope researchers will enjoy reading this book as much as I have enjoyed putting it together. I'd also like to give special thanks to John Walker for giving me this unique opportunity.

Declaration of Interest: Dr Nicole Stone is currently an employee at COMPASS Pathways plc. This work is unrelated to COMPASS Pathways plc.

Sheffield, UK *Nicole Stone*

Contents

Part IV Microfluidic and Chip Models of the BBB

Part V Models to Study Specific Pathologies at the BBB: Cancer, Neurodegeneration, and More

Contributors

ABRAHAM J. AL-AHMAD • *Department of Pharmaceutical Sciences, Jerry H. Hodge School of Pharmacy, Texas Tech University Health Sciences Center, Amarillo, TX, USA*

MANSOUR ALWJWAJ • *Stroke, Academic Unit of Mental Health and Clinical Neuroscience, School of Medicine, The University of Nottingham, Nottingham, UK*

ANUSKA V. ANDJELKOVIC • *Department of Pathology, University of Michigan Medical School, Ann Arbor, MI, USA; Department of Neurosurgery, University of Michigan Medical School, Ann Arbor, MI, USA*

MERCEDES BALCELLS • *Institute for Medical Engineering and Science, Massachusetts Institute of Technology, Cambridge, MA, USA; IQS School of Engineering, Universitat Ramon Llull, Barcelona, Spain*

ULVI BAYRAKTUTAN • *Stroke, Academic Unit of Mental Health and Clinical Neuroscience, School of Medicine, The University of Nottingham, Nottingham, UK*

ADITYA BHALERAO • *Department of Biological and Biomedical Sciences, Oakland University, Rochester, MI, USA*

MATTHEW CAMPBELL • *Smurfit Institute of Genetics, Trinity College Dublin, Dublin, Ireland; FutureNeuro SFI Research Centre, Royal College of Surgeons in Ireland, Dublin, Ireland*

CAMILLA CERUTTI • *School of Life, Health and Chemical Sciences, The Open University, Milton Keynes, UK; Department of Experimental Oncology, European Institute of Oncology IRCCS, Milan, Italy*

MINHWAN CHUNG • *Department of Mechanical Engineering, Seoul National University, Seoul, Republic of Korea*

LUCA CUCULLO • *Department of Foundational Medical Studies, Oakland University William Beaumont School of Medicine, Rochester, MI, USA*

LUCIE DEHOUCK • *Université d'Artois, UR 2465, Laboratoire de la Barrière Hémato-Encé phalique (LBHE), Lens, France*

MARIE-PIERRE DEHOUCK • *Université d'Artois, UR 2465, Laboratoire de la Barrière Hé mato-Encéphalique (LBHE), Lens, France*

ERIC R. ESPINAL • *Department of Biological Sciences, University of Alabama, Tuscaloosa, AL, USA*

MARY GOODWIN-TROTMAN • *Stroke Research Group, Department of Clinical Neurosciences, University of Cambridge, Cambridge, UK*

FABIEN GOSSELET • *Université d'Artois, UR 2465, Laboratoire de la Barrière Hémato-Encé phalique (LBHE), Lens, France*

ALESSANDRA GRANATA • *Stroke Research Group, Department of Clinical Neurosciences, University of Cambridge, Cambridge, UK*

CHRIS GREENE • *Smurfit Institute of Genetics, Trinity College Dublin, Dublin, Ireland*

PAUL M. HOLLOWAY • *Radcliffe Department of Medicine, University of Oxford, Oxford, UK*

NOO LI JEON • *Interdisciplinary Program for Bioengineering, Seoul National University, Seoul, Republic of Korea; Department of Mechanical Engineering, Seoul National University, Seoul, Republic of Korea; Institute of Advanced Machinery and Design, Seoul National University, Seoul, Republic of Korea*

RAIS RESKIAWAN A. KADIR • *Stroke, Academic Unit of Mental Health and Clinical Neuroscience, School of Medicine, The University of Nottingham, Nottingham, UK*
RICHARD F. KEEP • *Department of Neurosurgery, University of Michigan Medical School, Ann Arbor, MI, USA; Department of Molecular and Integrative Physiology, University of Michigan Medical School, Ann Arbor, MI, USA*
BRANDON J. KIM • *Department of Biological Sciences, University of Alabama, Tuscaloosa, AL, USA*
YOUNGMI KOO • *Department of Chemical, Biological, and Bioengineering, North Carolina Agricultural and Technical State University, Greensboro, NC, USA*
ELIANA LAURANZANO • *Humanitas Clinical and Research Center, IRCCS, Rozzano, Italy*
SOMIN LEE • *Interdisciplinary Program for Bioengineering, Seoul National University, Seoul, Republic of Korea*
LUMEI LIU • *Department of Chemical, Biological, and Bioengineering, North Carolina Agricultural and Technical State University, Greensboro, NC, USA*
SHAHRIMA MAHARUBIN • *Industrial Manufacturing and Systems Engineering, Texas Tech University, Lubbock, TX, USA*
CLÁUDIA MARTINS • *Instituto de Investigação e Inovação em Saúde (i3S), Universidade do Porto, Porto, Portugal; Instituto de Engenharia Biomédica (INEB), Universidade do Porto, Porto, Portugal; Instituto Ciências Biomédicas Abel Salazar (ICBAS), Universidade do Porto, Porto, Portugal*
JORDI MARTORELL • *IQS School of Engineering, Universitat Ramon Llull, Barcelona, Spain*
MICHELA MATTEOLI • *Humanitas Clinical and Research Center, IRCCS, Rozzano, Italy; CNR Institute of Neuroscience, Milano, Italy*
HEATHER L. MCCONNELL • *Department of Neurology, Jungers Center for Neurosciences Research, Oregon Health & Science University, Portland, OR, USA; Office of Academic Development, Houston Methodist Research Institute, Houston, TX, USA*
PATRICIA MIRANDA-AZPIAZU • *Leeds Institute of Cardiovascular and Metabolic Medicine, Faculty of Medicine and Health, University of Leeds, Leeds, UK*
ANUSHA MISHRA • *Department of Neurology, Jungers Center for Neurosciences Research, Oregon Health & Science University, Portland, OR, USA; Knight Cardiovascular Institute, Oregon Health & Science University, Portland, OR, USA*
GEMMA MOLINS GUTIÉRREZ • *Institute for Medical Engineering and Science, Massachusetts Institute of Technology, Cambridge, MA, USA*
CAROLINE MYSIOREK • *Université d'Artois, UR 2465, Laboratoire de la Barrière Hémato-Encéphalique (LBHE), Lens, France*
BEHNAM NOORANI • *Department of Pharmaceutical Sciences, Texas Tech University Health Sciences Center, Amarillo, TX, USA*
BIRGIT OBERMEIER • *Biogen, Cambridge, MA, USA*
ADJANIE PATABENDIGE • *Department of Biology, Edge Hill University, Ormskirk, UK; School of Biomedical Sciences & Pharmacy, The University of Newcastle, Callaghan, NSW, Australia*
KRUSHANGI PATEL • *Stroke Research Group, Department of Clinical Neurosciences, University of Cambridge, Cambridge, UK*
IQRA PERVAIZ • *Department of Pharmaceutical Sciences, Jerry H. Hodge School of Pharmacy, Texas Tech University Health Sciences Center, Amarillo, TX, USA*
CHELSEA M. PHILLIPS • *Graduate Program in Neuroscience, University of Michigan Medical School, Ann Arbor, MI, USA*

MARCO RASILE • *Humanitas Clinical and Research Center, IRCCS, Rozzano, Italy; Department of Biomedical Sciences, Humanitas University, Pieve Emanuele, Italy*
SNEHAL RAUT • *Department of Foundational Medical Studies, Oakland University William Beaumont School of Medicine, Rochester, MI, USA*
IGNACIO A. ROMERO • *School of Life, Health and Chemical Sciences, The Open University, Milton Keynes, UK*
TEAL RUSSELL • *Department of Chemical, Biological, and Bioengineering, North Carolina Agricultural and Technical State University, Greensboro, NC, USA*
SIKHA SAHA • *Leeds Institute of Cardiovascular and Metabolic Medicine, Faculty of Medicine and Health, University of Leeds, Leeds, UK*
ANTONIO G. SALAZAR-MARTIN • *Institute for Medical Engineering and Science, Massachusetts Institute of Technology, Cambridge, MA, USA*
BRUNO SARMENTO • *Instituto de Investigação e Inovação em Saúde (i3S), Universidade do Porto, Porto, Portugal; Instituto de Engenharia Biomédica (INEB), Universidade do Porto, Porto, Portugal; Instituto de Investigação e Formação Avançada em Ciências e Tecnologias da Saúde (CESPU), Gandra, Portugal*
S. JEROD SHARP • *Department of Biological Sciences, University of Alabama, Tuscaloosa, AL, USA; Jacksonville State University, Jacksonville, AL, USA*
DILSHAN SOORIYAARACHCHI • *Industrial Manufacturing and Systems Engineering, Texas Tech University, Lubbock, TX, USA*
SIMONA FEDERICA SPAMPINATO • *Department of Biomedical and Biotechnological Sciences, Section of Pharmacology, University of Catania, Catania, Italy; Department of Scienza e Tecnologia del Farmaco, University of Turin, Turin, IT, Italy*
SVETLANA M. STAMATOVIC • *Department of Pathology, University of Michigan Medical School, Ann Arbor, MI, USA*
YUKIO TAKESHITA • *Department of Neurology and Clinical Neuroscience, Yamaguchi University Graduate School of Medicine, Ube, Yamaguchi, Japan*
GEORGE Z. TAN • *Industrial Manufacturing and Systems Engineering, Texas Tech University, Lubbock, TX, USA*
YEOHEUNG YUN • *Department of Chemical, Biological, and Bioengineering, North Carolina Agricultural and Technical State University, Greensboro, NC, USA*

Part I

Introduction to Blood-Brain Barrier (BBB) Physiology

Chapter 1

Cells of the Blood–Brain Barrier: An Overview of the Neurovascular Unit in Health and Disease

Heather L. McConnell and Anusha Mishra

Abstract

The brain is endowed with highly specialized vasculature that is both structurally and functionally unique compared to vasculature supplying peripheral organs. The blood–brain barrier (BBB) is formed by endothelial cells of the cerebral vasculature and prevents extravasation of blood products into the brain to protect neural tissue and maintain a homeostatic environment. The BBB functions as part of the neurovascular unit (NVU), which is composed of neurons, astrocytes, and microglia in addition to the specialized endothelial cells, mural cells, and the basement membrane. Through coordinated intercellular signaling, these cells function as a dynamic unit to tightly regulate brain blood flow, vascular function, neuroimmune responses, and waste clearance. In this chapter, we review the functions of individual NVU components, describe neurovascular coupling as a classic example of NVU function, and discuss archetypal NVU pathophysiology during disease.

Key words Neurovascular unit, Blood–brain barrier, Cerebral blood flow, Astrogliosis, Neurovascular coupling

1 Introduction

Vasculature in the central nervous system (CNS) is structurally and functionally unique compared to vasculature that supplies the peripheral organs. To protect neural tissue, which is particularly sensitive to changes in the composition of its interstitial fluid, the cerebral vasculature has evolved to tightly regulate the entry and exit of compounds between the blood and the brain. This blood–brain barrier (BBB) is composed of several different cell types and encompasses a physical barrier as well as selectively regulated transport mechanisms. Highly specialized endothelial cells line the lumenal surface of blood vessels and abut an ablumenal basement membrane that contains embedded mural cells—pericytes or vascular smooth muscle cells (VSMCs). Astrocytes project endfeet processes that completely ensheathe these components, while their fine processes contact neuronal synapses in the neuropil.

Nicole Stone (ed.), *The Blood-Brain Barrier: Methods and Protocols*,
Methods in Molecular Biology, vol. 2492, https://doi.org/10.1007/978-1-0716-2289-6_1,

Motile microglia are also present within this biological interface. Functionally, these components interact to regulate vascular permeability, neuroimmune responses, and cerebral blood flow (CBF) to maintain a homeostatic CNS environment.

With an evolving understanding of the dynamic signaling interplay that occurs between the BBB components and neurons to support neuronal homeostasis and information processing, the BBB is now often considered in the context of a functional neurovascular unit (NVU) [1]. The NVU is a fascinating and clinically relevant disease target. Neurovascular dysfunction is present in numerous neurological diseases, yet whether it is a cause or an effect of a particular disease state is not well understood [2]. The cells of the CNS are specified and differentiated during early development. Thereafter, with few exceptions, these cells neither turn over nor regenerate effectively. This cell longevity renders the CNS susceptible to factors from the periphery that can disrupt cellular function and survival. Furthermore, as a highly energy-dependent organ with minimal energy stores, the brain relies on a near-constant supply of oxygen, glucose, and other nutrients to drive adenosine triphosphate (ATP) production and maintain metabolic homeostasis. Indeed, no brain cell is located further than 25 μm from a capillary [3]. Thus, both from a cell survival and metabolic standpoint, the NVU is critical for CNS health, and NVU dysfunction is poised to exacerbate or even initiate neurological disease. Herein, we describe the function of NVU components and the role of the NVU in health and disease.

2 Components of the Neurovascular Unit

NVU components exist along a three-dimensional vascular network of pial and penetrating arterioles, capillaries, ascending venules, and pial veins. This vascular network densely permeates all brain tissue to ensure adequate nutrient delivery [1, 4] and clearance of metabolic waste products [5, 6]. To maintain CNS homeostasis and meet the fluctuating metabolic demands of the brain tissue, NVU components have functional properties that vary by brain region and vascular zone. Although the NVU is present throughout the vascular tree, we focus largely on the capillary NVU as this is where nutrient exchange occurs and the strongest BBB properties are manifest.

We describe non-neuronal components of the NVU radially, beginning with endothelial cells and progressing outward through the layers of basement membranes (noncellular component), mural cells, astrocytes, and finally microglia.

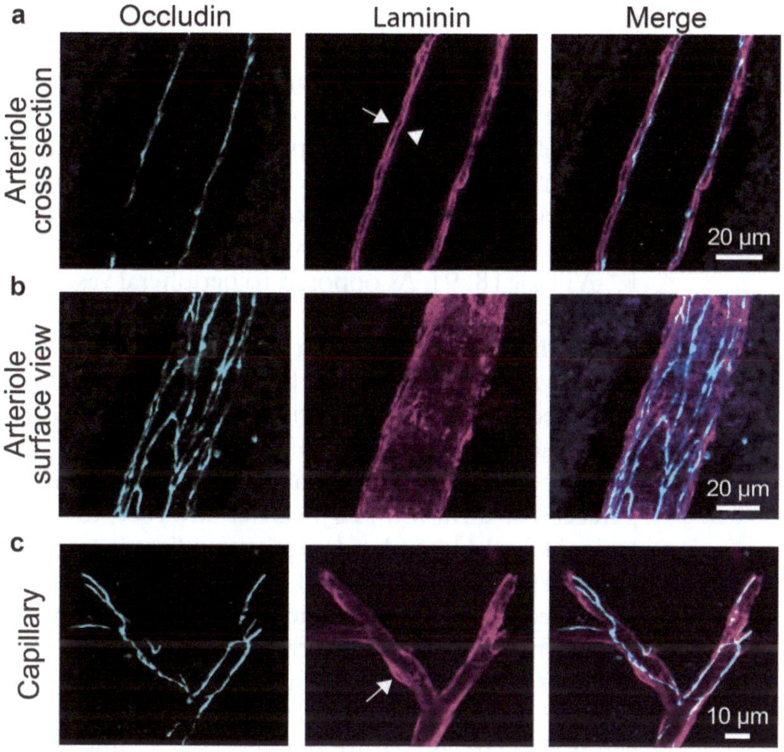

Fig. 1 Distribution of occludin and laminin blood–brain barrier proteins in the cerebral vasculature. Double immunofluorescence confocal microscopy performed with anti-occludin (cyan) and anti-laminin (magenta) antibodies on formalin-fixed 50 μm-thick adult rat brain sections. (**a**) Arteriolar cross section showing occludin-positive tight junctions between endothelial cells arranged circumferentially around the vascular lumen. Laminin-containing basement membrane circumscribes the occludin staining and delineates both the endothelial-secreted vascular basement membrane (arrowhead) and the astrocyte-secreted parenchymal basement membrane layer (arrow). Vascular smooth muscle cells occupy the space between the layers of laminin (not labeled). (**b**) Surface view of an arteriole shows occludin-positive tight junctions outlining the contiguous and elongated endothelial cells. Laminin staining reveals basement membrane coverage along the entire vessel. (**c**) View of a capillary branch point shows occludin tight junction staining between adjacent endothelial cells. The laminin-positive vascular and parenchymal basement membranes are fused along most regions along a capillary and indistinguishable except in regions encasing pericytes (arrow). Merged images are shown on the right

2.1 Endothelial Cells

A continuous monolayer of specialized endothelial cells lines the blood vessels of the brain. Unlike the endothelial cells lining the peripheral vasculature, brain vascular endothelial cells are completely confluent and interconnected circumferentially by transmembrane tight junction proteins that prevent the passive transfer of cells and molecules between the blood and the brain based on size, surface electrical charge, and lipid solubility (Fig. 1). These tight junctions create a strict seal throughout vascular lumenal surfaces, giving rise to in vivo transendothelial electrical resistances of up to ~1800 $\Omega \cdot cm^2$ and preventing even the flow of water or small ions across the endothelium [7].

Besides this physical barrier, brain vascular endothelial cells maintain a dynamic regulatory system that facilitates bidirectional transport of substances between the blood and the brain. The transport functions of these cells are mediated and maintained by polarized expression of various luminal and abluminal membrane receptors and transporters, such as HCO_3^- exchangers and Na^+/K^+-ATPase [8, 9]. As opposed to peripheral vascular systems where substances in the blood are more or less indiscriminately moved across the endothelial lining into the tissue via transcytosis, these pathways are suppressed in brain endothelial cells, reducing their non-specific transport capacity [10, 11]. Thus, depending on the chemical entity, transport of substances across the BBB must be actively regulated. This occurs via several mechanisms, including paracellular diffusion [12], carrier and receptor-mediated transcytosis [13, 14], and adsorptive-mediated transcytosis (Fig. 2) [15, 16]. These specialized properties of brain endothelial cells are reflected in their unique gene expression patterns, a finding highlighted in recent transcriptomic studies [17–19]. For example, expression of Mfsd2a, the transcytosis-suppressing protein responsible for lipid transport regulation, is high in endothelial cells of the cerebral microvasculature but very low in lung and heart endothelial cells [18, 19]. Although this regulation effectively maintains the delicate ionic milieu and extracellular homeostasis of neural tissue, it also creates a pharmacological bottleneck for CNS drug delivery [15, 20].

Brain vascular endothelial cells are phenotypically heterogeneous and have numerous location-specific functional roles that vary with both their position in the vascular hierarchy and location in the brain [18, 21, 22]. For example, the circumventricular organs and hypothalamic regions contain specialized endothelial cells that make their vasculature more permeable than the rest of the brain [21, 22]. In addition to their barrier functions, brain vascular endothelial cells dynamically modify permeability in response to inflammatory and immunological stimuli [19] or in a rhythmic, circadian fashion [23–25]. Further, endothelial cells also help mediate neurovascular coupling (NVC) [26–28] and regulate the spread of vascular responses via endothelial gap junctions, which allow propagation of signals along the vascular tree to mediate changes in regional CBF (Fig. 4) [29, 30]. The increased mitochondrial load of brain vascular endothelial cells reflects their active role in NVU regulation but also makes them more sensitive to changes in oxidative stress [31].

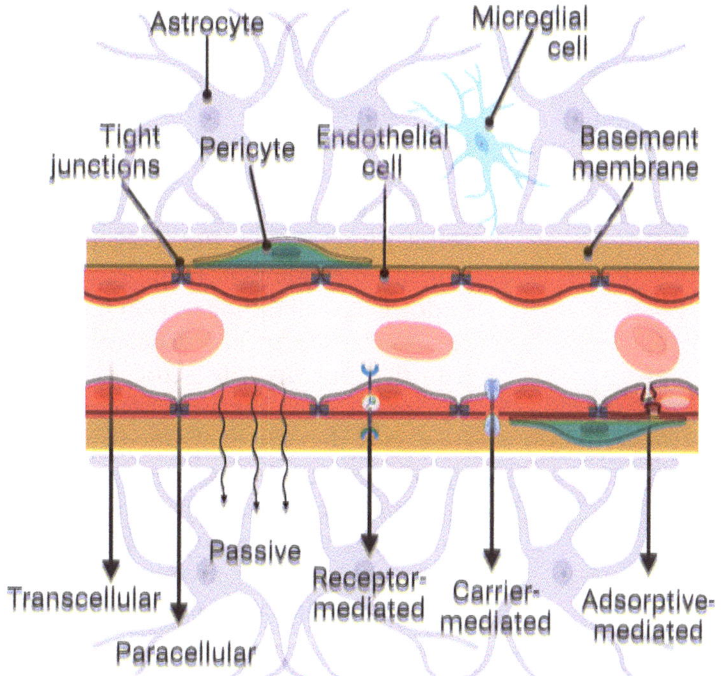

Fig. 2 Specialized brain vascular endothelial cells regulate transport at the BBB. A contiguous monolayer of vascular endothelial cells lines the brain vasculature. These cells are interconnected by tight junctions, creating a physical barrier with a remarkably high transendothelial electrical resistance. Transport across the BBB is mediated by three energy-independent diffusion mechanisms and three energy-dependent transport mechanisms. The energy-independent mechanisms include transcellular diffusion, paracellular diffusion, and passive diffusion. The energy-dependent mechanisms encompass receptor-mediated transcytosis, carrier-mediated transport, and adsorptive-mediated transport. Transport is shown only unidirectionally for simplicity. (Created with BioRender.com)

2.2 Basement Membrane

The cerebrovascular endothelial tube is ensheathed by a noncellular basement membrane composed of five major proteins—collagens, laminins, nidogens, perlecan, and agrin—arranged in a compact lattice meshwork. This vascular basement membrane merges with a parenchymal basement membrane in areas along the vasculature lacking mural cell coverage, particularly in capillary regions not covered by pericytes (Fig. 1). Vascular basement membrane proteins are secreted largely by endothelial cells while parenchymal basement membrane components are secreted by astrocytes via their vascular endfeet [32]. Bidirectional signaling between the cells of the NVU—endothelial cells, pericytes, and astrocytes—and the basement membrane dynamically regulates BBB permeability. For example, pericyte recruitment to endothelial tubes during development is an important stimulus that helps generate the basement membrane [33], but absence of the basement membrane protein laminin can, in turn, reduce pericyte coverage of the vasculature [34].

Structurally, basement membrane proteins form a sheet-like extracellular matrix that stabilizes the BBB, anchors cells, facilitates signal transduction, and enhances vascular barrier properties [35]. The composition and thickness of the basement membrane are dynamic, vary regionally throughout the brain, and can affect cellular interactions and receptor activation states [35, 36]. Functionally, the basement membrane contributes to BBB development, endothelial cell tight junction formation, and maintenance of trans-endothelial electrical resistance [35]. It also helps recruit and establish the polarity of astrocyte endfeet, including anchoring of aquaporin 4 to the vessel-facing cell membrane of the endfoot, functions that are likely mediated via interactions of laminin with its receptor dystroglycan in astrocyte endfeet [34]. However, as astrocytes are an important source of the parenchymal basement membrane components, astrocyte dysfunction can also result in a weakened basement membrane [37], underscoring the importance of bidirectional crosstalk between BBB components.

2.3 Mural Cells: Pericytes

Mural cells called pericytes enwrap capillaries in all vascular beds. Capillaries in the CNS are associated with the highest pericyte coverage in the body, exhibiting a pericyte to endothelial cell ratio of 1:4 [38, 39] compared to, for example, a ratio of 1:10 or more in skeletal muscle [40, 41]. This suggests an important role of pericytes in cerebral circulation. Indeed, brain pericytes play critical roles during both embryogenesis and adulthood by establishing and maintaining BBB integrity [42], orchestrating vascular development [43], and regulating capillary blood flow [44]. During brain development, pericytes migrate to vessels and, in concert with endothelial cells and other NVU components, contribute to vascular pruning and angiogenesis [43, 45]. A balance of these processes effectively shapes and organizes the cerebral vasculature to ensure an optimal angioarchitecture that facilitates nutrient delivery to brain tissue. Like the endothelial cells and basement membrane, pericytes tune the polarization of astrocyte endfeet along the capillary wall [46, 47]. Embedded between the endothelial tube and astrocyte endfeet, pericytes are often described as having a "bump on a log" morphology (Fig. 3) [48]. This description accurately describes the cell body but may not adequately convey the extensive coverage of the vasculature by pericyte processes, which can contact up to 90% of brain vasculature in mice [49]. Although pericytes are fixed in place and do not migrate after development under physiologically healthy conditions, they are structurally dynamic and can extend their processes and remodel to compensate for loss of nearby pericytes or endothelial cell coverage in areas along the vasculature [50]. In adult brains, pericytes exhibit contractile responses to vasoactive compounds and therefore control capillary diameter [44, 51, 52]. However, not all

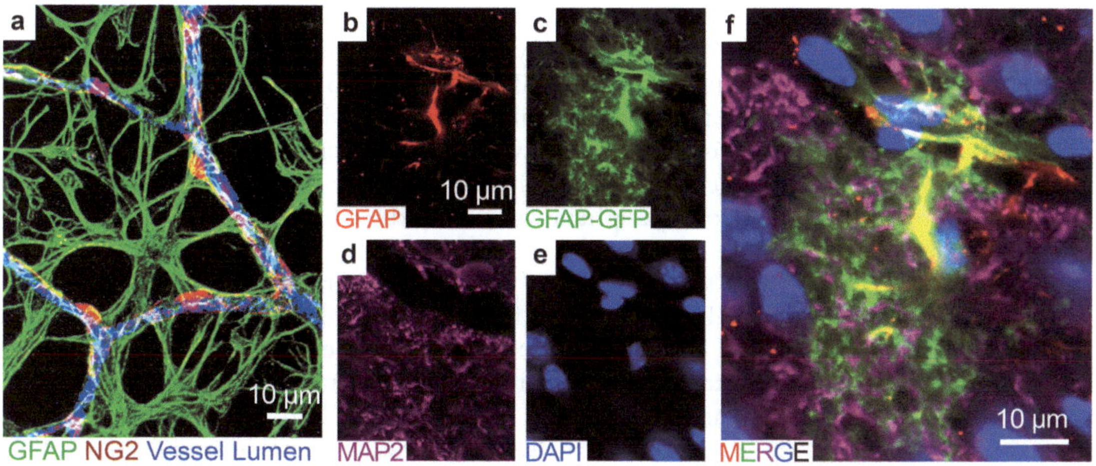

Fig. 3 Astrocyte ensheathment of the cerebral vasculature. (**a**) A capillary network in a mouse retina showing the relationship of astrocytes to pericytes. Astrocytes are labeled with an anti-glial fibrillary acidic protein (GFAP) antibody (green) in a retina whole mount from an NG2-dsRed mouse, wherein the red fluorescent protein dsRed is expressed in NG2-positive pericytes. The vascular lumen is labeled with an anti-mouse secondary (blue) that labels natural mouse serum IgGs present in blood vessels (retina post-fixed following live dissection without perfusion). Astrocyte endfeet encapsulate the capillary wall and contact all pericytes. (**b–f**) Astrocyte endfeet around a cortical capillary with the surrounding neuropil. (**b**) An astrocyte labeled with anti-GFAP antibody (red). (**c**) The same astrocyte expressing virally induced cytosolic green fluorescent protein (GFP) under the truncated GFAP promoter, GfaABC$_1$D (green). (**d**) The surrounding neuropil labeled with the neuronal marker, microtubule-associated protein 2 (MAP2; magenta). (**e**) Nuclei are labeled with DAPI (blue). (**f**) Merged image showing GFAP, GFP, MAP2, and DAPI. Though GFAP labeling shows only the cytoskeleton of the astrocyte, including the endfoot processes, GFP expression reveals its extensive fine processes permeating the neuropil (where it contacts synapses) and its complete ensheathment of the capillary via its endfoot

pericytes are contractile [53–55], and they may be relatively less contractile than the VSMCs on larger vessels. It has been proposed that pericytes may maintain basal CBF equilibrium at the capillary level via slow time-scale regulation [56] while arterial VSMCs are responsible for larger, more rapid CBF fluctuations [50].

Pericytes display broad heterogeneity in function, gene expression profile, and appearance. They are morphologically variable along the vascular tree, showing different structures and vessel coverage at capillary, pre-capillary, and post-capillary vascular zones. Therefore, depending on their location and morphology, pericytes may exert different functions in their regulation of the BBB and CBF at different levels of the vascular tree [48, 57]. Pericytes located at branch points between arterioles and capillaries may act as "pre-capillary sphincters" to direct flow into or away from a region [51, 58]. There is also strong evidence that pericytes on pre-capillary arterioles and capillaries closer to arterioles participate in NVC [44] and regulate CBF [50, 59]. Gap-junctional connections between neighboring pericytes and between pericytes and endothelial cells propagate signals along the vascular tree [30], such that a contractile response initiated in one region quickly spreads up- and downstream to increase blood flow through that region [60, 61]. Pericytes in mid-capillary branches may be more important for basal CBF regulation [56, 62] and maintenance of BBB integrity and stability in concert with endothelial cells [48, 57, 63]. Very little is known about the pericytes on cerebral venules, but they are hypothesized to regulate immune cell trafficking, as they do in peripheral vascular beds [64, 65].

2.4 Mural Cells: Vascular Smooth Muscle Cells

VSMCs are a distinct contractile mural cell type from pericytes [66]. They are embedded in the arterial wall where they form several concentric layers around these larger vessels. They have important roles in maintaining and generating basal vascular tone, vascular reactivity, and autoregulation [67, 68]. VSMCs express a variety of ion channels, are electrically excitable, and display robust calcium signals and consequent contractile responses; hence, they were traditionally thought to be the principal site of functional hyperemia in the brain [1, 67] (but also see: Sect. 2.4 on pericyte-mediated regulation and Sect. 3 on NVC). Furthermore, VSMCs are gap junctionally coupled to each other [69] and also to endothelial cells lining the vasculature, which allows evoked vasomotor responses to propagate along vessels [70] over long distances as calcium waves [71]. The polarity and magnitude of these calcium signals determine the degree of evoked vasoconstriction or dilation. Fluctuations between relaxation and contraction states in cerebral VSMCs gives rise to vasomotion, which is postulated to be a mechanism of perivascular drainage to clear waste from the brain interstitium [72].

2.5 Astrocytes

Astrocytes are the most numerous glial cell type in the CNS with a complex, polarized morphology. On the one hand, they have vascular endfeet processes, which almost completely encapsulate the vasculature of the brain, including the endothelial tube and surrounding mural cells (Fig. 3) [73]. On the other hand, they have numerous fine, ramified processes extending from a stellate cell body which permeate the neuropil (Fig. 3). Like neurons, astrocytes are heterogeneous in structure, function, and distribution throughout the brain. Traditionally, astrocytes were classified into two groups: protoplasmic astrocytes found in the well-vascularized gray matter and fibrous astrocytes found in the less vascular white matter. In both gray and white matter areas, astrocytes physically separate synapses or nodes of Ranvier, contact neurons, and cover the vasculature [74]. More recent work has revealed considerable heterogeneity even within gray matter astrocytes, with pronounced regional or even subregional differences [75–79], some of which are regulated by signals from the surrounding NVU components, including neurons [80] and endothelial cells [81, 82]. This heterogeneity manifests as variations in astrocyte functions, their interactions with neurons [83, 84], and their response to injuries [85].

Astrocytes play a multifaceted role in maintaining CNS homeostasis. They spatially tile the brain parenchyma and are interconnected by gap junction proteins to form a functional syncytium [86], which allows them to signal over long ranges via calcium wave propagation [87]. Astrocytes also express high levels of the inwardly rectifying potassium channel Kir4.1, which is pronounced in their peri-synaptic processes and vascular endfeet. This allows astrocytes to take up potassium after bouts of neuronal activity and spatially buffer potassium along the syncytium [86]. At synapses, astrocytes are active participants in tuning the information transfer between neurons by taking up neurotransmitters at the cleft to spatially and temporally refine synaptic signaling [74].

Astrocyte endfeet processes terminate on and enwrap all capillaries and arterioles in the CNS to form a contiguous gliovascular interface. Astrocytes extend other fine processes to a variety of targets within the CNS, including synapses, nerve cell bodies, and nodes of Ranvier. This polarized morphology positions astrocytes centrally within the NVU to dynamically mediate many regulatory functions, including maintenance of BBB integrity, mechanical and metabolic support, endothelial transport, innate immune responses, and cerebrovascular regulation [88, 89]. Via their endfeet processes, astrocytes secrete basement membrane proteins such as laminins [90] to stabilize the BBB. They also signal to pericytes [90] and endothelial cells [91] to maintain the CNS-specific properties of these vascular cells within the BBB. Specialized membrane proteins anchored to astrocyte endfeet processes regulate the transport of nutrients and neurotransmitters after they cross the BBB, including glucose and glutamate, which are necessary for healthy brain function [81, 92]. Further, selective expression of channels such as Kir4.1 and aquaporin 4 at the endfeet polarize astrocytes and facilitate the directional flow of potassium and water molecules to maintain ion homeostasis and neuropil volume [93–95].

Astrocytes serve important roles as sensors of neuronal activity and stabilizers of the extracellular environment. They are also active players in information processing within the brain. In response to synaptic activity, numerous intracellular signaling pathways are activated in astrocytes, the most studied so far being calcium-dependent. Calcium increases can be transient and localized within microdomains, spread through an entire process, spread throughout the astrocyte territory, or even be sustained and propagated across the astrocyte syncytium [87, 96]. This activation of astrocytes results in the release of gliotransmitters (to the spatial extent dictated by their calcium signals), which can then modulate activity ranging from anywhere between individual synapses to whole circuits [25, 87, 97–99]. Given that astrocytes also express virtually all neuromodulator receptors, they have been proposed to be the cellular substrates underlying coordinated long-range neuromodulatory networks [100]. With their endfeet on vessels for nutrient uptake and their gap junction-coupled fine neuropil processes, astrocytes also form a crucial metabolic supply network that not only regulates glucose delivery to active regions [101] but may also modulate circuit properties that regulate plasticity [102] and behaviors such as sleep [103].

Within the NVU, astrocytes are key regulators of CBF [104]. Astrocytes contribute to the resting tone of arterioles through release of vasoactive agents such as prostaglandin E2 [105] and ATP [106, 107]. They also convey signals from active neurons to the vasculature to mediate NVC. This intercellular signaling pathway—from neurons to astrocytes to the vascular mural cells—is the primary pathway regulating capillary NVC [29, 52, 108], but it appears to be less important in regulating arteriolar NVC, where it is engaged only upon sustained or strong activation (Fig. 4) [109]. Even so, recruitment of astrocytes results in a threefold enhancement of functional hyperemia in vivo, suggesting an important role for astrocytes in activity-dependent energy supply to the brain [110].

Finally, astrocytes serve an important role in glymphatic waste clearance. During sleep, aquaporin 4 water transport channels on the vessel-facing endfeet processes of astrocytes facilitate the exchange of perivascular cerebrospinal fluid (CSF) with interstitial fluid at capillaries [111]. This exchange results in the drainage of CNS waste products into the CSF, which is then removed at CSF clearance sites. Both sleep deprivation [112] and aging [113] result in glymphatic dysfunction by causing mislocalization of aquaporin 4 on astrocyte endfeet, potentially contributing to the accumulation of toxic waste products in the brain.

2.6 Microglia

Microglia are yolk sac-derived myeloid cells that seed the brain early in development [114]. During this early period, microglia stimulate angiogenesis, increase vascular complexity, and promote endothelial tip cell fusion within capillary beds [115–117]. Conversely, the vascular network may also provide microglia with a migration "highway" to populate the brain during development [118]. Adult microglia are the tissue-resident macrophages of the brain. These highly ramified, motile cells act as phagocytic sentinels, surveilling and probing their environment for any alterations or injury by extending and retracting their processes. Microglia are scattered evenly throughout the brain, in a manner similar to the tiling of astrocytes, and occasionally extend a process directly to the vasculature at locations containing pericytes [73]. They are exquisitely sensitive to changes in the neuropil environment or in nearby perivascular homeostasis. Upon sensing insult, microglia take on an ameboid morphology and undergo dynamic transcriptomic and phenotypic changes that are characterized along a context-dependent spectrum of pro-inflammatory to anti-inflammatory activated

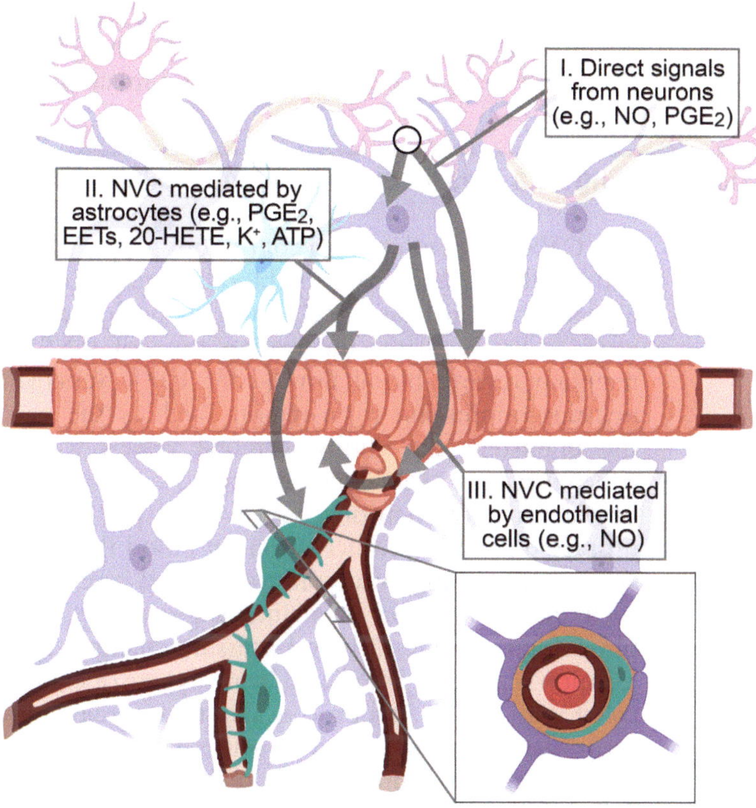

Fig. 4 Neurovascular coupling (NVC) is the phenomenon by which active neurons increase local blood flow to satisfy their energy demands. During synaptic activity (circled), post-synaptic neurons (magenta) can directly release vasoactive substances onto vessels to change the luminal diameter (I). Synaptic activity can also stimulate astrocytes (purple) to release vasoactive substances via their endfeet (II). Vascular endothelial cells (red) may also respond to signals from astrocytes, and perhaps neurons, to release vasoactive substances onto vessels (III). Regulation of lumenal diameter at arterioles occurs via signaling from neurons and endothelial cells to arteriolar smooth muscle cells (pink), while capillary responses depend upon signaling from astrocytes to capillary pericytes (green). Astrocytes may also modulate the response of arteriolar smooth muscle cells. Further, endothelial cells, pericytes, and vascular smooth muscle cells can also propagate the spread of vascular responses via gap junctions to mediate regional changes in CBF. Through this coordinated intercellular signaling, NVC components function as a dynamic functional unit to tightly regulate brain blood flow. Inset shows positioning of astrocytes, pericytes, basement membrane (brown), and endothelial cells within a capillary cross section. (Created with BioRender.com)

states. These activated microglia serve a critical role in phagocytosing debris [119, 120], signaling to peripheral immune cells [120], and regulating reactive astrogliosis [121, 122]. Microglia may also modulate the BBB, sometimes enhancing BBB properties and sometimes contributing to BBB breakdown and NVU dysfunction [123–125], depending on the type, extent, and chronicity of the injury [126–128]. As with any other component of the NVU, microglial properties are also dynamically modulated by surrounding NVU cells such as astrocytes and endothelial cells

[129]. Together, astroglial and microglial activation may optimize innate immune responses in the CNS by engaging in coordinated gliotransmitter and cytokine release and metabolite uptake [130].

3 Neurovascular Coupling as a Classic Example of Intercellular NVU Signaling

The CNS is dependent on a continuous supply of energy substrates and other nutrients from the blood to maintain healthy function. Signaling within the NVU, particularly between astrocytes and mural cells, regulates the resting tone of CNS vessels to help maintain resting blood flow [105, 107]. Subcortical projection neurons originating from regions such as the basal ganglia and the brainstem can endow cerebral vessels with vascular tone, which may occur via direct signaling to vascular cells, via astrocytes, or via local interneurons (reviewed in [29]). The brain is also protected from large fluctuations in blood flow due to changes in systemic physiology via a process termed cerebral autoregulation, such that blood flow is maintained relatively constant within a physiological range of blood pressure changes. Although autoregulation is generally mediated by myogenic or endothelial factors, a role for astrocytes in a process akin to autoregulation was recently identified, whereby astrocytes sense pressure changes and release vasoconstrictors onto mural cells to maintain vascular tone [106]. Layered upon these processes is NVC, an intercellular signaling mechanism that engages almost all cellular components of the NVU to increase blood supply to regions of increased neuronal activity in a spatially and temporally coordinated manner (Fig. 4) [104], and gives rise to functional hyperemia in the brain. Neuronal activity and blood flow changes are coupled so tightly that this phenomenon has been exploited by various functional neuroimaging techniques, such as blood oxygenation level-dependent functional magnetic resonance imaging [29, 131, 132], positron emissions tomography [133, 134], and the recently developed functional ultrasound [135] to image brain activity in real time.

The study of pathways regulating NVC has blossomed in the last two decades, revealing complex intercellular interactions at the NVU interface. NVC is initiated by neuronal activity and can be mediated by vasoactive signals released directly from neurons or indirectly via astrocytes, which trigger a change in vascular lumen diameter (Fig. 4) (reviewed in [1, 29, 67]). Vasodilatory signals such as nitric oxide (NO), prostaglandins, potassium, and epoxyeicosatrienoic acids released by neurons and/or astrocytes can relax mural cells, thereby dilating blood vessels and increasing CBF. Conversely, vasoconstrictive signals such as ATP and 20-hydroxytetraenoic acids from neurons/astrocytes cause contraction of mural cells, thereby constricting blood vessels and decreasing CBF. Though traditional views held that NVC only involved direct signaling from neurons to the VSMCs to increase CBF at the arteriolar level, recent studies have found significant evidence for a role of pericytes in capillary flow regulation and for astrocytes in mediating signals to capillaries (Fig. 5) [52, 108]. Indeed, with their fine processes apposed to numerous synapses and their endfeet ensheathing the entire cerebral vasculature (including arterioles and capillaries) astrocytes are ideally positioned to be NVC intermediaries. Current evidence suggests that astrocytes play an obligatory role in capillary NVC [52, 108], while their role in arteriolar NVC may be more modulatory [109, 110, 136]. Ultimately, signals from both neurons and astrocytes must alter the contractile state of vascular mural cells—pericytes on capillaries or VSMCs on arterioles—for NVC to successfully produce functional hyperemia. Furthermore, a role for endothelial cells in this intercellular signaling pathway, wherein endothelial cells respond to signals from astrocytes and in turn secrete vasodilators onto VMSCs, has also been discovered (Fig. 4) [28]. Thus, functional interactions between NVU components are crucial regulators of blood flow and hence energy supply to the brain.

4 The Neurovascular Unit in Disease

Neurovascular dysfunction is a central theme of numerous brain disease states, though whether it is a cause or an effect of a particular disease state is not always well defined. Broadly, neurovascular dysfunction encompasses a range of pathological states that can be attributed to individual NVU components or the unit as a whole. In lieu of a particular disease focus, we instead describe elements of NVU dysfunction that are common to many brain diseases. Critically, NVU dysfunction occurs along a spectrum with variable pathophysiological features depending on the disease state and individual.

As a major function of the NVU is to maintain BBB characteristics and function, a general consequence of NVU dysfunction is often BBB disruption, though this term can reflect distinct and nuanced structural, functional, cellular, and molecular deficits depending on the particular disease state. For example, NVU cellular responses to hypoxia differ from those occurring in neurodegeneration or traumatic injury. On a structural level, BBB disruption reflects a loss of integrity and an increase in permeability due to altered endothelial cell expression of tight junction proteins [137, 138] and transporters [139], enlarged perivascular spaces [140], or even degeneration of NVU components including endothelial cells [141], pericytes [142–144], astrocytes [145–147], or basement membranes [148]. Pathologically impaired waste clearance may also compromise BBB integrity [149]. A functional consequence of increased BBB permeability is permissive infiltration of

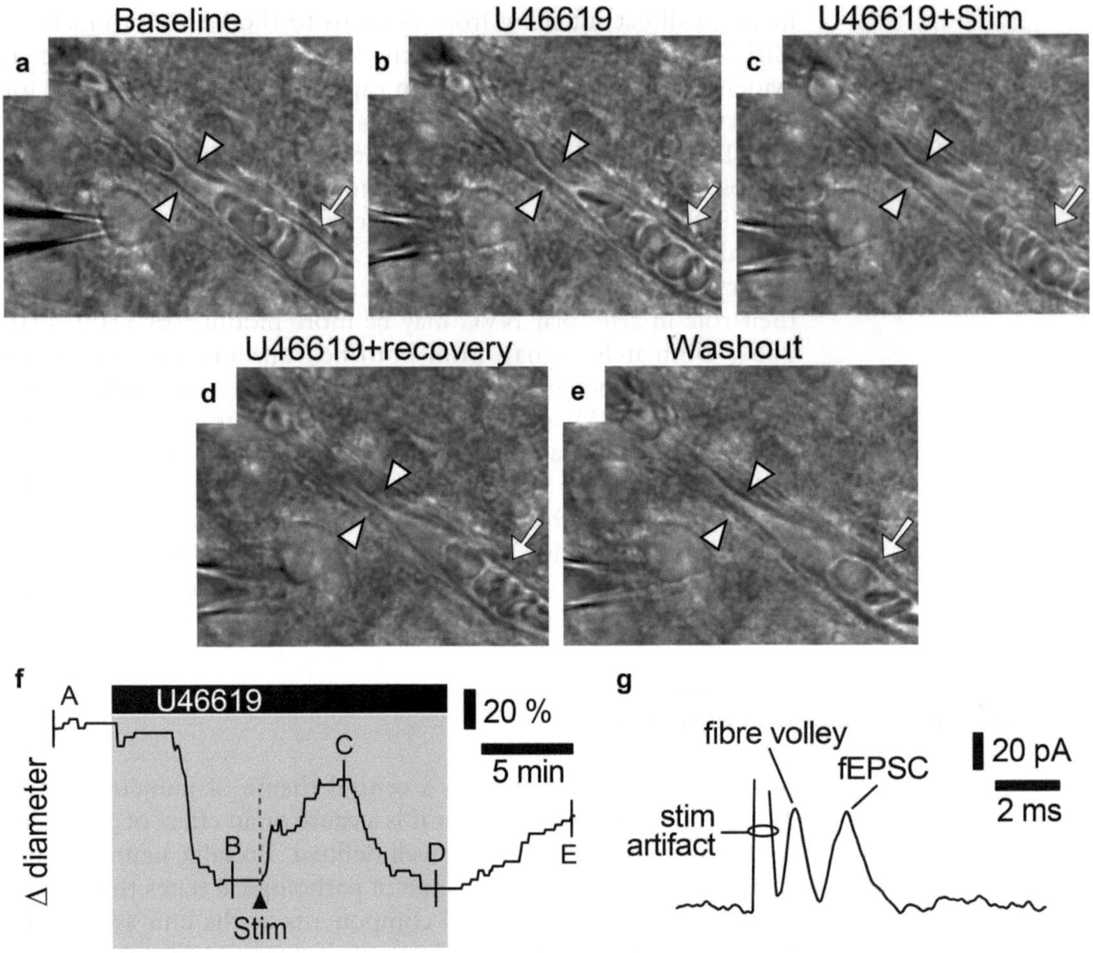

Fig. 5 Neuronal stimulation evokes dilation of capillaries in the cortex. (**a**–**e**) A 300 μm-thick acute rat brain slice preparation showing a cortical capillary and the surrounding neuropil. A recording electrode for measuring field potentials is also visible. Successive images show the response of the capillary to the preconstrictor U46619 (**b**), electrical stimulation of cortical layer I/II axons (**c**), recovery from stimulation-evoked response (**d**), and after washout of U46619 (**e**). (**f**) The diameter of the capillary region marked by arrowheads in **a**–**e** is plotted over time, demonstrating a strong constriction in response to the thromboxane analog U46619 (200 nM), followed by a brief dilation in response to neuronal stimulation. Washing out U46619 reverses the constricted tone (end of trace). Notably, the changes in capillary diameter occur only in a restricted segment while the neighboring region remains undisturbed (arrow). Such a response is characteristic of pericyte-mediated diameter changes. This response is blocked by inhibitors of action potentials and post-synaptic activity, indicating that it is dependent on neuronal activity (data not shown, see [52] for more details). (**g**) A field recording of neuronal activity evoked by electrical stimulation of layer 1/II in the same experiment (detected by the recording electrode visible in **a**). The fiber volley (compound action potentials) and the field excitatory post-synaptic current are clearly visible. A 3 s, 20 Hz stimulus train was applied to induce the capillary response shown in **c**, **f**

blood products through the vulnerable barrier, including but not limited to peripheral immune cell infiltrates.

Under conditions that deviate from a normal physiological range, astrocytes often become "reactive" and alter their morphologic and transcriptomic properties. Reactive astrogliosis is also induced following BBB breakdown via signaling from blood-borne proteins such as albumin, fibronectin, and thrombin [2, 150–152]. Reactive astrocytes are characterized by gain of new functions or loss of existing functions, which together alter the overall functions of the region affected [2, 153]. They can also release various cytokines that can alter the properties of nearby neurons, microglia, and vascular cells [154] and perhaps even attract or modulate peripheral immune cells to the rescue of the CNS, with deleterious or beneficial effects [155–157]. The current consensus is that reactive astrogliosis is a context-dependent process engaged to re-establish homeostasis and protect neurons from further damage [122, 158–160]; however, in some contexts, they acquire detrimental properties and may amplify injury [121, 161, 162]. The central positioning of astrocytes in neurovascular signaling also poises them to subtly or dramatically impair energy homeostasis in the CNS due to an "uncoupling" of neuronal activity and vascular responses during both acute and chronic NVU impairment [2, 104]. Indeed, CBF alterations are commonly seen in many brain disease states and can manifest as decreased perfusion to deep brain structures, loss of NVC, and impaired vascular reactivity to flow and pressure changes [163–167].

5 Conclusion

The mammalian cerebrovascular system is characterized by structurally and functionally unique vessels endowed with a strong biological barrier that selectively separates blood and brain components. Distinct functional features characterize individual NVU components, though they maintain many intimate functional relationships with one another. Further, cells of the NVU can change both functionally and phenotypically, depending on microenvironmental signaling cues.

The presence of the BBB has important implications for neuronal function during physiology and pathology. It protects the CNS extracellular environment from peripheral signals that could destabilize neuronal function. Further, complex intercellular signals between NVU components allows these cells to sense and respond to changes in regional neuronal activity to regulate CBF and meet the energy demands of neurons. The recent development and application of novel and sophisticated neuroimaging paradigms that allow real-time imaging of these physiological processes offers exciting prospects for delineating the contributions of individual NVU components to regional hemodynamic changes.

Finally, NVU dysfunction can precede or exacerbate CNS disease states. With their multifaceted functional roles in physiological processes, dysfunctional NVU components are poised to adversely influence a broad range of metabolic, hemodynamic, and immunologic regulatory responses in the brain. The NVU, therefore, represents a viable therapeutic target to stall disease and promote health. Compelling evidence thus far supports this view; however, considerable research will be required to form a complete picture of the NVU's role in neurological diseases.

Acknowledgments

The authors thank Dr. Leslie Muldoon, Dr. Sasha Pejerrey, and Adrienne Winston for helpful comments on the manuscript and gratefully acknowledge Dr. Zhihua Wang for assistance with the immunohistochemistry shown in Fig. 1. H.L.M was supported by a Ruth L. Kirschstein National Research Service Award T32HL094294 (NIH). A.M.'s lab is supported by NIH grants R01NS110690 (NINDS), R01DA047237 (NIMH; PI Andrew Adey) and a Development Project grant under P30AG066518-01; Oregon Alzheimer Disease Research Center (NIA; PI Jeffrey Kaye).

References

1. Iadecola C (2017) The neurovascular unit coming of age: a journey through neurovascular coupling in health and disease. Neuron 96:17–42
2. Mcconnell HL, Li Z, Woltjer RL et al (2019) Astrocyte dysfunction and neurovascular impairment in neurological disorders: correlation or causation? Neurochem Int 128:70–84
3. Lovick TA, Brown LA, Key BJ (1999) Neurovascular relationships in hippocampal slices: physiological and anatomical studies of mechanisms underlying flow-metabolism coupling in intraparenchymal microvessels. Neuroscience 92:47–60
4. Blinder P, Tsai PS, Kaufhold JP et al (2013) The cortical angiome: an interconnected vascular network with noncolumnar patterns of blood flow. Nat Neurosci 16:889–897
5. Xie L, Kang H, Xu Q et al (2013) Sleep drives metabolite clearance from the adult brain. Science 342:373–377
6. Kisler K, Nelson AR, Montagne A et al (2017) Cerebral blood flow regulation and neurovascular dysfunction in Alzheimer disease. Nat Rev Neurosci 18:419–434
7. Butt AM, Jones HC, Abbott NJ (1990) Electrical resistance across the blood-brain barrier in anaesthetized rats: a developmental study. J Physiol 429:47–62
8. Pluimer BR, Colt M, Zhao Z (2020) G protein-coupled receptors in the mammalian blood-brain barrier. Front Cell Neurosci 14:139
9. Cornford EM, Hyman S (2005) Localization of brain endothelial luminal and abluminal transporters with immunogold electron microscopy. NeuroRx 2:27–43
10. Andreone BJ, Chow BW, Tata A et al (2017) Blood-brain barrier permeability is regulated by lipid transport-dependent suppression of caveolae-mediated transcytosis. Neuron 94:581–594 e585
11. Chow BW, Gu C (2017) Gradual suppression of transcytosis governs functional blood-retinal barrier formation. Neuron 93:1325–1333 e1323
12. Keaney J, Walsh DM, O'malley T et al (2015) Autoregulated paracellular clearance of amyloid-beta across the blood-brain barrier. Sci Adv 1:e1500472
13. Urayama A, Grubb JH, Sly WS et al (2004) Developmentally regulated mannose 6-phosphate receptor-mediated transport of a lysosomal enzyme across the blood-brain barrier. Proc Natl Acad Sci U S A 101:12658–12663
14. Pardridge WM, Triguero D, Yang J et al (1990) Comparison of in vitro and in vivo

models of drug transcytosis through the blood-brain barrier. J Pharmacol Exp Ther 253:884–891

15. Lombardo SM, Schneider M, Tureli AE et al (2020) Key for crossing the bbb with nanoparticles: the rational design. Beilstein J Nanotechnol 11:866–883
16. Matsumoto J, Stewart T, Sheng L et al (2017) Transmission of alpha-synuclein-containing erythrocyte-derived extracellular vesicles across the blood-brain barrier via adsorptive mediated transcytosis: another mechanism for initiation and progression of Parkinson's disease? Acta Neuropathol Commun 5:71
17. Daneman R, Zhou L, Agalliu D et al (2010) The mouse blood-brain barrier transcriptome: a new resource for understanding the development and function of brain endothelial cells. PLoS One 5:e13741
18. He L, Vanlandewijck M, Mae MA et al (2018) Single-cell RNA sequencing of mouse brain and lung vascular and vessel-associated cell types. Sci Data 5:180160
19. Jambusaria A, Hong Z, Zhang L et al (2020) Endothelial heterogeneity across distinct vascular beds during homeostasis and inflammation. elife 9:e51413
20. Lochhead JJ, Yang J, Ronaldson PT et al (2020) Structure, function, and regulation of the blood-brain barrier tight junction in central nervous system disorders. Front Physiol 11:914
21. Ganong WF (2000) Circumventricular organs: definition and role in the regulation of endocrine and autonomic function. Clin Exp Pharmacol Physiol 27:422–427
22. Rodriguez EM, Blazquez JL, Guerra M (2010) The design of barriers in the hypothalamus allows the median eminence and the arcuate nucleus to enjoy private milieus: the former opens to the portal blood and the latter to the cerebrospinal fluid. Peptides 31:757–776
23. Cuddapah VA, Zhang SL, Sehgal A (2019) Regulation of the blood-brain barrier by circadian rhythms and sleep. Trends Neurosci 42:500–510
24. Zhang SL, Yue Z, Arnold DM et al (2018) A circadian clock in the blood-brain barrier regulates xenobiotic efflux. Cell 173:130–139 e110
25. Gomez-Gonzalez B, Hurtado-Alvarado G, Esqueda-Leon E et al (2013) Rem sleep loss and recovery regulates blood-brain barrier function. Curr Neurovasc Res 10:197–207
26. Atochin DN, Huang PL (2011) Role of endothelial nitric oxide in cerebrovascular regulation. Curr Pharm Biotechnol 12:1334–1342
27. Hogan-Cann AD, Lu P, Anderson CM (2019) Endothelial NMDA receptors mediate activity-dependent brain hemodynamic responses in mice. Proc Natl Acad Sci U S A 116:10229–10231
28. Stobart JL, Lu L, Anderson HD et al (2013) Astrocyte-induced cortical vasodilation is mediated by d-serine and endothelial nitric oxide synthase. Proc Natl Acad Sci U S A 110:3149–3154
29. Howarth C, Mishra A, Hall CN (2021) More than just summed neuronal activity: how multiple cell types shape the bold response. Philos Trans R Soc Lond Ser B Biol Sci 376:20190630
30. Kovacs-Oller T, Ivanova E, Bianchimano P et al (2019) Dynamic connectivity maps of pericytes and endothelial cells mediate neurovascular coupling in health and disease. bioRxiv:830398
31. Davidson SM, Duchen MR (2007) Endothelial mitochondria: contributing to vascular function and disease. Circ Res 100:1128–1141
32. Castro Dias M, Mapunda JA, Vladymyrov M et al (2019) Structure and junctional complexes of endothelial, epithelial and glial brain barriers. Int J Mol Sci 20:5372
33. Stratman AN, Malotte KM, Mahan RD et al (2009) Pericyte recruitment during vasculogenic tube assembly stimulates endothelial basement membrane matrix formation. Blood 114:5091–5101
34. Menezes MJ, Mcclenahan FK, Leiton CV et al (2014) The extracellular matrix protein laminin alpha2 regulates the maturation and function of the blood-brain barrier. J Neurosci 34:15260–15280
35. Xu L, Nirwane A, Yao Y (2019) Basement membrane and blood-brain barrier. Stroke Vasc Neurol 4:78–82
36. Ceafalan LC, Fertig TE, Gheorghe TC et al (2019) Age-related ultrastructural changes of the basement membrane in the mouse blood-brain barrier. J Cell Mol Med 23:819–827
37. Robel S, Mori T, Zoubaa S et al (2009) Conditional deletion of beta1-integrin in astroglia causes partial reactive gliosis. Glia 57:1630–1647
38. Bonkowski D, Katyshev V, Balabanov RD et al (2011) The cns microvascular pericyte: Pericyte-astrocyte crosstalk in the regulation of tissue survival. Fluids Barriers CNS 8:8

39. Smyth LCD, Rustenhoven J, Scotter EL et al (2018) Markers for human brain pericytes and smooth muscle cells. J Chem Neuroanat 92: 48–60
40. Armulik A, Genove G, Betsholtz C (2011) Pericytes: developmental, physiological, and pathological perspectives, problems, and promises. Dev Cell 21:193–215
41. Geevarghese A, Herman IM (2014) Pericyte-endothelial crosstalk: implications and opportunities for advanced cellular therapies. Transl Res 163:296–306
42. Daneman R, Zhou L, Kebede AA et al (2010) Pericytes are required for blood-brain barrier integrity during embryogenesis. Nature 468: 562–566
43. Gaengel K, Genove G, Armulik A et al (2009) Endothelial-mural cell signaling in vascular development and angiogenesis. Arterioscler Thromb Vasc Biol 29:630–638
44. Hall CN, Reynell C, Gesslein B et al (2014) Capillary pericytes regulate cerebral blood flow in health and disease. Nature 508:55–60
45. Korn C, Augustin HG (2015) Mechanisms of vessel pruning and regression. Dev Cell 34: 5–17
46. Armulik A, Genove G, Mae M et al (2010) Pericytes regulate the blood-brain barrier. Nature 468:557–561
47. Gundersen GA, Vindedal GF, Skare O et al (2014) Evidence that pericytes regulate aquaporin-4 polarization in mouse cortical astrocytes. Brain Struct Funct 219: 2181–2186
48. Attwell D, Mishra A, Hall CN et al (2016) What is a pericyte? J Cereb Blood Flow Metab 36:451–455
49. Grant RI, Hartmann DA, Underly RG et al (2019) Organizational hierarchy and structural diversity of microvascular pericytes in adult mouse cortex. J Cereb Blood Flow Metab 39:411–425
50. Berthiaume AA, Hartmann DA, Majesky MW et al (2018) Pericyte structural remodeling in cerebrovascular health and homeostasis. Front Aging Neurosci 10:210
51. Gonzales AL, Klug NR, Moshkforoush A et al (2020) Contractile pericytes determine the direction of blood flow at capillary junctions. Proc Natl Acad Sci U S A 117:27022–27033
52. Mishra A, Reynolds JP, Chen Y et al (2016) Astrocytes mediate neurovascular signaling to capillary pericytes but not to arterioles. Nat Neurosci 19:1619–1627
53. Alarcon-Martinez L, Yilmaz-Ozcan S, Yemisci M et al (2019) Retinal ischemia induces alpha-SMA-mediated capillary pericyte contraction coincident with perivascular glycogen depletion. Acta Neuropathol Commun 7:134
54. Bandopadhyay R, Orte C, Lawrenson JG et al (2001) Contractile proteins in pericytes at the blood-brain and blood-retinal barriers. J Neurocytol 30:35–44
55. Nehls V, Drenckhahn D (1991) Heterogeneity of microvascular pericytes for smooth muscle type alpha-actin. J Cell Biol 113:147–154
56. Hartmann DA, Berthiaume AA, Grant RI et al (2021) Brain capillary pericytes exert a substantial but slow influence on blood flow. Nat Neurosci 24: 633–645
57. Arango-Lievano M, Boussadia B, De Terdonck LDT et al (2018) Topographic reorganization of cerebrovascular mural cells under seizure conditions. Cell Rep 23:1045–1059
58. Grubb S, Cai C, Hald BO et al (2020) Precapillary sphincters maintain perfusion in the cerebral cortex. Nat Commun 11:395
59. Montagne A, Nikolakopoulou AM, Zhao Z et al (2018) Pericyte degeneration causes white matter dysfunction in the mouse central nervous system. Nat Med 24:326–337
60. Cai C, Fordsmann JC, Jensen SH et al (2018) Stimulation-induced increases in cerebral blood flow and local capillary vasoconstriction depend on conducted vascular responses. Proc Natl Acad Sci U S A 115:E5796–E5804
61. Rungta RL, Chaigneau E, Osmanski BF et al (2018) Vascular compartmentalization of functional hyperemia from the synapse to the pia. Neuron 99:362–375 e364
62. Nelson AR, Sagare MA, Wang Y et al (2020) Channelrhodopsin excitation contracts brain pericytes and reduces blood flow in the aging mouse brain in vivo. Front Aging Neurosci 12:108
63. Profaci CP, Munji RN, Pulido RS et al (2020) The blood-brain barrier in health and disease: Important unanswered questions. J Exp Med 217:e20190062
64. Proebstl D, Voisin MB, Woodfin A et al (2012) Pericytes support neutrophil subendothelial cell crawling and breaching of venular walls in vivo. J Exp Med 209:1219–1234
65. Sims DE, Miller FN, Donald A et al (1990) Ultrastructure of pericytes in early stages of histamine-induced inflammation. J Morphol 206:333–342
66. Damisah EC, Hill RA, Tong L et al (2017) A fluoro-Nissl dye identifies pericytes as distinct vascular mural cells during in vivo brain imaging. Nat Neurosci 20:1023–1032
67. Attwell D, Buchan AM, Charpak S et al (2010) Glial and neuronal control of brain blood flow. Nature 468:232–243

68. Iadecola C, Nedergaard M (2007) Glial regulation of the cerebral microvasculature. Nat Neurosci 10:1369–1376
69. Christ GJ, Spray DC, El-Sabban M et al (1996) Gap junctions in vascular tissues. Evaluating the role of intercellular communication in the modulation of vasomotor tone. Circ Res 79:631–646
70. Welsh DG, Tran CHT, Hald BO et al (2018) The conducted vasomotor response: function, biophysical basis, and pharmacological control. Annu Rev Pharmacol Toxicol 58: 391–410
71. Seppey D, Sauser R, Koenigsberger M et al (2010) Intercellular calcium waves are associated with the propagation of vasomotion along arterial strips. Am J Physiol Heart Circ Physiol 298:H488–H496
72. Aldea R, Weller RO, Wilcock DM et al (2019) Cerebrovascular smooth muscle cells as the drivers of intramural periarterial drainage of the brain. Front Aging Neurosci 11:1
73. Mathiisen TM, Lehre KP, Danbolt NC et al (2010) The perivascular astroglial sheath provides a complete covering of the brain microvessels: an electron microscopic 3d reconstruction. Glia 58:1094–1103
74. Sofroniew MV, Vinters HV (2010) Astrocytes: biology and pathology. Acta Neuropathol 119:7–35
75. Batiuk MY, Martirosyan A, Wahis J et al (2020) Identification of region-specific astrocyte subtypes at single cell resolution. Nat Commun 11:1220
76. Bayraktar OA, Bartels T, Holmqvist S et al (2020) Astrocyte layers in the mammalian cerebral cortex revealed by a single-cell in situ transcriptomic map. Nat Neurosci 23: 500–509
77. Chai H, Diaz-Castro B, Shigetomi E et al (2017) Neural circuit-specialized astrocytes: transcriptomic, proteomic, morphological, and functional evidence. Neuron 95: 531–549 e539
78. John Lin CC, Yu K, Hatcher A et al (2017) Identification of diverse astrocyte populations and their malignant analogs. Nat Neurosci 20: 396–405
79. Tsai HH, Li H, Fuentealba LC et al (2012) Regional astrocyte allocation regulates cns synaptogenesis and repair. Science 337: 358–362
80. Farmer WT, Abrahamsson T, Chierzi S et al (2016) Neurons diversify astrocytes in the adult brain through sonic hedgehog signaling. Science 351:849–854
81. Lee ML, Martinez-Lozada Z, Krizman EN et al (2017) Brain endothelial cells induce astrocytic expression of the glutamate transporter glt-1 by a notch-dependent mechanism. J Neurochem 143:489–506
82. Mi H, Haeberle H, Barres BA (2001) Induction of astrocyte differentiation by endothelial cells. J Neurosci 21:1538–1547
83. Hu X, Qin S, Huang X et al (2019) Region-restrict astrocytes exhibit heterogeneous susceptibility to neuronal reprogramming. Stem Cell Rep 12:290–304
84. Ziemens D, Oschmann F, Gerkau NJ et al (2019) Heterogeneity of activity-induced sodium transients between astrocytes of the mouse hippocampus and neocortex: mechanisms and consequences. J Neurosci 39: 2620–2634
85. Rusnakova V, Honsa P, Dzamba D et al (2013) Heterogeneity of astrocytes: from development to injury - single cell gene expression. PLoS One 8:e69734
86. Scemes E, Spray DC (2003) The astrocyte syncytium non-neural cells in the nervous system. Elsevier, pp 165–179
87. Araque A, Carmignoto G, Haydon PG et al (2014) Gliotransmitters travel in time and space. Neuron 81:728–739
88. Boulay AC, Saubamea B, Adam N et al (2017) Translation in astrocyte distal processes sets molecular heterogeneity at the gliovascular interface. Cell Discov 3:17005
89. Li L, Acioglu C, Heary RF et al (2021) Role of astroglial toll-like receptors (TLRs) in central nervous system infections, injury and neurodegenerative diseases. Brain Behav Immun 91:740–755
90. Yao Y, Chen ZL, Norris EH et al (2014) Astrocytic laminin regulates pericyte differentiation and maintains blood brain barrier integrity. Nat Commun 5:3413
91. Garcia CM, Darland DC, Massingham LJ et al (2004) Endothelial cell-astrocyte interactions and TGF beta are required for induction of blood-neural barrier properties. Brain Res Dev Brain Res 152:25–38
92. Kacem K, Lacombe P, Seylaz J et al (1998) Structural organization of the perivascular astrocyte endfeet and their relationship with the endothelial glucose transporter: a confocal microscopy study. Glia 23:1–10
93. Amiry-Moghaddam M, Williamson A, Palomba M et al (2003) Delayed k+ clearance associated with aquaporin-4 mislocalization: phenotypic defects in brains of alpha-syntrophin-null mice. Proc Natl Acad Sci U S A 100:13615–13620

94. Ishii M, Horio Y, Tada Y et al (1997) Expression and clustered distribution of an inwardly rectifying potassium channel, kab-2/kir4.1, on mammalian retinal Muller cell membrane: their regulation by insulin and laminin signals. J Neurosci 17:7725–7735
95. Manley GT, Fujimura M, Ma T et al (2000) Aquaporin-4 deletion in mice reduces brain edema after acute water intoxication and ischemic stroke. Nat Med 6:159–163
96. Bazargani N, Attwell D (2016) Astrocyte calcium signaling: the third wave. Nat Neurosci 19:182–189
97. Covelo A, Araque A (2018) Neuronal activity determines distinct gliotransmitter release from a single astrocyte. elife 7:e32237
98. Mederos S, Sanchez-Puelles C, Esparza J et al (2021) Gabaergic signaling to astrocytes in the prefrontal cortex sustains goal-directed behaviors. Nat Neurosci 24:82–92
99. Shigetomi E, Jackson-Weaver O, Huckstepp RT et al (2013) Trpa1 channels are regulators of astrocyte basal calcium levels and long-term potentiation via constitutive d-serine release. J Neurosci 33:10143–10153
100. Pacholko AG, Wotton CA, Bekar LK (2020) Astrocytes-the ultimate effectors of long-range neuromodulatory networks? Front Cell Neurosci 14:581075
101. Rouach N, Koulakoff A, Abudara V et al (2008) Astroglial metabolic networks sustain hippocampal synaptic transmission. Science 322:1551–1555
102. Murphy-Royal C, Johnston AD, Boyce AKJ et al (2020) Stress gates an astrocytic energy reservoir to impair synaptic plasticity. Nat Commun 11:2014
103. Clasadonte J, Scemes E, Wang Z et al (2017) Connexin 43-mediated astroglial metabolic networks contribute to the regulation of the sleep-wake cycle. Neuron 95:1365–1380 e1365
104. Mishra A (2017) Binaural blood flow control by astrocytes: listening to synapses and the vasculature. J Physiol 595:1885–1902
105. Rosenegger DG, Tran CH, Wamsteeker Cusulin JI et al (2015) Tonic local brain blood flow control by astrocytes independent of phasic neurovascular coupling. J Neurosci 35:13463–13474
106. Kim KJ, Iddings JA, Stern JE et al (2015) Astrocyte contributions to flow/pressure-evoked parenchymal arteriole vasoconstriction. J Neurosci 35:8245–8257
107. Kur J, Newman EA (2014) Purinergic control of vascular tone in the retina. J Physiol 592: 491–504
108. Biesecker KR, Srienc AI, Shimoda AM et al (2016) Glial cell calcium signaling mediates capillary regulation of blood flow in the retina. J Neurosci 36:9435–9445
109. Institoris A, Rosenegger DG, Gordon GR (2015) Arteriole dilation to synaptic activation that is sub-threshold to astrocyte endfoot ca2+ transients. J Cereb Blood Flow Metab 35:1411–1415
110. Gu X, Chen W, Volkow ND et al (2018) Synchronized astrocytic ca(2+) responses in neurovascular coupling during somatosensory stimulation and for the resting state. Cell Rep 23:3878–3890
111. Iliff JJ, Wang M, Liao Y et al (2012) A paravascular pathway facilitates CSF flow through the brain parenchyma and the clearance of interstitial solutes, including amyloid beta. Sci Transl Med 4:147ra111
112. Liu DX, He X, Wu D et al (2017) Continuous theta burst stimulation facilitates the clearance efficiency of the glymphatic pathway in a mouse model of sleep deprivation. Neurosci Lett 653:189–194
113. Kress BT, Iliff JJ, Xia M et al (2014) Impairment of paravascular clearance pathways in the aging brain. Ann Neurol 76: 845–861
114. Thion MS, Ginhoux F, Garel S (2018) Microglia and early brain development: an intimate journey. Science 362:185–189
115. Coelho-Santos V, Shih AY (2020) Postnatal development of cerebrovascular structure and the neurogliovascular unit. Wiley Interdiscip Rev Dev Biol 9:e363
116. Rymo SF, Gerhardt H, Wolfhagen Sand F et al (2011) A two-way communication between microglial cells and angiogenic sprouts regulates angiogenesis in aortic ring cultures. PLoS One 6:e15846
117. Fantin A, Vieira JM, Gestri G et al (2010) Tissue macrophages act as cellular chaperones for vascular anastomosis downstream of VEGF-mediated endothelial tip cell induction. Blood 116:829–840
118. Mondo E, Becker SC, Kautzman AG et al (2020) A developmental analysis of juxtavascular microglia dynamics and interactions with the vasculature. J Neurosci 40: 6503–6521
119. Tanaka T, Ueno M, Yamashita T (2009) Engulfment of axon debris by microglia requires p38 MAPK activity. J Biol Chem 284:21626–21636
120. Fenn AM, Hall JC, Gensel JC et al (2014) Il-4 signaling drives a unique arginase+/il-1beta+ microglia phenotype and recruits

macrophages to the inflammatory cns: consequences of age-related deficits in il-4ralpha after traumatic spinal cord injury. J Neurosci 34:8904–8917
121. Liddelow SA, Guttenplan KA, Clarke LE et al (2017) Neurotoxic reactive astrocytes are induced by activated microglia. Nature 541: 481–487
122. Shinozaki Y, Shibata K, Yoshida K et al (2017) Transformation of astrocytes to a neuroprotective phenotype by microglia via p2y1 receptor downregulation. Cell Rep 19: 1151–1164
123. Haruwaka K, Ikegami A, Tachibana Y et al (2019) Dual microglia effects on blood brain barrier permeability induced by systemic inflammation. Nat Commun 10:5816
124. Jolivel V, Bicker F, Biname F et al (2015) Perivascular microglia promote blood vessel disintegration in the ischemic penumbra. Acta Neuropathol 129:279–295
125. Halder SK, Milner R (2019) A critical role for microglia in maintaining vascular integrity in the hypoxic spinal cord. Proc Natl Acad Sci U S A 116:26029–26037
126. Boche D, Perry VH, Nicoll JA (2013) Review: activation patterns of microglia and their identification in the human brain. Neuropathol Appl Neurobiol 39:3–18
127. Ronaldson PT, Davis TP (2020) Regulation of blood-brain barrier integrity by microglia in health and disease: a therapeutic opportunity. J Cereb Blood Flow Metab :271678X20951995
128. Thurgur H, Pinteaux E (2019) Microglia in the neurovascular unit: blood-brain barrier-microglia interactions after central nervous system disorders. Neuroscience 405:55–67
129. Xing C, Li W, Deng W et al (2018) A potential gliovascular mechanism for microglial activation: differential phenotypic switching of microglia by endothelium versus astrocytes. J Neuroinflammation 15:143
130. Osipova ED, Semyachkina-Glushkovskaya OV, Morgun AV et al (2018) Gliotransmitters and cytokines in the control of blood-brain barrier permeability. Rev Neurosci 29: 567–591
131. Logothetis NK (2003) The underpinnings of the bold functional magnetic resonance imaging signal. J Neurosci 23:3963–3971
132. Raichle ME (1998) Behind the scenes of functional brain imaging: a historical and physiological perspective. Proc Natl Acad Sci U S A 95:765–772
133. Fox PT, Mintun MA, Raichle ME et al (1986) Mapping human visual cortex with positron emission tomography. Nature 323:806–809
134. Frackowiak RS, Friston KJ (1994) Functional neuroanatomy of the human brain: positron emission tomography--a new neuroanatomical technique. J Anat 184(Pt 2):211–225
135. Mace E, Montaldo G, Osmanski BF et al (2013) Functional ultrasound imaging of the brain: theory and basic principles. IEEE Trans Ultrason Ferroelectr Freq Control 60: 492–506
136. Schulz K, Sydekum E, Krueppel R et al (2012) Simultaneous bold fMRI and fiber-optic calcium recording in rat neocortex. Nat Methods 9:597–602
137. Jiao H, Wang Z, Liu Y et al (2011) Specific role of tight junction proteins claudin-5, occludin, and zo-1 of the blood-brain barrier in a focal cerebral ischemic insult. J Mol Neurosci 44:130–139
138. Yang Y, Estrada EY, Thompson JF et al (2007) Matrix metalloproteinase-mediated disruption of tight junction proteins in cerebral vessels is reversed by synthetic matrix metalloproteinase inhibitor in focal ischemia in rat. J Cereb Blood Flow Metab 27: 697–709
139. Sweeney MD, Sagare AP, Zlokovic BV (2018) Blood-brain barrier breakdown in Alzheimer disease and other neurodegenerative disorders. Nat Rev Neurol 14:133–150
140. Li Y, Li M, Yang L et al (2019) The relationship between blood-brain barrier permeability and enlarged perivascular spaces: a cross-sectional study. Clin Interv Aging 14: 871–878
141. Krueger M, Bechmann I, Immig K et al (2015) Blood-brain barrier breakdown involves four distinct stages of vascular damage in various models of experimental focal cerebral ischemia. J Cereb Blood Flow Metab 35:292–303
142. Halliday MR, Rege SV, Ma Q et al (2016) Accelerated pericyte degeneration and blood-brain barrier breakdown in apolipoprotein e4 carriers with Alzheimer's disease. J Cereb Blood Flow Metab 36:216–227
143. Kisler K, Nelson AR, Rege SV et al (2017) Pericyte degeneration leads to neurovascular uncoupling and limits oxygen supply to brain. Nat Neurosci 20:406–416
144. Watson AN, Berthiaume AA, Faino AV et al (2020) Mild pericyte deficiency is associated with aberrant brain microvascular flow in aged PDGFRbeta(+/−) mice. J Cereb Blood Flow Metab 40:2387–2400

145. Chen A, Akinyemi RO, Hase Y et al (2016) Frontal white matter hyperintensities, clasmatodendrosis and gliovascular abnormalities in ageing and post-stroke dementia. Brain 139: 242–258
146. Early AN, Gorman AA, Van Eldik LJ et al (2020) Effects of advanced age upon astrocyte-specific responses to acute traumatic brain injury in mice. J Neuroinflammation 17: 115
147. Yamashita T, Kamiya T, Deguchi K et al (2009) Dissociation and protection of the neurovascular unit after thrombolysis and reperfusion in ischemic rat brain. J Cereb Blood Flow Metab 29:715–725
148. Fukuda S, Fini CA, Mabuchi T et al (2004) Focal cerebral ischemia induces active proteases that degrade microvascular matrix. Stroke 35:998–1004
149. Zlokovic BV (2008) The blood-brain barrier in health and chronic neurodegenerative disorders. Neuron 57:178–201
150. Ivens S, Kaufer D, Flores LP et al (2007) Tgf-beta receptor-mediated albumin uptake into astrocytes is involved in neocortical epileptogenesis. Brain 130:535–547
151. Piao CS, Holloway AL, Hong-Routson S et al (2019) Depression following traumatic brain injury in mice is associated with down-regulation of hippocampal astrocyte glutamate transporters by thrombin. J Cereb Blood Flow Metab 39:58–73
152. Schachtrup C, Ryu JK, Helmrick MJ et al (2010) Fibrinogen triggers astrocyte scar formation by promoting the availability of active TGF-beta after vascular damage. J Neurosci 30:5843–5854
153. Sofroniew MV (2020) Astrocyte reactivity: subtypes, states, and functions in cns innate immunity. Trends Immunol 41:758–770
154. Dong Y, Benveniste EN (2001) Immune function of astrocytes. Glia 36:180–190
155. Kim RY, Hoffman AS, Itoh N et al (2014) Astrocyte CCL2 sustains immune cell infiltration in chronic experimental autoimmune encephalomyelitis. J Neuroimmunol 274: 53–61
156. Moreno M, Bannerman P, Ma J et al (2014) Conditional ablation of astroglial CCL2 suppresses CNS accumulation of M1 macrophages and preserves axons in mice with MOG peptide EAE. J Neurosci 34: 8175–8185
157. Cekanaviciute E, Dietrich HK, Axtell RC et al (2014) Astrocytic TGF-beta signaling limits inflammation and reduces neuronal damage during central nervous system toxoplasma infection. J Immunol 193:139–149
158. Anderson MA, Burda JE, Ren Y et al (2016) Astrocyte scar formation aids central nervous system axon regeneration. Nature 532: 195–200
159. Faulkner JR, Herrmann JE, Woo MJ et al (2004) Reactive astrocytes protect tissue and preserve function after spinal cord injury. J Neurosci 24:2143–2155
160. Herrmann JE, Imura T, Song B et al (2008) Stat3 is a critical regulator of astrogliosis and scar formation after spinal cord injury. J Neurosci 28:7231–7243
161. Clarke LE, Liddelow SA, Chakraborty C et al (2018) Normal aging induces a1-like astrocyte reactivity. Proc Natl Acad Sci U S A 115: E1896–E1905
162. Zamanian JL, Xu L, Foo LC et al (2012) Genomic analysis of reactive astrogliosis. J Neurosci 32:6391–6410
163. Brown WR, Thore CR (2011) Review: cerebral microvascular pathology in ageing and neurodegeneration. Neuropathol Appl Neurobiol 37:56–74
164. Iturria-Medina Y, Sotero RC, Toussaint PJ et al (2016) Early role of vascular dysregulation on late-onset Alzheimer's disease based on multifactorial data-driven analysis. Nat Commun 7:11934
165. Korte N, Nortley R, Attwell D (2020) Cerebral blood flow decrease as an early pathological mechanism in Alzheimer's disease. Acta Neuropathol 140:793–810
166. Salinet AS, Robinson TG, Panerai RB (2015) Effects of cerebral ischemia on human neurovascular coupling, CO2 reactivity, and dynamic cerebral autoregulation. J Appl Physiol (1985) 118:170–177
167. Vermeer SE, Hollander M, Van Dijk EJ et al (2003) Silent brain infarcts and white matter lesions increase stroke risk in the general population: the Rotterdam scan study. Stroke 34: 1126–1129

Chapter 2

In Vitro Models of the Blood–Brain Barrier

Snehal Raut, Aditya Bhalerao, Behnam Noorani, and Luca Cucullo

Abstract

Traditional in vitro models can replicate many essential features of drug transport/permeability across the blood-brain barrier (BBB) but are not entirely projecting in vivo central nervous system (CNS) uptake. Species differences fail to translate experimental therapeutics from the research laboratory to the clinic. Improved in vitro modeling of human BBB is vital for both CNS drug discovery and delivery. High-end human BBB models fabricated by microfluidic technologies offer some solutions to this problem. BBB's complex physiological microenvironment has been established by increasing device complexity in terms of multiple cells, dynamic conditions, and 3D designs. It is now possible to predict the therapeutic effects of a candidate drug and identify new druggable targets by studying multicellular interactions using the advanced in vitro BBB models. This chapter reviews the current as well as an ideal in vitro model of the BBB.

Key words Cerebrovascular, Alternative, Barrier, TEER, Endothelium, iPSC, Models, Artificial, Cells, Shear stress

1 Introduction: What Is the Blood–Brain Barrier?

The blood–brain barrier (BBB) is a highly efficient and dynamic interface that envelops blood vessels in most parts of the brain (Fig. 1) [1]. It serves as a highly selective barrier between the circulatory system and the brain, thus preventing unwanted blood–borne substances from crossing over [2]. The BBB's structural and functional units are specialized endothelial cells (EC), which form the interior surface of cerebral blood vessels [3]. Blood vessels in other parts of the body have fenestrations or small openings between ECs that allow the exchange of molecules across tissues. However, cerebral ECs are characterized by tight junctions that closely regulate diffusion across the blood vessel [4].

The primary components of the blood-brain barrier (BBB) are the brain microvascular endothelial cells (ECs), astrocytes, and pericytes, as shown in Fig. 2 [5]. Other cellular elements like neurons and microglia play an essential role in maintaining BBB phenotype and function. All these cells together form a

Nicole Stone (ed.), *The Blood-Brain Barrier: Methods and Protocols*,
Methods in Molecular Biology, vol. 2492, https://doi.org/10.1007/978-1-0716-2289-6_2,

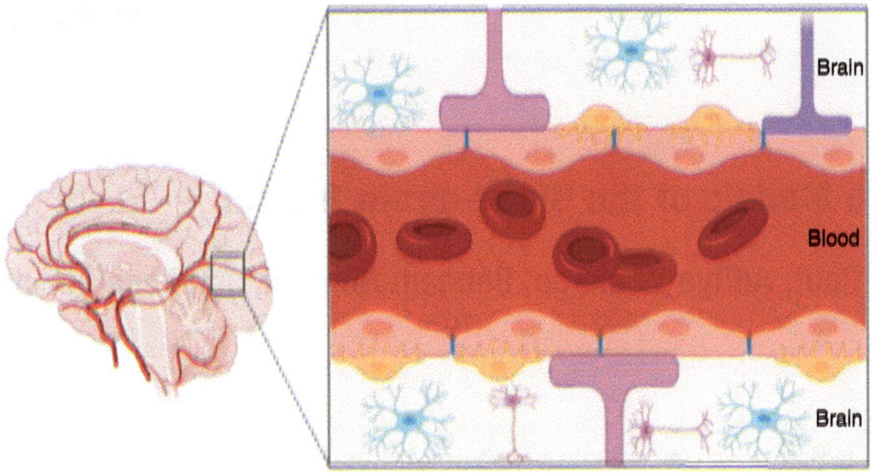

Fig. 1 A cross section of brain microvessel: The capillary basement membrane, as well as endothelial cells, astrocytic endfeets, and brain pericytes abluminal to endothelial cells, share a continuous basement membrane

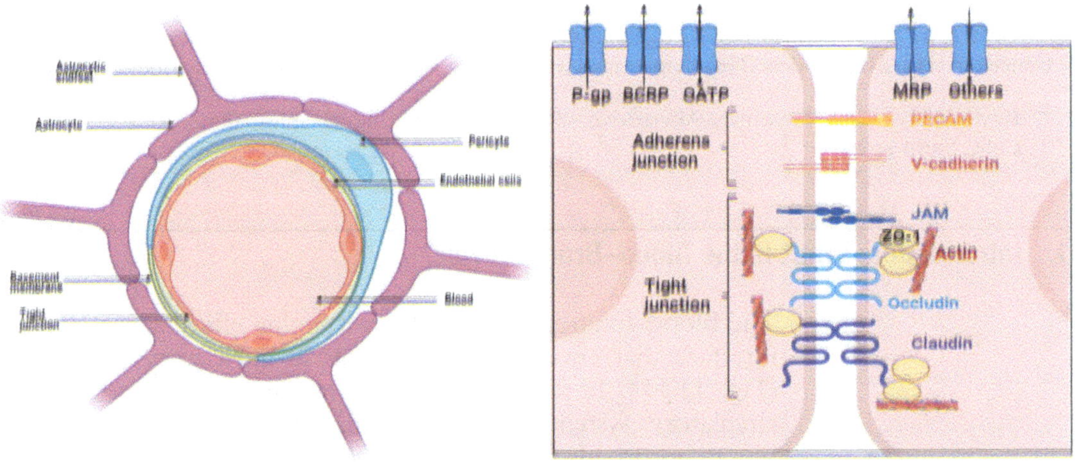

Fig. 2 Schematic view of a multimodal barrier system on the blood-brain barrier (BBB): Many adhesive molecules are integrated into cellular components, which are functionally regulated by ligand–receptor systems

neurovascular unit (NVU). The BBB ECs are surrounded by the basement membrane (BM), composed of various proteins and proteoglycans. The BM is thought to play an essential role in maintaining barrier function; however, the exact mechanisms are unknown.

Pericytes are embedded within the basement membrane (BM), covering approximately 22–32% of the endothelium. They act as chemical sensors to facilitate communication between neurons and the CNS vasculature to meet the brain's energy requirements. Pericytes also regulate endothelial cell proliferation and play an essential role in angiogenesis, inflammatory processes [6]. Astrocytes are star-shaped glial cells with astrocytic endfeet covering a significant part of the endothelial surface [7]. Astrocytes have a unique function in ionic, amino acid, and water homeostasis of the brain. Another essential component of the BBB is the extracellular matrix (ECM), which is the extracellular space that comprises 15–30% of the brain volume. It is mainly composed of brain interstitial fluid and hyaluronic acid-derived matrix. It provides structural support and facilitates multicellular dynamics. Apart from these cellular and biochemical components, several biomechanical factors such as shear stress, 3D microenvironment, flow rate are essential for BBB integrity [8].

2 Why Is There a Need for Modeling the BBB?

Disorders of the central nervous system (CNS), including Alzheimer's, Parkinson's, and brain tumors, affect millions and make day-to-day life challenging for patients and caregivers. Existing treatments for these disorders can only relieve symptoms temporarily and do not cure the disease itself [9]. Why haven't effective therapies been developed for CNS diseases? The peculiar nature of BBB is the answer to this question. This natural barrier also restricts, among other molecules, the entry of drugs to treat these diseases [10]. There is no perfect system for screening drugs for brain diseases. Drug development for CNS diseases requires drug molecules to cross the BBB and show therapeutic effects at the action site. The complex structure of the brain results in longer drug development times and an increased risk of failure.

In vivo, BBB modeling using laboratory mice provided significant advances in obtaining physiological information on BBB transport mechanisms within the cerebrovascular microenvironment [11]. The entire experiment takes place under a natural environment and can generate consistent data. However, animal models used for lab research are expensive and animal experimentation is subject to ethical concerns. Moreover, drug candidates that were first tested based on animal studies repeatedly failed to translate in humans due to the interspecies differences between the BBB in animals and humans [12].

Several in vitro BBB models were established and tested for their accurate functioning [13]. The human in vitro BBB models are critical for studying pathological and physiological mechanisms of the BBB and identify therapeutic formulations (e.g., drugs, antibodies, nanoparticles) that can cross this biological barrier [14, 15].

In vitro models have several essential advantages:

- Maintain microenvironment throughout the experiment and reproducible.
- Allow detailed mechanistic analyses.
- Accelerate pharmaceutical procedures.
- Limit ethical constraints save time and money needed for ethical approval procedures.

The most important question is how efficiently in vitro models mimic the complexity of in vivo. In vitro BBB models that are currently available are listed in Table 1 (Fig. 3) [16].

3 Modeling the BBB In Vitro: What Components Are Needed?

CNS drug discovery or development studies must be mindful of the intricacies associated with the BBB. To attain clinically relevant CNS exposure levels to putative drug candidates, it is crucial to determine the BBB permeability of drugs within the CNS, which is usually very low [17]. Selecting the best BBB modeling methodology is typically a compromise between capacity, cost, time, and how closely a model needs to mimic in vivo conditions [18].

There are significant variations between the various BBB cell culture models. Still, there are specific requirements common to all that must be addressed for the model to be as functional and precise, as well as being as close to the in vivo BBB phenotype as possible. Thus, a "good" in vitro BBB model should:

- Display functional tight junctions between ECs.
- Have low paracellular permeability resulting from high transendothelial electrical resistance (TEER).
- Exhibit selective permeability to ions (Na^+, K^+, etc.) and other molecules.
- Express specific BBB phenotypic transporters and transcytotic activity [19].
- Display physiologically relevant morphology with all the cellular components of the BBB, etc.
- Have ease of culture and low cost.
- Availability of the desired cell type, including primary and stem cell-differentiated cultures.

3.1 Primary and Stem Cell-Based Cultures in BBB Modeling

Ferenc Joo and his co-workers developed isolated brain capillaries from rat brains as an in vitro BBB model system [20]. Brain microvessels were isolated by various filtration and/or density gradient centrifugation techniques, mechanical homogenization, and enzymatic digestion. The preferential sources of primary cerebral endothelial cells were rats, dogs, bovines, pigs, and primates [21]. This

Table 1
Current 2D and 3D in vitro models of BBB/NVU with their advantages and disadvantages [78, 79]

2D models		BBB models	3D models	
BMEC monolayer	Transwell co-culture	DIV-BBB	Microfluidic model	BBB-on-a-chip
		Advantages		
Drug transport HTS assay Molecular cell analysis	Improved phenotype, cell–cell interaction, barrier properties Ease of establishing cultures, moderate scalability, and low cost	Physiological cell distribution Presence of flow and shear stress Low permeability to intraluminal polar molecules High TEER Expression of specialized transporters, ion channels, and efflux systems	Microvascular channels Physiological shear stress More realistic BBB dimensions and geometries Real-time imaging and analysis Flexibility in the design, low-cost fabrication Built-in multicellular network	Microvascular channels Built-in multicellular network Real-time analysis Whole organ-on-a-chip Better replicates the in vivo microenvironment
		Disadvantages		
Difficult isolation protocols Poor viability No flow and shear stress Unable to replicate key characteristics of the BBB	Lack of a three-dimensional structure No flow and shear stress Unable to replicate key characteristics of the BBB Reduced polarized transport	Large vascular bed Invisible cell-cell interaction Limited transmigration Not intended for HTS studies Requires more time and technical skills, high cell load The complex design does not allow visualization	No standardized parameters and quantification of critical experimental factors such as: luminal shear stress, TEER, selective permeability The complicated design of BBB models Limited scalability	No standardization for optimal designs and protocols

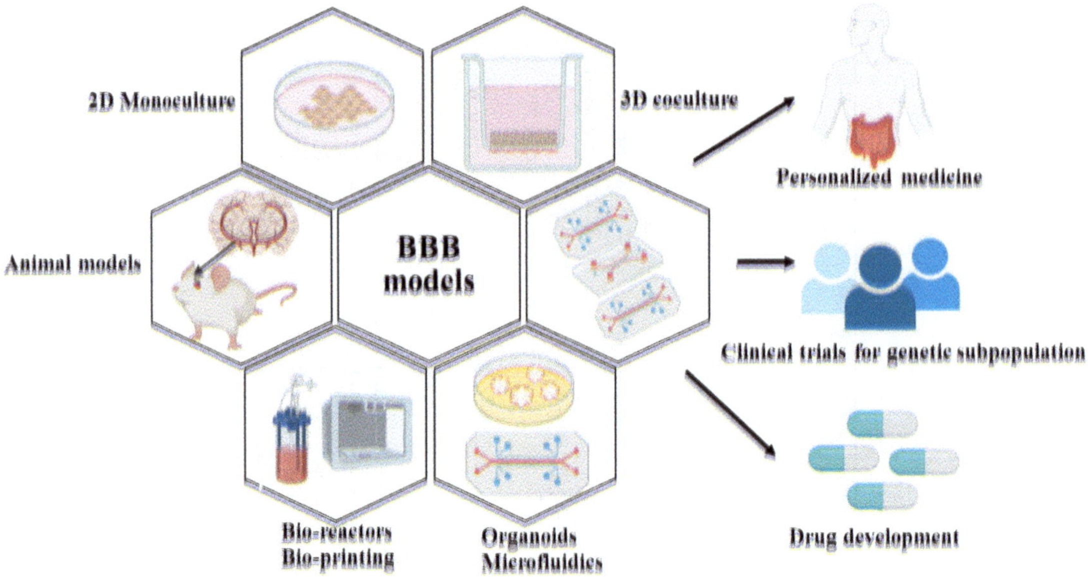

Fig. 3 Various in vitro BBB models and their application

model system provided typical BBB-specific features (cerebral endothelial receptors, transporters), leading to the discovery of many molecular events underlying endothelial cell function [22, 23]. Isolated microvessels and ECs grown out of brain capillaries are used for binding, uptake, or efflux studies and also provide valuable data on the transcriptome/proteome of BBB [24]. However, these models are not appropriate for transcellular transport studies [25] and fail to translate in humans because of interspecies differences.

The use of primary cultures of human brain ECs is ideal given the complexity and specificity of BBB transporters. However, the availability and ease of obtaining human specimens can be challenging due to ethical constraints. Moreover, preparation of primary cultures is expensive, time-consuming, and needs expertise. Primary isolation protocols do not yield pure cultures of primary ECs, microvessel viability is often low, and nonenzymatic preparation's success depends mainly on the EC's division ability in tissue culture. In general, primary cells are used at low passages to avoid downregulation of BBB characteristics. This limitation, together with low capacity, makes this experimental system less suited for BBB permeability screening in the pharmaceutical industry. Stem cell engineering may solve the problem of limited cell lines, while maintaining a phenotype close to primary human cells.

Pluripotent stem cells (PSCs) (embryonic stem (ES) cells and induced pluripotent stem cells (iPSCs) solve the problem of interspecies differences. Stem cell-based in vitro BBB models are widely used for CNS drug screening [26]. Induced pluripotent stem cells (iPSCs) technology allows somatic cells from patients, such as skin fibroblasts, to be converted into pluripotent stem cells capable of forming many different differentiated types of cells [27, 28]. The iPSCs can be differentiated into brain microvascular endothelial cells (BMECs), neurons, astrocytes, etc., thus allowing the barrier activities to be compared in vitro [29]. BBB modeling using iPSCs has significantly impacted neuroscience research and has been explored for various neurological diseases [30]. The stem cell-derived BBB model provides a scalable and reproducible human cell source, which can accomplish barrier properties like in vivo. Moreover, iPSC-based BBB models enable the use of isogenic co-cultures and create models close to in vivo [31].

Even though the iPSC-based in vitro BBB models offer significant advantages over primary cells, the longer differentiation time and dependency upon various factors make them difficult to use. Modeling using patient-derived iPSCs is hindered by the inadequate supply of patient-derived iPSC lines, matched controls, and maintenance. Moreover, derivation by genome editing approach comprises a significant problem in translating such outcomes into treatments. Current techniques such as CRISPR/Cas 9 offer exciting opportunities in CNS modeling using iPSC cell lines. A significant problem of in vitro BBB modeling is the absence of three-dimensional structures observed in vivo. Brain "organoids" provide a solution in the close representation of the brain. In organoids, BMECs, pericytes, and astrocytes can freely self-assemble to ball-shaped cellular aggregates without scaffolding. These aggregates have several shortcomings, like the absence of an extracellular matrix and lack of neuronal projections and tracts.

4 What We Got So Far: An Overview of Available In Vitro BBB Platforms

Initial studies of the BBB were performed in vivo to establish various drug molecules' permeability across the brain endothelial layer. This approach presented vital information about the BBB's structure and function. However, the BBB's further characterization, specifically at the cellular and molecular level, was hindered due to the complex natural environment present in vivo. To date, there is no perfect in vitro BBB model as none of these models imitate entirely the in vivo conditions [32]. The in vitro BBB model is carefully chosen according to the study's requirements and efficiently interprets the data. Here, we have discussed the most commonly used in vitro BBB models, including the recently established microfluidic models [33]. These models are categorized into static and dynamic BBB models based on shear stress [34]. Their characteristics and application to gain in-depth knowledge of cellular and molecular mechanisms are discussed in the following sections.

4.1 Static BBB Models

Commonly used static in vitro BBB models do not offer the shear stress typically generated in vivo by the intravascular blood flow [35]. Based on the BBB design's cell types (Table 2), these are further divided into monolayer and co-culture models [36].

4.2 Mono and Two-Dimensional In Vitro BBB Models

The most common in vitro model of the BBB created today is growing a monolayer of endothelial cells in the transwell system (Fig. 4a). This apparatus was designed based on a simplified view of the BBB. The transwell system consists of a semipermeable microporous membrane (polycarbonate or polyester) submerged in the culture medium. The bottom chamber of the transwell insert mimics the abluminal side, whereas the top surface of the insert mimics the abluminal side. The porous membrane (0.2–0.4 μm) acts as a tight BBB restricting transport between the two compartments. The attractive features of this model are its simplicity and the ability to perform many experiments simultaneously. However, BBB forming properties of ECs in vivo are induced by the surrounding environment (shear stress, astrocytes, pericytes, blood cells, etc.). The monolayer model is static and shows high permeability to sucrose with very low TEER. Another aspect to be thought of is that only the luminal side of ECs is exposed to serum proteins in vivo, whereas the abluminal side is exposed to cerebrospinal fluid (CSF). ECs grown on monodimensional porous membranes are exposed to serum on both sides. This non-physiological condition may enhance the de-differentiation process. Also, ECs grown in static monoculture generally have a short lifespan compared to flow-based culture models.

BMECs are co-cultured with other CNS cells that directly contribute to the induction, maintenance, and regulation of barrier properties. The presence of other brain cells increases transporter expression and induces cell polarity in BMECs as observed physiologically. The co-culture can be established on the transwell apparatus to facilitate cell–cell contact via astrocytic endfeet (Fig. 4b) (astrocytes and ECs are seeded on opposite sides of the porous membrane) or in a contactless set up by seeding the astrocytes at the bottom of the well and the ECs on the luminal (upper) side of the membrane (Fig. 4c) [37, 38]. The co-culture model is valuable for studying the BBB functionality and transport processes [39]. The bidimensional model establishes an environment close to in vivo compared to the monodimensional model. BMEC–astrocyte–pericyte co-culture model has also been developed and termed as a triple-culture system. This arrangement contains endothelial cells seeded on the support's upper surface, pericytes on the lower body, and astrocytes seeded on the bottom of the culture wells (Fig. 4d). A triple co-culture model is a more reliable in vitro BBB model due to the higher TEER and lower permeability [40]. However, maintaining different cell types together in a small transwell area is challenging and demands extra attention.

Table 2
Cells/cell lines used in static in vitro BBB models

Cell lines	Origin	Advantages	Disadvantages
Isolated brain capillaries	Rat Dog Bovine Pig Primate Human	Typical microvascular endothelial cells' specific features Discovery of many molecular events underlying endothelial cell function Assessment of transport mainly from the abluminal to the luminal side	The availability and ease of obtaining human specimens can be challenging The isolation of microvessels is labor-intensive and complicated Isolation protocol gives a heterogeneous mixture of ECs Microvessels' viability is often low Not suited for BBB permeability screening in the pharmaceutical industry
Immortalized brain endothelial cell lines [80] hCMEC/D3 RBE4 bEnd.3 TY08 HMEC-1 A735 C6 HMO6 HBPC/ci37 NT2 SH-SY5Y	Human Mouse Rat Human Human Human Rat Human Human Human Human	Reduces primary cells' usage and expands at will [81] Decrease cost and labor	Immortalization causes possible loss of contact inhibition and results in an immature BBB characteristic [82] Do not form rigid, tight junctions and are unsuitable for permeability studies [83]
Cell lines of non-cerebral origin MDCK (Madin-Darby canine kidney)	Canine kidney	Low paracellular permeability to sucrose, easy to culture	Significant differences in morphology, transport properties, etc., between MDCK and brain ECs [34, 84] The typology of TJs and transporters in MDCK cell lines can differ from their human counterpart
Caco-2	Human intestine	Epithelial cell line models showed suitable paracellular tightness	Significant disparities in drug permeability between the intestine and brain-derived cells
HUVECs	Human	Human origin and relatively low cost compared to other primary vascular human cells	Failed to replicate complete physiological conditions

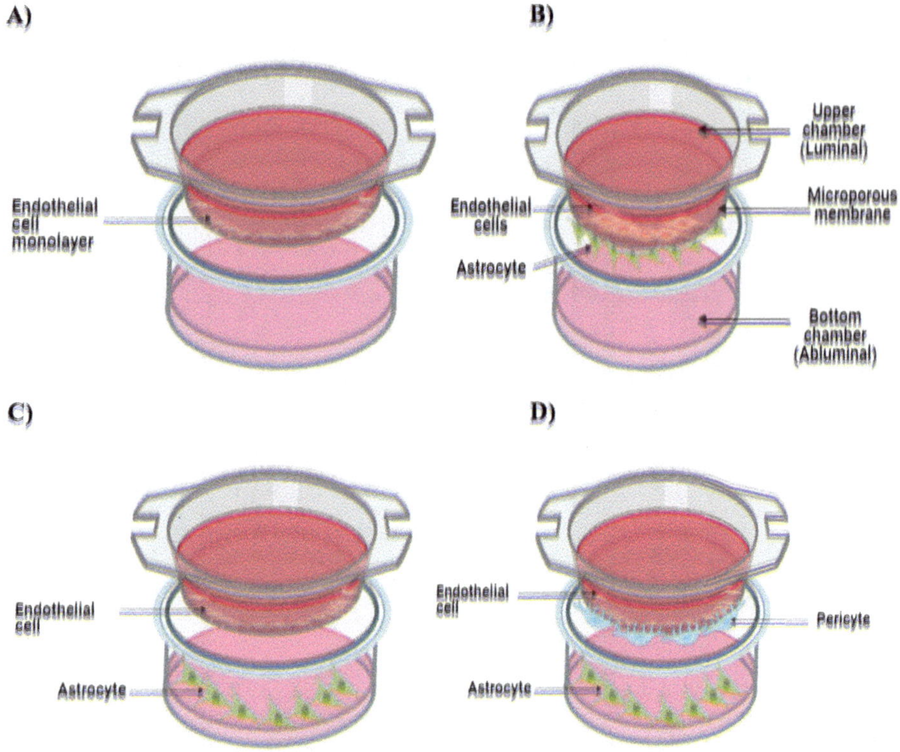

Fig. 4 Mono- and two-dimensional in vitro BBB models on the transwell apparatus: (**a**) Monoculture of endothelial cells, (**b**) co-culture of endothelial cells and astrocytes (cell–cell contact), (**c**) co-culture of endothelial cells and astrocytes (no contact), (**d**) triple-co-culture system of endothelial cells, pericytes, and astrocytes

Also, co-culture models still lack shear stress, which plays a substantial role in promoting and maintaining the EC differentiation into a BBB phenotype [41].

5 Importance of Intravascular Flow and Shear Stress in BBB Endothelial Physiology

Shear stress (SS) is a mechanical force that arises due to the shear flow of the fluid and is decided by the flow rate, vessel radius, and viscosity of the fluid [42]. Blood pressure exerts a force on the vessels' walls that imposes circumferential stress, whereas frictional hemodynamic force (blood flow) or shear stress acts parallel to the endothelial layer.

These stresses play a critical role in maintaining vascular homeostasis and vary with time, magnitude, the direction of blood flow, and anatomy [43]. ECs actively respond to shear stress by becoming polarized, migrating, and go through morphological changes, cell division, etc. [44, 45].

For an ideal Newtonian fluid, the shear stress (τ) under constant flow is given by the Poiseuille equation:

$$\tau = 4\mu Q/\pi r^3$$

where μ is the fluid viscosity, Q is the flow rate, and r is the vessel radius. Thus, the high flow rate and small diameter of a vessel expose EC cells to significant shear stress. Fluid SS ranges from 5 to 23 dyne/cm^2 in microcapillaries of the human brain vascular network. Physiological SS augments BMEC barrier function through improving the tight junctions, induces the endothelial expression of drug transporters, and strictly regulates the exchange of substances into the brain circulation [46].

6 Dynamic BBB Models

The modern opinion of the BBB has shifted from a solely anatomic concept to a more physiological and dynamic vision. This change was brought about by the proof of active regulation of transport across BBB. ECs are continuously exposed to shear stress produced by the blood flow across their apical surfaces in physiological conditions. Shear stress affects the EC structure and function. The dynamic BBB models with shear stress have been classified in three types: the cone–plate, dynamic, and microfluidic-based models.

The cone–plate apparatus was initially used to construct shear force. The model lacks astrocytes and pericytes; therefore, its application and reliability are low (Fig. 5a). Taking these shortcomings into consideration, new dynamic in vitro model was developed. The dynamic in vitro model of BBB (DIV-BBB) combines both the components, i.e., shear stress and different NVU cell types (Fig. 5b). DIV-BBB model facilitates the real-time monitoring of BBB function by TEER measurement across the barrier via electrodes introduced in the luminal and abluminal compartments. However, the model requires a relatively high cell number and enormous media volume for proper functioning, limiting its applicability. Direct imagining of the endothelial morphology is also difficult due to the cylindrical shape of the apparatus, and the DIV-BBB model is not optimized for high-throughput screening.

The more recent generation of DIV-BBBs requires fewer cells for the initial setup (approximately one-fourth of the cell number used in the standard model [47]). It is not limited to ECs and astrocyte culture since other relevant cell types can also be seeded on the abluminal chamber bottom. However, the membrane wall thickness of 150 μm is still suboptimal, and the platform still does not allow for high throughput screening.

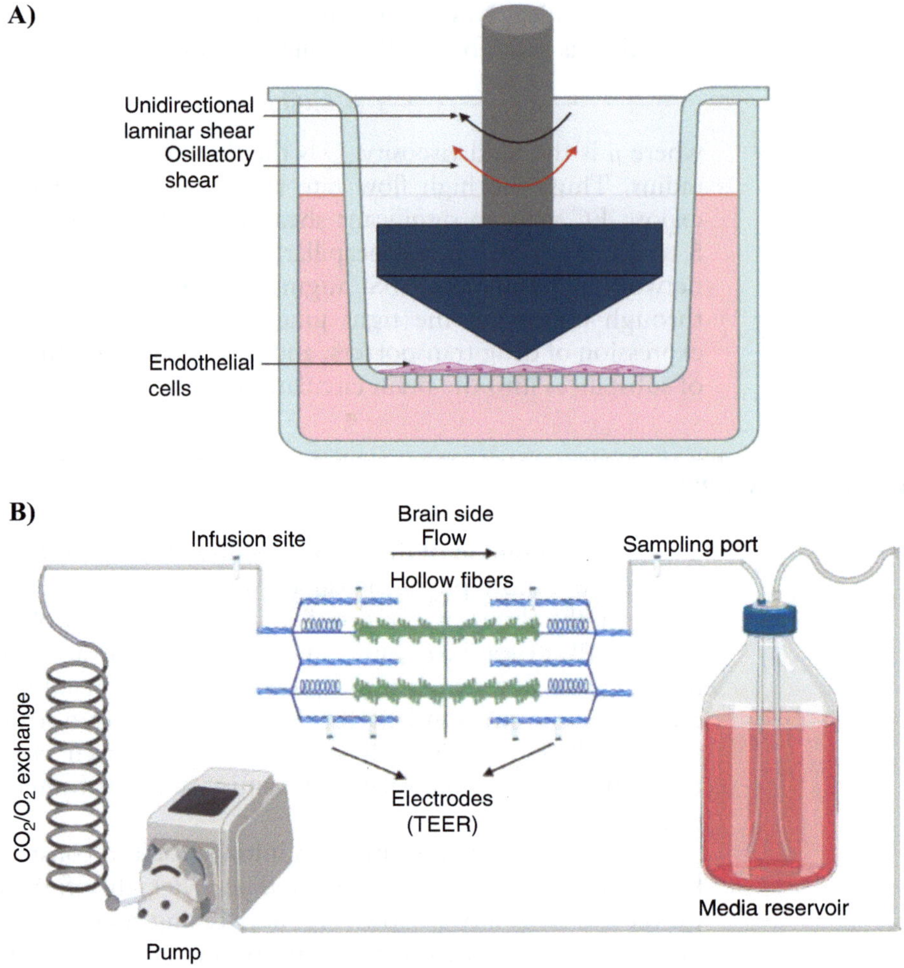

Fig. 5 Various designs of dynamic in vitro BBB models: (**a**) Cone–plate BBB apparatus, (**b**) DIV-BBB model

6.1 Microfluidic Platforms

The BBB models mentioned above have several critical shortcomings: they do not provide an ideal BBB environment, they lack of physiological size or hemodynamic shear stress, and they do not provide real-time imaging. These shortcomings prompted the development of microfluidic 3D in vitro models. These systems are produced using advanced micro-electro-mechanical systems (MEMS) technology and imitate the unique biological microenvironment observed in vivo [48]. Microfluidic designs/chips require microengineering technologies to construct channels (microvessels), chambers (luminal, abluminal), and valves on silicon, glass, quartz, or macromolecule polymer material to simulate the BBB microenvironment [48]. These channels are designed with similar width scales as brain microstructure. Microfluidics aims to recapitulate in vivo elements within scalable in vitro models by manipulating fluids at the micrometer scale [49]. These models also provide clinically relevant insights into drug discovery, molecular transport, and disease modeling [50].

6.2 Chip Fabrication

Different materials can be used to manufacture microfluidic chips. However, an appropriate choice of the fabrication material can reduce experimental cost, improve stability, sensitivity, and accuracy. Silicon is the first-generation material processed by MEMS technology to construct microfluidic structures. However, the fabrication process is usually not cost-effective, is lengthy, and requires special environmental conditions [51]. Nowadays, polydimethylsiloxane (PDMS)-based systems are preferred because they are cheaper and easier to produce [52]. PDMS is a silicon-based organic polymer that is an optically transparent, nontoxic, non-flammable, gas–water permeable, and stiff elastomer (Young's modulus of ~0.5 MPa). It is mainly known for its unusual flow properties. PDMS's transparent nature is convenient for real-time monitoring and imaging, while gas–water permeability is essential for cell culture in the microfluidic chip environment. The replica molding process (REM) is used to fabricate PDMS-based microfluidic chips and allow mass production by duplicating a mold's shape and structure.

The REM process, the form of soft lithography techniques, is a primary channel creation method in replica molding [53]. Silicon wafer, which acts as a master for the microfluidic device, is fabricated by rapid prototyping (Fig. 6). Firstly, it is photo masked with photoresist, such as SU8, and then spin-coated to cover the entire surface. The photoresist–wafer complex is then baked or UV exposed to cross-link both with silicon–oxygen bonds forming (Si–O–Si). PDMS go through hydrophobic treatment using a chemical (perfluoroalkyl trichlorosilane) and then poured onto the master mold. It prevents sticking PDMS to the wafer, which otherwise would have been ruined. It is then heated in the oven at a temperature between 60 and 70 °C to allow solidification (PDMS curing). The PDMS chips which are peeled off contain a negatively imprinted image of the microstructure. PDMS surfaces are further modified by plasma treatment to enhance cell adhesion. The microfluidic devices of different shapes are created by replica molding, which helped to take research to another level [54].

In microfluidic models, porous membranes are used for compartmentalization present in various organs, for example, the brain. These membranes separate the luminal–abluminal layers and offer the platform for different co-culturing of cells present in the NVU. Membranes such as polycarbonate (PC), polyester (PE), polyethylene terephthalate, and polytetrafluoroethylene (PTFE) based on the purpose of the study/microstructure of the organ [55]. Pore density, porosity, and membrane thickness of the porous membrane may affect cell attachment (static, dynamic conditions) and cell–cell communication on both sides of the membrane. Most membranes used in the BBB models are about 10 μm-thick with 0.2 or 0.4 μm pore size. Though PC membranes are optically translucent and compromise real-time visualization, they are still most widely used in microfluidic models.

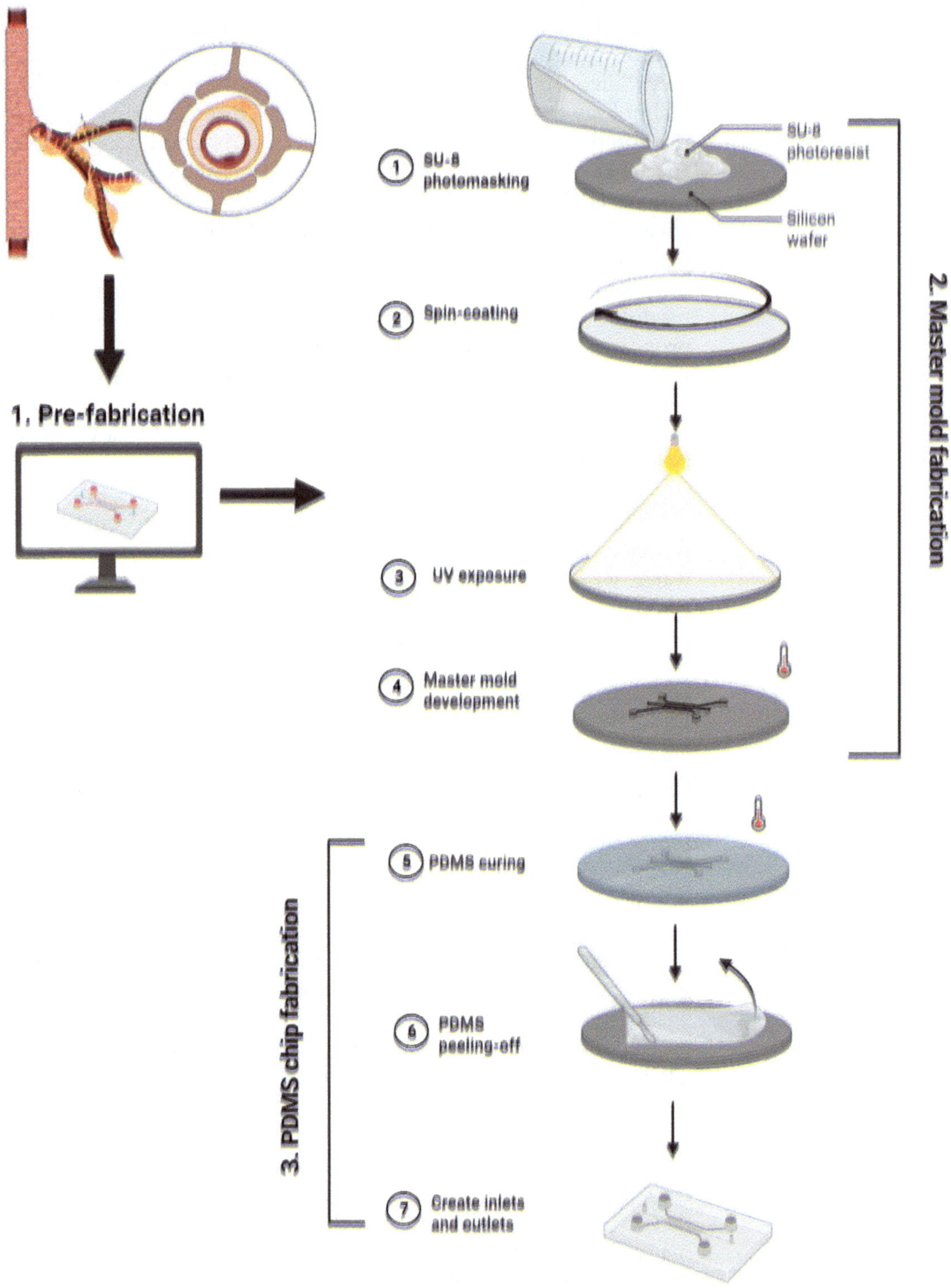

Fig. 6 Representation of the entire microfabrication process for a microfluidic PDMS chip

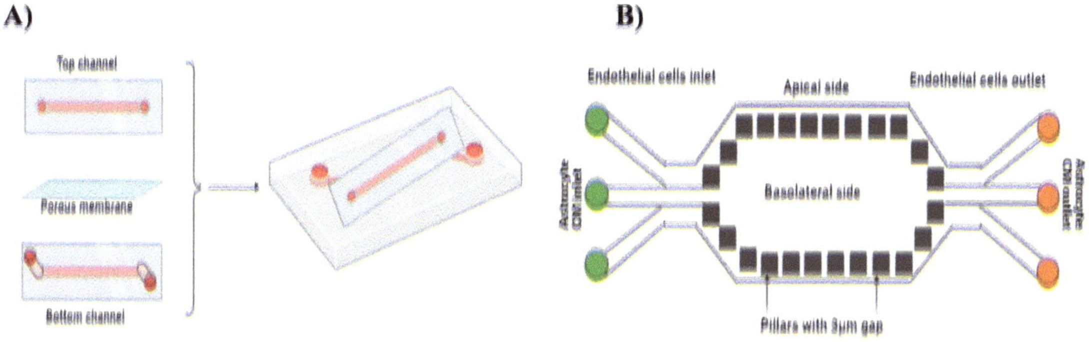

Fig. 7 Schematic illustrations of various microfluidic-based BBB designs: (**a**) Sandwich design, (**b**) SyM-BBB model

6.3 Shear Stress Generation in Microfluidic BBB Designs

Media flow creates shear stress in a microfluidic chip, as is the case for biological fluids in vivo. Studies have shown that shear stress at physiological levels affects cell growth, morphology, and cell function and regulates gene expression in a microfluidic chip [43, 56]. With rapid advances in microfluidics, the generation of shear stress with microfabricated channels has become increasingly popular. Several parallel channels with different cross-sectional measurements are connected to a single inlet for the high throughput studies. After the chip fabrication and perfusion of cells, the microfluidic device is switched to the FSS mode by circulating the culture medium. Flow rates in microfluidic models vary from 0.01 to 120 μL/min because of fluid viscosity, flow rates in different channel geometry, and the flow profile. Various microfluidic-based BBB designs (Fig. 7) with shear stress fabricated are as follows [57, 58]:

1. Sandwich design is the first μBBB model developed from the conventional transwell design which contains upper and lower PDMS channels. These channels are separated by porous polycarbonate membranes (0.2–3 μm pore diameter), like a sandwich's structure.
2. Parallel design is another evolved μBBB model, where an array of PDMS microchannels horizontally separates two aligned channels. This design replaces the polycarbonate membrane from the sandwich design with a PDMS-based microchannel membrane. The PDMS-based membrane significantly improves the assembly of μBBB devices. However, the thickness of PDMS membranes (~50 μm) is still not equivalent to the basal membrane's thickness (BM).
3. 3D tubular structure design consists of an improved collagen gel-based 3D microvascular tube structure constructed using microneedles of different diameters (75–150 μm) and was later applied in μBBB devices. This design ensured shear stress with the uniform flow and enabled co-culture with astrocytes or pericytes embedded in a collagen matrix.

4. Vasculogenesis design incorporates collagen/fibrin matrix in a channel close to ECs, enabling sprouting and new blood vessel formation. ECs gradually form vascular networks inside the fibrin hydrogel and establishes contacts with astrocytes and neurons.

6.4 Organ-on-a-Chip (OOC)

The biggest challenge in drug development is to go from a bench discovery to the lengthy approval process required for animal studies. Apart from the ethical issues, the real problem is that these preclinical animal models often do not predict humans' clinical responses. Many therapeutic antibodies and crRNA therapies only recognize human targets causing a desperate need for preclinical human models. Another health challenge is that many of the emerging diseases are zoonotic diseases where animals are often the original disease sources/vectors, for example, bats for SARS-Cov-2 [59]. The fast spread in humans and reoccurrence of the infection results in culling livestock, substantial economic costs, and significant ethical concerns. It is not even clear if the measures taken to solve the problems mentioned above are effective or not. Also, many diverse species can be the source of an infectious viral variant, making it impossible to maintain animal husbandry and housing for all relevant species. There is even a need for preclinical models that must apply to multiple different species.

To solve the challenges mentioned earlier, "human organs on chips" are developed, consisting of engineered microchips incorporating living human cells and reconstitute organ-level functions, not just cell or tissue [60–63] (Table 3). So, the ultimate goals are to replace animal testing and accelerate drug development, recapitulate human anatomy and diseases states, predict human drug responses using clinical dose exposures (pharmacokinetic profiles), develop personalized disease models, and create animal organ-on-chips that replicate species–species drug responses.

6.5 Human BBB-on-Chip

An enhanced human BBB model incorporates human iPSC-derived BMECs interfaced with primary pericytes and astrocytes (Fig. 8) [64, 65]. PDMS was used to create an upper CNS microchannel separated from a parallel vascular microchannel by a porous PET membrane (2 μm thick) coated on both sides with an ECM composed of collagen type IV, fibronectin. Astrocytes and pericytes are seeded on the PET membrane in the upper channel under static conditions in the ratio of 7:3. Human iPS-BMVECs were cultured under hypoxic conditions plated on the bottom surface of the lower channel's porous membrane [66]. The flow through the vascular channel was controlled to maintain physiological levels of shear stress (6 dyne cm^{-2} at 100 $\mu L\,h^{-1}$) and viscosity modified by adding 3.5% dextran to the medium (same as blood, 3–4 cP). With the help of microfluidic technology, the research in the neuroscience field

Table 3
Various organ-on-a-chip disease models and their application

Organ-on-a-chip	Disease condition	Drug testing
Lung-on-a-chip [85]	Lung cancer (NSCLC)	Tyrosine kinase inhibitor
Liver-on-a-chip [86–88]	Non-alcoholic fatty liver disease (NAFLD)	Pioglitazone, metformin
	Potential drug toxic effects	Acetaminophen (APAP) [89]
	Drug-induced liver injury (DILI)	Troglitazone
	Mitochondrial dysfunction	Troglitazone, rotenone
Kidney-on-a-chip [90]	Albuminuria	Adriamycin
	CysA-induced damage	Cyclosporine A
	Chronic kidney disease, urothelial cancer	Aristolochic acid I
Gut-on-a-chip [91]	Drug absorption	SN-38 (7-ethyl-10-hydroxycamptothecin)
	Gut radiation injury	DMOG (dimethyl oxaloylglycin)
	Biologically responsive to exogenous stimuli	Tumor necrosis factor-α, IFN-γ
CNS- and PNS-on-a-chip [92]	Molecular toxicology	Acetaminophen, 5-fluorouracil, retinoic acid, doxorubicin, pitavastatin
	Familial Alzheimer's disease	β-secretase inhibitor
	Motor neuron disease	Motor neuron progenitor
BBB-on-a-chip [92]	CNS disorders	Caffeine, cimetidine, doxorubicin
	Inflammatory stimulation	Lipopolysaccharide
	Brain tumor	Paclitaxel, bortezomib
Heart-on-a-chip [93–95]	Dilated cardiomyopathy	Isoproterenol, E-4031, verapamil, and metoprolol
	Low blood pressure, heart failure [96]	Norepinephrine
	Cardiac cell outgrowth correlation of pharmacological compounds	Doxorubicin, endothelin-1, isoproterenol, phenylephrine, amiodarone
Muscle-on-a-chip [97]	Duchenne muscular dystrophy	Stem cell therapy
	Multi-organ toxicity	Doxorubicin, atorvastatin, valproic acid, acetaminophen, N-acetyl-m-aminophenol
	Cardiovascular diseases	Hemodynamic force, mechanical stimuli

and related drug discovery has advanced significantly over the last 10–15 years to the point where we can use these preclinical models to obtain relevant clinical data.

6.6 Human-Derived Multi-organs-on-a-Chip

Nowadays, pharmaceutical companies spend more and more on clinical trials of drug candidate screening. Regardless of remarkable progress in drug candidate screening with computational models, traditional in vitro models, animal models, most recent humanized

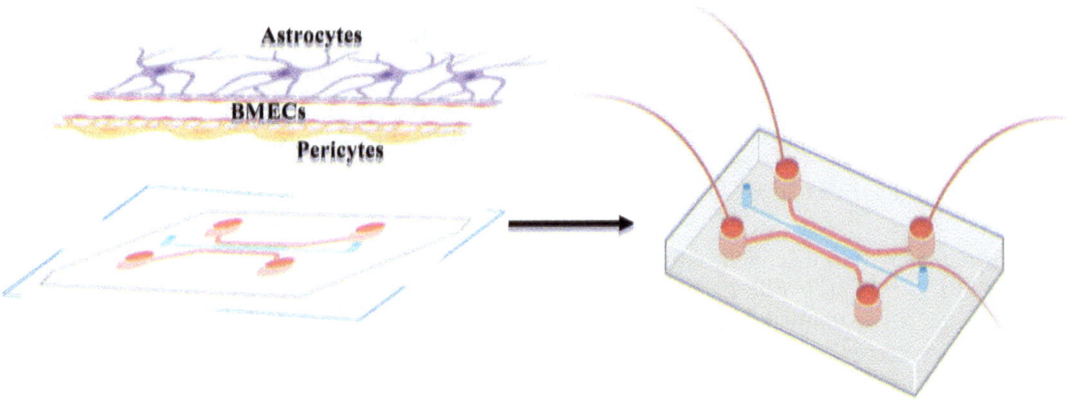

Fig. 8 BBB-on-a-chip with astrocytes and pericytes

animals, and 3D one-organ microfluidic models, there is still a significant gap in our ability to foresee drug response in patients' groups. The overall failure rates of clinical studies from phase I through phase IV remain well above the required percentage. Organ-on-a-chip platforms have demonstrated high potential in providing incredible robustness in drug screening by employing microengineering techniques. Organ-on-a-chip technology has successfully created many microfluidic chips that can mimic organ function (liver, lungs, gut, etc., and even tumors) on a chip. Human iPSC utilization to develop tissue or organ models has made drug development more reliable for human use and correct genetic disorders. In time, organs-on-a-chip could allow for the development of personalized treatments beyond preclinical testing.

More complex humanized models encompassing multiple organ systems are being developed. Integrating multiple organs-on-chips in a single design mimicking the human body increases the possibilities for a more sophisticated representation of absorption, distribution, metabolism, excretion (ADME), and toxicity process. It would also help understanding drug interactions within multiple organ systems to predict drug efficacy and safety and fast-track the drug development process (Fig. 9). As of now, up to ten organs can be integrated on a single multichip—skin, heart, liver, gut, kidney, lung, etc. By generating organoids from a particular patient, a diseased body can be emulated on a chip.

7 Advantages of Organs-on-a-Chip Platform

Trials with a wide range of drug concentrations to check for its efficacy is made possible due to the low manufacturing cost of OCC. During new drug development, the initial tests can be conducted several times without causing a financial burden. Moreover, there would be no ethical problems encountered by animal testing,

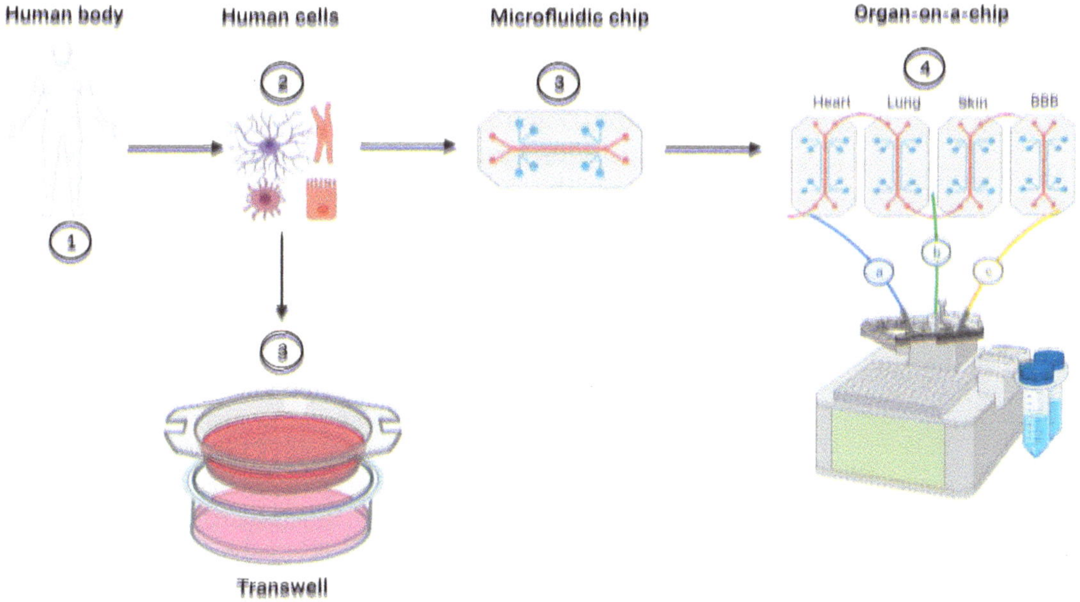

Fig. 9 Schematic illustrations of an individualized organs-on-chips for personalized medicine (genetic subpopulations)

which is a growing social concern. This allows OCC technology to substantially fast-track scientific research. The 3D structure inside a chip better replicates the human organ's microenvironment than in a Petri dish, and it is more reliable [67, 68]. Diverse designs created with microfluidics technology make this platform exciting and innovative for research. OCC's are easy to use and portable, and multiple chips can now be incorporated on a single chip due to their small size to study the toxicity of a drug throughout the human body [69].

8 Disadvantages of Organs-on-a-Chip Platform

Microfluidic chips have some significant advantages compared to the other available in vitro–in vivo models. However, some drawbacks remain. Due to the micrometer dimensions of OCC, surface effects widely take over volume effects. It can cause the adsorption of products of interest on the chip's inner surface and affect the data analysis [70]. The Reynolds number will always be very small (<1), and as a result, the flow into the chips will remain laminar. It allows for precise control of experimental conditions but lacks mixing capabilities. The drawbacks are common to all organs on chips since the nature of the device links them. Each of them also has minor flaws. However, advances in this technology are primarily compensated for, especially compared to the existing alternatives.

9 The Future of BBB Modeling

The intricate architecture of the BBB and NVU is still a challenge to replicate in vitro. Despite the availability of numerous in vitro BBB models, most studies have concentrated on practical validation or toxicity testing with a low output. These models are incapable of assessing various experimental stimuli in parallel, thereby restraining readout possibilities. Out of all the currently available BBB models, none can be considered as being ideal. Preclinical and clinical studies account for 33% and 63% of drug development's overall cost, respectively. The need for novel CNS pharmaceutical strategies and the necessity to limit experimental studies' increasing costs, drug development, and testing continuously push researchers to develop improved in vitro BBB models [71].

The ideal BBB model should comprise a dynamic system responding to fluid dynamic (blood), inflammatory (viruses, bacteria, etc.), and pharmacological stimuli (drugs, antibodies, etc.) [72]. The unmet need for more accurate in vitro BBB models simulating in vivo anatomical or disease conditions requires researchers' constant efforts [73]. Pharmaceutical companies prioritize cost-effective and high-throughput systems to speed-up drug development projects [74]. OOC models to be implemented in preclinical studies, parallelization, and computerization with high output screening of variables are required. In silico models could be the best choice to cut short the cost–time and foresee CNS bioavailability. However, the chemical structure of a drug could mislead its many physicochemical, biological parameters. The NVU-on-a-chip or BBB-on-a-chip system is developed to close this gap. BBB-on-a-chip can recapitulate the complex microarchitecture and physiochemical microenvironment of human BBB by incorporating patient-derived iPSCs and facilitating predictive, personalized medicine applications [75]. Microfluidic organ-on-a-chip model of human BBB that recapitulates in vivo barrier functions offers a new preclinical tool for developing brain-targeting therapeutics. The system also provides opportunities to combine multi-organ-on-chips on a single platform for drug screening and study pK parameters in the human body [76].

Even with all the limitations of current in vitro models, these models have deduced some of the critical processes involved in the BBB/NVU. Further improvements in BBB model fabrication are necessary to "humanize" models and generate human iPSCs that best mimic the BBB phenotype [12, 77]. Personalized models of particular cerebrovascular diseases where a specific donor's iPSCs create different NVU cell types will be a step forward [75]. The development of improved BBB-on-a-chip should be enhanced to facilitate discoveries of underlying mechanisms involved in BBB dysfunction in various cerebrovascular diseases. It will limit species differences and open a path for more translational research (personalized medicine and drug testing) in vitro.

10 Conclusions

Despite the rapid technological progress in BBB modeling, current in vitro platforms cannot fully recapitulate the features and physiological responses of the BBB/NVU in vivo. Therefore, in vivo studies in animals or retrospectively in human patients will stay the main course in this research field. In vitro solution will play a complementary role to continue assisting basic, translational, and pharmaceutical research. However, the breath and levels of details that in vitro technologies can muster will continue to expand significantly and allow us to assess and dissect out more complex physiological conundrum with increased relevance and translational significance than ever before. In this respect, knowing that no BBB model "rules them all," proper planning should be in place to choose the best model according to the specific research paradigms and study objectives. The strive to develop BBB models to provide more predictable results while remaining fully scalable, customizable, and well suited for extensive research or industrial use will likely continue for the foreseeable future. With the novel prospects of stem cell research and microengineering, we move in a direction where new BBB models will become competent and highly translational. The development of efficient high-throughput in vitro BBB models will transition from individual research labs to the industry, opening new prospects for personalized drug discovery.

Acknowledgments

This work was also supported by the National Institutes of Health/National Institute on Drug Abuse 2R01DA029121-01A1, 1R01DA049737-01, and National Institute of Neurological Disorders and Stroke 1R01NS117906-01 to Dr. Luca Cucullo. Figures created with BioRender.com.

References

1. Abbott NJ, Patabendige AA, Dolman DE et al (2010) Structure and function of the blood–brain barrier. Neurobiol Dis 37:13–25
2. Ballabh P, Braun A, Nedergaard M (2004) The blood–brain barrier: an overview: structure, regulation, and clinical implications. Neurobiol Dis 16:1–13
3. Pardridge WM (1999) Blood-brain barrier biology and methodology. J Neurovirol 5: 556–569
4. Huber JD, Egleton RD, Davis TP (2001) Molecular physiology and pathophysiology of tight junctions in the blood-brain barrier. Trends Neurosci 24:719–725

5. Goldstein GW, Betz AL (1986) The blood-brain barrier. Cold Spring Harb Perspect Biol 255:74–83
6. Brown LS, Foster CG, Courtney JM et al (2019) Pericytes and neurovascular function in the healthy and diseased brain. Front Cell Neurosci 13:282
7. Abbott NJ, Rönnbäck L, Hansson E (2006) Astrocyte-endothelial interactions at the blood-brain barrier. Nat Rev Neurosci 7:41–53
8. Begley DJ, Brightman MW (2003) Structural and functional aspects of the blood-brain barrier. Prog Drug Res 61:39–78
9. Jeffrey P, Summerfield S (2010) Assessment of the blood–brain barrier in CNS drug discovery. Neurobiol Dis 37:33–37
10. Pardridge WM (2001) Crossing the blood–brain barrier: are we getting it right? Drug Discov Today 6:1–2
11. Harilal S, Jose J, Parambi DGT et al (2020) Revisiting the blood-brain barrier: a hard nut to crack in the transportation of drug molecules. Brain Res Bull 160:121–140
12. Stanimirovic DB, Bani-Yaghoub M, Perkins M et al (2015) Blood–brain barrier models: in vitro to in vivo translation in preclinical development of CNS-targeting biotherapeutics. Expert Opin Drug Discov 10:141–155
13. Bhalerao A, Sivandzade F, Archie SR et al (2020) In vitro modeling of the neurovascular unit: advances in the field. Fluids Barriers CNS 17:1–20
14. Cecchelli R, Berezowski V, Lundquist S et al (2007) Modelling of the blood–brain barrier in drug discovery and development. Nat Rev Drug Discov 6:650–661
15. Cecchelli R, Dehouck B, Descamps L et al (1999) In vitro model for evaluating drug transport across the blood-brain barrier. Adv Drug Deliv Rev 36:165–178
16. Naik P, Cucullo L (2012) In vitro blood–brain barrier models: current and perspective technologies. J Pharm Sci 101:1337–1354
17. Chowdhury EA, Noorani B, Alqahtani F et al (2021) Understanding the brain uptake and permeability of small molecules through the BBB: a technical overview. J Cereb Blood Flow Metab 41:0271678X20985946
18. Hatherell K, Couraud P-O, Romero IA et al (2011) Development of a three-dimensional, all-human in vitro model of the blood–brain barrier using mono-, co-, and tri-cultivation Transwell models. J Neurosci Methods 199:223–229
19. de Lange EC (2012) The physiological characteristics and transcytosis mechanisms of the blood-brain barrier (BBB). Curr Pharm Biotechnol 13:2319–2327
20. Joo F (1973) A procedure for the isolation of capillaries from rat brain. Cytobios 8:41–48
21. Deli MA, Ábrahám CS, Kataoka Y et al (2005) Permeability studies on in vitro blood–brain barrier models: physiology, pathology, and pharmacology. Cell Mol Neurobiol 25:59–127
22. Joó F (1985) The blood-brain barrier in vitro: ten years of research on microvessels isolated from the brain. Neurochem Int 7:1–25
23. Joo F (1993) The blood-brain barrier in vitro: the second decade. Neurochem Int 23:499–521
24. Enerson BE, Drewes LR (2006) The rat blood—brain barrier transcriptome. J Cereb Blood Flow Metab 26:959–973
25. Bickel U (2005) How to measure drug transport across the blood-brain barrier. NeuroRx 2:15–26
26. Aday S, Cecchelli R, Hallier-Vanuxeem D et al (2016) Stem cell-based human blood-brain barrier models for drug discovery and delivery. Trends Biotechnol 34:382–393
27. Lippmann ES, Al-Ahmad A, Palecek SP et al (2013) Modeling the blood-brain barrier using stem cell sources. Fluids Barriers CNS 10:2
28. Raut S, Patel R, Al-Ahmad AJ (2021) Presence of a mutation in PSEN1 or PSEN2 gene is associated with an impaired brain endothelial cell phenotype in vitro. Fluids Barriers CNS 18:1–16
29. Appelt-Menzel A, Cubukova A, Günther K et al (2017) Establishment of a human blood-brain barrier co-culture model mimicking the neurovascular unit using induced pluri- and multipotent stem cells. Stem Cell Rep 8:894–906
30. Page S, Patel R, Raut S et al (2020) Neurological diseases at the blood-brain barrier: stemming new scientific paradigms using patient-derived induced pluripotent cells. Biochim Biophys Acta Mol Basis Dis 1866:165358
31. Delsing L, Herland A, Falk A et al (2020) Models of the blood-brain barrier using iPSC-derived cells. Mol Cell Neurosci 107:103533
32. Grant GA, Abbott NJ, Janigro D (1998) Understanding the physiology of the blood-brain barrier: in vitro models. News Physiol Sci 13:287–293
33. Sivandzade F, Cucullo L (2018) In-vitro blood–brain barrier modeling: a review of

modern and fast-advancing technologies. J Cereb Blood Flow Metab 38:1667–1681

34. Garberg P, Ball M, Borg N et al (2005) In vitro models for the blood–brain barrier. Toxicol In Vitro 19:299–334
35. Helms HC, Abbott NJ, Burek M et al (2016) In vitro models of the blood–brain barrier: an overview of commonly used brain endothelial cell culture models and guidelines for their use. J Cereb Blood Flow Metab 36:862–890
36. Gomes MJ, Mendes B, Martins S, Sarmento B (2016) 3.9 – Cell-based in vitro models for studying blood–brain barrier (BBB) permeability. In: Sarmento B, (Ed) Concepts and models for drug permeability studies: Woodhead Publishing, pp 169–188
37. Demeuse P, Kerkhofs A, Struys-Ponsar C et al (2002) Compartmentalized coculture of rat brain endothelial cells and astrocytes: a syngenic model to study the blood–brain barrier. J Neurosci Methods 121:21–31
38. Nakagawa S, Deli MA, Kawaguchi H et al (2009) A new blood–brain barrier model using primary rat brain endothelial cells, pericytes and astrocytes. Neurochem Int 54:253–263
39. Gaillard PJ, Voorwinden LH, Nielsen JL et al (2001) Establishment and functional characterization of an in vitro model of the blood–brain barrier, comprising a co-culture of brain capillary endothelial cells and astrocytes. Eur J Pharm Sci 12:215–222
40. Thomsen LB, Burkhart A, Moos T (2015) A triple culture model of the blood-brain barrier using porcine brain endothelial cells, astrocytes and pericytes. PLoS One 10:e0134765
41. Santaguida S, Janigro D, Hossain M et al (2006) Side by side comparison between dynamic versus static models of blood–brain barrier in vitro: a permeability study. Brain Res 1109:1–13
42. Della-Morte D, Rundek T (2015) The role of shear stress and arteriogenesis in maintaining vascular homeostasis and preventing cerebral atherosclerosis. Brain Circ 1:53
43. Cucullo L, Hossain M, Puvenna V et al (2011) The role of shear stress in Blood-Brain Barrier endothelial physiology. BMC Neurosci 12:1–15
44. Campinho P, Vilfan A, Vermot J (2020) Blood flow forces in shaping the vascular system: a focus on endothelial cell behavior. Front Physiol 11:552
45. Davies PF (2009) Hemodynamic shear stress and the endothelium in cardiovascular pathophysiology. Nat Clin Pract Cardiovasc Med 6: 16–26
46. Wang X, Xu B, Xiang M et al (2020) Advances on fluid shear stress regulating blood-brain barrier. Microvasc Res 128:103930
47. Cucullo L, McAllister MS, Kight K et al (2002) A new dynamic in vitro model for the multidimensional study of astrocyte–endothelial cell interactions at the blood–brain barrier. Brain Res 951:243–254
48. Oddo A, Peng B, Tong Z et al (2019) Advances in microfluidic blood–brain barrier (BBB) models. Trends Biotechnol 37:1295–1314
49. De Jong J, Lammertink RG, Wessling M (2006) Membranes and microfluidics: a review. Lab Chip 6:1125–1139
50. Wolff A, Antfolk M, Brodin B et al (2015) In vitro blood–brain barrier models—an overview of established models and new microfluidic approaches. J Pharm Sci 104:2727–2746
51. van der Helm MW, Odijk M, Frimat JP et al (2017) Fabrication and validation of an organ-on-chip system with integrated electrodes to directly quantify transendothelial electrical resistance. J Vis Exp (127):56334
52. Palchesko RN, Zhang L, Sun Y et al (2012) Development of polydimethylsiloxane substrates with tunable elastic modulus to study cell mechanobiology in muscle and nerve. PLoS One 7:e51499
53. Qin D, Xia Y, Whitesides GM (2010) Soft lithography for micro-and nanoscale patterning. Nat Protoc 5:491–502
54. Anderson JR, Chiu DT, Jackman RJ et al (2000) Fabrication of topologically complex three-dimensional microfluidic systems in PDMS by rapid prototyping. Anal Chem 72: 3158–3164
55. Esch MB, Sung JH, Yang J et al (2012) On chip porous polymer membranes for integration of gastrointestinal tract epithelium with microfluidic 'body-on-a-chip' devices. Biomed Microdevices 14:895–906
56. Wang YI, Abaci HE, Shuler ML (2017) Microfluidic blood–brain barrier model provides in vivo-like barrier properties for drug permeability screening. Biotechnol Bioeng 114:184–194
57. Booth R, Kim H (2012) Characterization of a microfluidic in vitro model of the blood-brain barrier (μBBB). Lab Chip 12:1784–1792
58. Prabhakarpandian B, Shen M-C, Nichols JB et al (2013) SyM-BBB: a microfluidic blood brain barrier model. Lab Chip 13:1093–1101
59. Buzhdygan TP, DeOre BJ, Baldwin-Leclair A et al (2020) The SARS-CoV-2 spike protein alters barrier function in 2D static and 3D microfluidic in-vitro models of the human

blood–brain barrier. Neurobiol Dis 146: 105131

60. Maoz BM, Herland A, FitzGerald EA et al (2018) A linked organ-on-chip model of the human neurovascular unit reveals the metabolic coupling of endothelial and neuronal cells. Nat Biotechnol 36:865–874
61. Ronaldson-Bouchard K, Vunjak-Novakovic G (2018) Organs-on-a-chip: a fast track for engineered human tissues in drug development. Cell Stem Cell 22:310–324
62. van der Helm MW, van der Meer AD, Eijkel JC et al (2016) Microfluidic organ-on-chip technology for blood-brain barrier research. Tissue Barriers 4:e1142493
63. Baker M (2011) Tissue models: a living system on a chip. Nature 471:661–665
64. Adriani G, Ma D, Pavesi A et al (2015) Modeling the blood-brain barrier in a 3D triple co-culture microfluidic system. Annu Int Conf IEEE Eng Med Biol Soc 2015:338–341
65. Griep LM, Wolbers F, de Wagenaar B et al (2013) BBB on chip: microfluidic platform to mechanically and biochemically modulate blood-brain barrier function. Biomed Microdevices 15:145–150
66. Park TE, Mustafaoglu N, Herland A et al (2019) Hypoxia-enhanced Blood-Brain Barrier Chip recapitulates human barrier function and shuttling of drugs and antibodies. Nat Commun 10:2621
67. Van Der Meer AD, Van Den Berg A (2012) Organs-on-chips: breaking the in vitro impasse. Integr Biol (Camb) 4:461–470
68. Huh D, Hamilton GA, Ingber DE (2011) From 3D cell culture to organs-on-chips. Trends Cell Biol 21:745–754
69. Moraes C, Mehta G, Lesher-Perez SC et al (2012) Organs-on-a-chip: a focus on compartmentalized microdevices. Ann Biomed Eng 40: 1211–1227
70. Mannino DM, Homa DM, Akinbami LJ et al (2002) Chronic obstructive pulmonary disease surveillance-United States, 1971–2000. Respir Care 47:1184–1199
71. Terasaki T, Ohtsuki S, Hori S et al (2003) New approaches to in vitro models of blood–brain barrier drug transport. Drug Discov Today 8: 944–954
72. Cucullo L, Aumayr B, Rapp E et al (2005) Drug delivery and in vitro models of the blood-brain barrier. Curr Opin Drug Discov Devel 8:89–99
73. Wilhelm I, Fazakas C, Krizbai IA (2011) In vitro models of the blood-brain barrier. Methods Mol Biol 71:113–128
74. Wevers NR, Kasi DG, Gray T et al (2018) A perfused human blood-brain barrier on-a-chip for high-throughput assessment of barrier function and antibody transport. Fluids Barriers CNS 15:23
75. Vatine GD, Barrile R, Workman MJ et al (2019) Human iPSC-derived blood-brain barrier chips enable disease modeling and personalized medicine applications. Cell Stem Cell 24:995–1005.e6
76. Polini A, Prodanov L, Bhise NS et al (2014) Organs-on-a-chip: a new tool for drug discovery. Expert Opin Drug Discov 9:335–352
77. Cucullo L, Hossain M, Rapp E et al (2007) Development of a humanized in vitro blood–brain barrier model to screen for brain penetration of antiepileptic drugs. Epilepsia 48:505–516
78. Logan S, Arzua T, Canfield SG et al (2011) Studying human neurological disorders using induced pluripotent stem cells: from 2D monolayer to 3D organoid and blood brain barrier models. Compr Physiol 9:565–611
79. Hajal C, Campisi M, Mattu C et al (2018) In vitro models of molecular and nano-particle transport across the blood-brain barrier. Biomicrofluidics 12:042213
80. Eigenmann DE, Xue G, Kim KS et al (2013) Comparative study of four immortalized human brain capillary endothelial cell lines, hCMEC/D3, hBMEC, TY10, and BB19, and optimization of culture conditions, for an in vitro blood–brain barrier model for drug permeability studies. Fluids Barriers CNS 10: 1–17
81. Rahman NA, Rasil ANHM, Meyding-Lamade U et al (2016) Immortalized endothelial cell lines for in vitro blood–brain barrier models: a systematic review. Brain Res 1642:532–545
82. Burek M, Salvador E, Förster CY (2012) Generation of an immortalized murine brain microvascular endothelial cell line as an in vitro blood brain barrier model. J Vis Exp 66:e4022
83. Kusch-Poddar M, Drewe J, Fux I et al (2005) Evaluation of the immortalized human brain capillary endothelial cell line BB19 as a human cell culture model for the blood–brain barrier. Brain Res 1064:21–31
84. Hellinger É, Veszelka S, Tóth AE et al (2012) Comparison of brain capillary endothelial cell-based and epithelial (MDCK-MDR1, Caco-2, and VB-Caco-2) cell-based surrogate blood–brain barrier penetration models. Eur J Pharm Biopharm 82:340–351

85. Huh D, Matthews BD, Mammoto A et al (2010) Reconstituting organ-level lung functions on a chip. Science 328:1662–1668
86. Bhise NS, Manoharan V, Massa S et al (2016) A liver-on-a-chip platform with bioprinted hepatic spheroids. Biofabrication 8:014101
87. Khetani SR, Bhatia SN (2008) Microscale culture of human liver cells for drug development. Nat Biotechnol 26:120–126
88. Yip D, Cho CH (2013) A multicellular 3D heterospheroid model of liver tumor and stromal cells in collagen gel for anti-cancer drug testing. Biochem Biophys Res Commun 433: 327–332
89. Prot J-M, Bunescu A, Elena-Herrmann B et al (2012) Predictive toxicology using systemic biology and liver microfluidic "on chip" approaches: application to acetaminophen injury. Toxicol Appl Pharmacol 259: 270–280
90. Jang K-J, Mehr AP, Hamilton GA et al (2013) Human kidney proximal tubule-on-a-chip for drug transport and nephrotoxicity assessment. Integr Biol (Camb) 5:1119–1129
91. Kim HJ, Huh D, Hamilton G et al (2012) Human gut-on-a-chip inhabited by microbial flora that experiences intestinal peristalsis-like motions and flow. Lab Chip 12:2165–2174
92. Taylor AM, Blurton-Jones M, Rhee SW et al (2005) A microfluidic culture platform for CNS axonal injury, regeneration and transport. Nat Methods 2:599–605
93. Agarwal A, Goss JA, Cho A et al (2013) Microfluidic heart on a chip for higher throughput pharmacological studies. Lab Chip 13:3599–3608
94. Klauke N, Smith G, Cooper JM (2007) Microfluidic systems to examine intercellular coupling of pairs of cardiac myocytes. Lab Chip 7:731–739
95. Martewicz S, Michielin F, Serena E et al (2012) Reversible alteration of calcium dynamics in cardiomyocytes during acute hypoxia transient in a microfluidic platform. Integr Biol (Camb) 4:153–164
96. Depre C, Vatner SF (2007) Cardioprotection in stunned and hibernating myocardium. Heart Fail Rev 12:307–317
97. Agrawal G, Aung A, Varghese S (2017) Skeletal muscle-on-a-chip: an in vitro model to evaluate tissue formation and injury. Lab Chip 17: 3447–3461

Part II

Using Pluripotent Stem Cells in Models of the BBB

Chapter 3

In Vitro Models of the Human Blood–Brain Barrier Utilising Human Induced Pluripotent Stem Cells: Opportunities and Challenges

Iqra Pervaiz and Abraham J. Al-Ahmad

Abstract

The blood–brain barrier (BBB) is a component of the neurovascular unit formed by specialized brain microvascular endothelial cells surrounded by astrocytes end-feet processes, pericytes, and a basement membrane. The BBB plays an important role in the maintenance of brain homeostasis and has seen a growing involvement in the pathophysiology of various neurological diseases. On the other hand, the presence of such a barrier remains an important challenge for drug delivery to treat such illnesses.

Since the pioneering work describing the isolation and cultivation of primary brain microvascular cells about 50 years ago until now, the development of an in vitro model of the BBB that is scalable, capable to form tight monolayers, and predictive of drug permeability in vivo remained extremely challenging.

The recent description of the use of induced pluripotent stem cells (iPSCs) as a modeling tool for neurological diseases raised momentum into the use of such cells to develop new in vitro models of the BBB. This chapter will provide an exhaustive description of the use of iPSCs as a source of cells for modeling the BBB in vitro, describe the advantages and limitations of such model, as well as describe their prospective use for disease modeling and drug permeability screening platforms.

Key words Blood–brain barrier, Induced pluripotent stem cells, Disease modeling, Brain microvascular endothelial cells

1 Introduction

Since the initial protocol of brain microvascular endothelial cells (BMECs) isolation by Ferenc Joo and colleagues almost 50 years ago [1], the use of in vitro models of the blood–brain barrier (BBB) based on primary cultures and immortalized cell lines have been essential in understanding its physiology, and its pathophysiology in various neurological disorders, by allowing us to address cellular and molecular mechanisms at a granularity not feasible with the current in vivo models [2, 3].

Nicole Stone (ed.), *The Blood-Brain Barrier: Methods and Protocols*,
Methods in Molecular Biology, vol. 2492, https://doi.org/10.1007/978-1-0716-2289-6_3,

However, a major issue encountered with such models is the rapid loss of their barrier phenotype following their isolation [4], in particular in terms of barrier function. Furthermore, the scarcity of access to human primary BMEC culture represents a major issue in terms of scalability needed to use such cells in drug discovery as a drug permeability and transport. These two major flaws observed with human BMECs, obtained from somatic cells, contributed to poor adoption of these cells in key applications in drug discovery (including their use as a permeability screening platform) and preferring the use of primary cultures of BMEC monolayers isolated from bovine and porcine sources [5–7].

The seminal work of Shusta and colleagues [8] provided a differentiation protocol to differentiate human pluripotent stem cells (hPSCs), using cell lines originated from both embryonic stem cells (hESCs) and induced pluripotent stem cells (iPSCs), which brought a noticeable momentum in the field. A decade after this pioneering work, a lot has been learned on the use of iPSC-derived BMECs (iBMECs) as an in vitro model of the human BBB, with both opportunities and challenges in using such cells.

In this chapter, we will provide a rapid overview of the feature and advantages of iPSCs in modeling neurological diseases, provide an overview of the differentiation protocol and the subsequent revisions brought to it, and finally highlight the opportunities and challenges inherent in the use of iBMECs to model the human BBB in vitro during health and diseases.

2 Induced Pluripotent Stem Cells: An Overall Overview

Somatic cells are by definition cells that underwent differentiation from an undifferentiated cell population and therefore engaged into an irreversible fate that leads them to programmed senescence and death.

In mammalian cells, such differentiation occurs rapidly during development, within hours following fertilization and division of the zygote into a multicellular structure. Totipotent cells quickly cleave into trophoblastic cells (that will arise into the formation of the trophoblast) and pluripotent stem cells that will form the blastocyst.

Up until recently, the existence of human pluripotent stem cells remained unique to their transient existence during embryogenesis.

Such limited existence requires the harvesting and use of human embryos obtained from in vitro fertilization, a developmental step that has an important ethical and cultural consideration to be taken. Such considerations as so controversial that on August 9, 2001, President Georges W. Bush introduced a moratorium, banning any federal funding on research projects involving newly created human embryonic stem cells from discarded embryos [9].

However, such moratorium did mildly affect the ability of Dr. Jamie Thompson and his colleagues to better understand the biology of hESCs [10] and their ability to be differentiated into various somatic cell types, as well as the use of other mammals (e.g., mouse, bovine, ovine, etc.), to further advance our understanding of stem cell biology.

However, the pioneering work of Takahashi and Yamanaka [11] opened new avenues and alternative routes to circumvent the roadblock set by President George W. Bush in 2001.

In this groundbreaking study, both authors observed that the apparent definitive differentiation of somatic cells could be reversed into a process named "derivation": a biological step consisting of the introduction of key transcription factors dubbed "*The Yamanaka factors*" (*KLF4*, *MYC*, *OCT4*, *SOX2*) was sufficient to allow some cells to revert their evolutionary clock counter-wise and display traits commonly found in embryonic stem cells (including immortality by harboring unlimited passaging, lack of senescence, ability to differentiate and grow teratoma when injected in animals). Such "derivation" occurs at the epigenetic level rather than at the genetic level (although the random insertion of these transcription factors into the genome may impair the quality and yield on the differentiation of such cells) [12], which is considered to be a process that removes most of the epigenetic signature found in somatic cells but also a very challenging procedure. As of today, this step remains a bottleneck in the generation of induced pluripotent stem cells (iPSCs) from patient somatic cells (e.g., fibroblasts, peripheral blood mononuclear cells, etc.), which usually yields less than 1% of iPSC clones from the initial amount of somatic cells used.

The second breakthrough in the field of iPSC biology was the adoption of the Yamanaka factors into human somatic cells by Thompson and colleagues in their 2007 study [13], which was the first reported derivation of human iPSCs from somatic cells.

iPSCs are derived from somatic cells and share high similarities with ESCs, the ability to undergo differentiation, self-renewal, and express markers of pluripotency. They have the robust potential of proliferation and differentiation into many cell types in vitro while retaining their developmental potential. Additionally, due to their effortlessly accessible somatic origin, they are easy to be produced from patient-specific pluripotent stem cells for toxicity testing, drug screening, and disease modeling. Hence, this factor directed reprogramming not only eliminates ethical limitations but also circumvents the immunological cross-reactivity associated with grafts and transplants as observed in ESC [14].

To realize the full potential of iPSCs, the process of derivation and expansion of iPSCs must be robust and efficient. Even after the successful derivation step, three major limitations of iPSC technology remain until now limiting: Firstly, the low yield of the reprogramming efficiency, which as of today remains relatively low (<1%). Secondly, the variability within clones obtained from the same somatic cell line used for derivation remains an inherent problem in the field [15].

Thirdly, the derivation process itself relies on viral vectors and the random insertion of the transcription factors. This issue, however, appears to be solvable soon via the use of episomal factors and other gene-editing techniques [14].

3 Modeling Neurological Diseases in a Dish: How Human Induced Pluripotent Stem Cells Became a Game-Changer

Neurological diseases are in their majority progressing and irreversible conditions that are mostly treated symptomatically using pharmacological and non-pharmacological interventions.

The peculiar nature of the human central nervous system (CNS), which is a complex tissue formed by excitable and post-mitotic cells, constitutes an important challenge in understanding the root cause of many of these diseases and by extension our ability to design and deliver drugs to target such conditions.

The use of in vitro models can only provide us with limited information on such diseases [16], whereas the use of animal models can only allow us to model human neurological and psychological conditions to an extent. Traditionally, the use of genetically engineered animal models (mostly mice) remained the mainstay for the modeling of neurological diseases. For example, mice genetically engineered to overexpress the human *APP* gene remains the most common in vivo model of Alzheimer's disease used to model the pathophysiology of the disease and identify novel therapeutic targets [17]. Yet these models, as useful as they are, fail to fill the divide between the bench to the bedside, with the blunt failure to translate our findings from these models into clinically efficacious treatment against such disease an important pitfall [18].

A major limitation of modeling human neurological diseases in vitro remains the very profound nature of neurons. These cells are by essence post-mitotic and virtually impossible to isolate from post-mortem tissues, except for neuron progenitor cells obtained from fetal brain tissues isolated from miscarried or aborted fetuses. On the other hand, few human neuron-like cells isolated from primary tumors such as neuroblastoma (e.g., SH-SY5Y, SK-N-SH, etc.) provide only a limited and faulty source of cells to model human neurological diseases in a Petri dish.

The pioneering work of Pr. Su-Chun Zhang in differentiating neurons from hESCs [19] opened the avenue to use hPSCs as an alternative source of human neurons but also to reconstitute other cell types found in the CNS (e.g., astrocytes, oligodendrocytes), in particular when applied to iPSCs.

The use of iPSCs, in place of hESCs, opened the possibility of using patient somatic cells as working material and by extent having a cellular model closer to the diseases studied. iPSC-derived neurons provide us the advantage of studying the implication of genes associated with a neurological disease, which function can be involved as early as during development (e.g., Rett syndrome [20]) or as late as in neurodegeneration (e.g., Parkinson's disease [21]).

Furthermore, iPSCs allows us to englobe a more holistic approach to such diseases by going beyond a neuron-centric approach, allowing us to explore how such mutations impact non-neuronal cells as well.

The derivation of patient-specific iPSCs has great potential to be a powerful tool for modeling neurological diseases [22]; however, there are still technical hurdles and limitations that remain to be addressed. For instance, the derivation step from somatic cells into differentiated remains a long and tortuous route that is still an important bottleneck in using iPSC for modeling diseases. Another issue also brought is the relative loss of the epigenetic signature inherent to aging or environmental exposure. One alternative to such hurdle is a direct reprogramming of patients' fibroblasts into neurons [23].

3.1 The Use of iPSCs to Model Neurodevelopmental Diseases

One of the major assets of iPSCs is their differentiation pattern which follows the same developmental lineage observed in mammalian development: development of a neuroepithelium, followed by the differentiation into neural stem cells, neural progenitor cells, and eventually differentiated neurons.

The first study documenting the use of iPSCs to document a neurodevelopmental (or juvenile neurological disorder) is the work of Ebert and colleagues [24] in which they documented the differentiation pattern of iPSCs isolated from a patient suffering from spinal muscular atrophy (SMA, an autosomal recessive neurodegenerative disease affecting motoneurons and associated with mutations in the *SMN1* gene). This study not only allowed to follow and provide a phenotype in vitro but also paved the way to personalized medicine in the aspect of prospective use of iPSCs as cell therapy. In that regard, the subsequent work of Chang and colleagues [25] demonstrated the ability to rescue the phenotype by inserting a functional *SMN1* gene into such iPSCs.

As of today, iPSCs have become a formidable tool to model various neurodevelopmental diseases [26] including autism spectrum disorders [27], Batten's disease [28], cerebellar ataxias [29], Down syndrome [30], Dravet syndrome [31], Fragile X syndrome [32], or Rett syndrome [33].

3.2 The Use of iPSCs to Model Neurodegenerative Diseases

Neurodegenerative diseases are a set of neurological disorders characterized by the loss of a particular neuronal population. Such diseases include (but not exclusively) Alzheimer's disease (AD), amyotrophic lateral sclerosis (ALS), Huntington's disease (HD), or Parkinson's disease (PD). These diseases are commonly classified into two major types: sporadic form (which represents the vast majority of the cases) and familial form (which represents a minor fraction).

Despite representing a minority fraction, familial forms of neurodegenerative diseases still represent an important asset to model such diseases in vitro, as they provide insights into genetic mutations associated with such forms of the disease and subsequently cellular and molecular mechanisms which can help to better understand such diseases.

Several studies have been reporting the use of iPSCs derived from patients suffering from a familial form of Alzheimer's disease (FAD) [34], amyotrophic lateral sclerosis [35], Huntington's disease [36], or Parkinson's disease [37].

4 In Vitro Model of the Human Blood–Brain Barrier Based on Stem Cells: Past and Current Protocols

In vitro models of the BBB are essential tools to better understand the cellular and molecular mechanisms underlying the physiology of it during development and aging, but also to better understand how diseases impair its function.

However, current in vitro models of the human BBB, based on primary cultures and immortalized cell lines [38], suffer from both scarcity and from poor barrier properties which render their use to limited applications.

Thus, the recent advances in stem cell biology and its application for modeling neurological diseases promoted an incentive for the establishment of differentiation protocols to obtain in vitro models of the human BBB using human pluripotent stem cells (hPSCs).

The first documented protocol in the differentiation of hPSCs was from Shusta and colleagues in 2012 [8]. In this protocol, the differentiation of cells was left unguided and occurred mostly by the removal of basic fibroblast growth factor (bFGF) from the cell culture media for 6 days, followed by a directed differentiation through incubation in endothelial cell medium supplemented with bFGF (used here as an angiogenic factor) for 2 days.

Under such conditions, two major cell populations rose: an early cell population harboring cell markers inherent to a neuronal cell lineage (nestin, βIII-tubulin) as early as day 4 of differentiation, emerging from the rims of iPSC colonies and ultimately matured into βIII-tubulin+ "tube-like" structures; followed by the late appearance of a $CD31^+$ population (indicative of an endothelial lineage) as early as day 4 in the middle of the former iPSC colonies, which ultimately represents about 60% of the cell population by day 8 of differentiation. This "endothelial cell" population was accompanied by the co-expression of GLUT1 (glucose transporter 1), a marker commonly associated with the brain endothelial cell lineage. Indeed, a flow cytometry analysis of the cell population identified that such $CD31^+$ cells could be divided into two major subpopulations: CD31/$GLUT1^{high}$ (which were labeled as "BMECs" by the authors) and CD31/GLUT1low (which were referred to as "non-BMECs" by the authors) populations.

Notably, the authors also reported that such differentiation of the BMEC population was under the influence of the canonical WNT pathway (WNT/β-catenin), a pathway playing an important role in brain angiogenesis and maturation of the developmental BBB [39, 40].

Upon purification, these BMEC monolayers showed a high purity, as almost 100% of the cells present expressed cell markers attributable to BMECs ($CD31^+$, $GLUT1^+$, claudin-5^+ cells). These cells were also able to form barrier tightness with values similar to primary rat BMECs, were capable of angiogenesis on Matrigel (R) assays, presented functional drug transporters (in particular presented functional efflux pumps), and were capable to induce their tightness upon co-cultures with astrocytes.

Although such a study pioneered the use of hPSCs as a source for BMECs, there was still room for improvement. Such monolayers were decently tight but remained sub-optimal when compared to in vivo [41], as well as to bovine and porcine in vitro models of the BBB [5, 7].

The documentation by Mizee and colleagues of the ability of retinoids, in particular, all-trans retinoic acid (RA), to elicit a barrier tightening of hCMEC/D3 monolayers [42], further motivated the improvement of the original protocol by the Shusta group.

The addition of RA in the differentiation protocol [43] from day 6 to day 9 not only allowed the additional expression of VE-cadherin (an endothelial-selective type of cadherin) but also significantly boosted the tightening of such monolayers as the authors reported transendothelial electrical resistance (TEER) values ranging from 1000 up to 5000 $\Omega.cm^2$, such induction being the consequence of the activation of one or several retinoid receptors and its signaling cascade [44].

The Shusta protocol was rapid, efficient, and simple. Yet, it was only documenting success in a select number of established hESC and iPSC lines, letting the use of other stem cells/progenitor cell sources undocumented.

Following the publication of the revised protocol by Shusta in 2014, it was quickly implemented and adapted with success in the differentiation of iBMECs from both BC1 cell lines (isolated hematopoietic stem cell line) by Searson and colleagues [45] but also encountered some success in the differentiation of endothelial progenitor cells (CD34+ cells) isolated from cord blood cells by Cecchelli and colleagues [46].

Follow-up studies came from Shusta and colleagues as optimization of the protocol rather than major revisions of it and include optimization of cell density needed to maintain a consistent quality output of the differentiated iBMEC monolayers [47], a directed differentiation through a mesoderm intermediate [48], the use of more defined cell culture media in the maintenance and differentiation of such hPSCs [49, 50], and more recently the addition of isogenic co-cultures of astrocytes, neurons [51, 52], and pericytes [53].

5 Features and Comparisons to Existing In Vitro Models Based on Immortalized Cell Lines and Primary Cultures

The main challenge encountered with the current in vitro models of the blood–brain barrier (BBB) is the ability of such models to form tight monolayers. The presence of tight monolayers is crucial for such models to be representative of the in vivo situation.

In vitro models of the BBB based on primary cell cultures or immortalized cell lines mostly suffer from poor barrier tightness (as measured by determination of transendothelial electrical resistance (TEER) and measurement of paracellular fluxes), with exception of bovine- and porcine-based models.

Based on the current literature [54], there is a biphasic relationship between TEER and paracellular permeability, with a minimum TEER of 500 $\Omega.cm^2$. Immortalized human brain microvascular endothelial cell lines (e.g., hCMEC/D3, bEnd.3, etc.) are usually suffering from low barrier tightness as reflected by their low transendothelial electrical resistance (TEER) below 100 $\Omega.cm^2$, whereas primary cultures usually display a better barrier tightness with TEER values ranging from 200 $\Omega.cm^2$ (rodents) to 1000–2000 $\Omega.cm^2$ as reported in bovine and porcine primary cultures.

A key feature that an iPSC-based in vitro model of the BBB should achieve is the formation of tight monolayers. The initial differentiation protocol by Shusta and colleagues [8] reported TEER values in various iPSC-derived brain microvascular brain endothelial cells (iBMECs) could achieve an average TEER value of 200 $\Omega.cm^2$, which is comparable to values reported in primary rodent cultures.

However, their first revision of this differentiation protocol [43], which included the addition of retinoic acid (RA) between day 6 and day 8 of differentiation, considerably increased the ability of such iBMECs to form tight monolayers, as such monolayers were capable to form monolayers exhibiting TEER values ranging from 1000 up to 5000 $\Omega.cm^2$.

Such tight monolayers appeared reproducible independently of the use of the iPS(IMR90)-c4 iPSC clone, as several groups (including our group) were capable to reproduce similar findings using other iPSC lines [45, 50–52, 54–60], suggesting that the current differentiation protocol can be applied uniformly across iPSC lines.

The second feature of these iBMECs is the expression of cell markers considered as enriched at the BBB. These cells are expressing BBB markers (GLUT1, claudin-5), but also other proteins involved in tight junction complexes (occludin, ZO1, tricellulin), but also a series of claudins recently identified as expressed at the human BBB [61] including claudin-1, claudin-2, claudin-3, claudin-4, claudin-6, claudin-9, claudin-11, claudin-12, claudin-15, claudin-17, claudin-20, claudin-22, claudin-23, claudin-25, and claudin-27 (unpublished data).

The third feature of iBMECs is the expression and the functionality of various solute carriers (SLCs) typically found at the human BBB including glucose transporters (GLUT1, GLUT3, and GLUT4) [62], monocarboxylate transporter 8 (MCT8) [58], organic anion transporter polypeptide 1A4 (OATP1A4) [63], as well as several members of the ATP-binding cassette (ABC) transporters super-family including P-glycoprotein (P-gp), breast cancer resistant protein, and MRP1 [8, 43, 52].

Finally, there are several instances that iBMECs were capable to respond to induction by surrounding non-endothelial cells part of the neurovascular unit (e.g., astrocytes, pericytes, or neurons). Induction of tight BMEC monolayers following co-cultures with other cellular partners found at the neurovascular unit is well-documented through the literature [64–71]. The ability of iBMECs to positively respond to surrounding cells have been documented in the literature by both our group and the Shusta group in several publications, including the use of differentiated neural progenitor cells (NPCs) [43], as well as astrocytes and neurons differentiated from the same iPSC lines [51, 52]. More recently, Shusta and colleagues reported a successful differentiation and co-culture of iPSC-derived pericytes co-cultured with iBMECs [53].

Such an approach is very interesting, as the use of iPSCs would allow generating isogenic co-cultures with all the different elements coming from the same patient cell line.

6 Use of Induced Pluripotent Stem Cell-Based Models as Modeling Diseases at the BBB

The use of iBMECs as a disease modeling tool can be divided into two major categories: diseases affecting the BBB following an environmental disruptor (including toxins and pathogens) and diseases involving a mutation in one or several genes.

6.1 Modeling Diseases at the Blood–Brain Barrier Associated with an Environmental Factor

The disruption of the BBB by extrinsic factors are multiple and can involve biological, chemical, and physical agents.

Various biological agents, in particular pathogens, have been reported to impact and compromise the BBB integrity in vivo. However, the use of in vitro models to understand the cellular and molecular mechanisms underlying such interactions have been limited.

Recent studies highlighted the ability of iBMEC monolayers to interact with various viruses or virus components, including SARS-CoV-2 [72, 73] or Zika virus [74].

More recently, Kim and colleagues have documented the ability of certain bacteria strains commonly involved with a certain form of meningitis (Group B Streptococcus (GBS), *Neisseria*, etc.) to interact with iBMECs [75, 76], including the ability to identify important signaling pathways involved in such pathogen–host interactions. Our group also identified gliotoxin (a fungal toxin produced by Aspergillus) as a potent opener of the BBB, as it elicited a loss of the barrier function following 24-hour treatment [77].

Finally, Engelhardt and colleagues also documented the ability of iBMEC monolayers to interact with immune cells, as such cells were able to express important cell adhesion molecules involved in leukocyte–endothelial cell interactions (VCAM-1, ICAM-1) [78].

In addition to biological agents, chemical agents and toxins capable of toxic effects on the central nervous systems are also an item of concern that is relevant to screen for. Several studies highlighted the use of iPSC-derived organoids to assess neurotoxicity; these models are yet to account for the permeability of such neurotoxins across the BBB.We have previously reported the use of iBMEC monolayers to determine the ability of glyphosate (and its metabolite AMPA (aminomethylphosphonic acid)) to interact and cross the BBB [79]. Our group also investigated the effects of known neurotoxic compounds (unpublished data), but as of today, the literature using iBMECs as a toxicology screening platform remains very limited.

Another use of the iBMEC model is to assess the effect of hypoxic/ischemic injury at the BBB. Hypoxia and hypoxemia (a condition marked by an inadequate oxygen supply compared to the oxygen need) is a condition commonly encountered in patients having impaired lung function or impaired oxygen availability from exposure to high altitude. However, when combined with a severe hypoperfusion state, hypoxia becomes a component of ischemic injury and a major contributor to ischemic stroke injury.

Hypoxic/ischemic injury at the BBB is commonly associated with an onset of cerebral edema which can occur during or after the onset of such insult. The response of the BBB to hypoxic/ischemic injury is commonly characterized by a loss of the barrier function (as reported through a decrease in TEER and increased permeability to paracellular tracers) [65, 80–88].

Our group has reported in several publications the ability of iBMECs to respond to hypoxic/ischemic injury in vitro similar to existing in vitro models [89, 90]. In particular, exposure of these cells to hypoxia resulted in a decreased barrier function (as measured by TEER) and increased permeability to fluorescein (a paracellular marker). As previously reported with immortalized cell lines or primary cultures [91–101], such increased permeability correlated with increased release of VEGF. Such observations and outcomes measured contribute to providing evidence of iBMECs showing more similarities to existing human brain endothelial cells (primary and immortalized), than differences.

6.2 Modeling Genetic Diseases at the Blood–Brain Barrier Using Patient-Derived Induced Pluripotent Stem Cells

The use of the iBMEC to model genetic diseases at the BBB is pretty novel and likely benefits the most from induced pluripotent stem cells, as it provides a unique approach that cannot be easily replicated by primary culture.

As of today, most of the studies published in the literature focused on assessing the impact of genetic diseases associated with a neurological phenotype and determine how such mutations may impact the BBB.

The first reported study using iBMECs to model how a genetic disorder affects the BBB is the study by Vatine and colleagues [58] in which iPSCs from patients suffering from Allan–Herndon–Dudley syndrome (AHDS, a disease marked by impaired T3 thyroid hormone uptake at the BBB) were differentiated into iBMECs and characterized. Notably, iBMECs from AHDS patients showed no major deficiencies in terms of phenotype; however, such iBMECs showed a significant decrease in T3 uptake and diffusion rate across the monolayers. The second reported use of iBMECs derived from patients iPSCs is related to Huntington's disease (HD) by Lim and colleagues [57]. In that study, the authors investigated the phenotype of iBMECs differentiated from iPSCs of patients suffering from HD and reported poor barrier function in such cells. One possible explanation of such detrimental phenotype was possible overactive angiogenesis and possibly a dysregulation in the WNT pathway.

More recently, Azarin and colleagues reported similar outcomes in iBMECs obtained from patients suffering from leukodystrophy [102], similar observation in the impairment of the phenotype was also reported by our group in an iPSC line obtained from a patient suffering from Batten's disease linked to a mutation in the *CLN3* gene [103].

Finally, Katt and colleagues provided a recent study assessing the phenotypic outcome on iBMECs derived from patients suffering from various familial forms of neurodegenerative diseases including amyotrophic lateral sclerosis (ALS), Parkinson's disease (PD), Alzheimer's disease (AD), or Huntington's disease [59]. In such study, the authors reported an overall impaired barrier phenotype (as measured by TEER and/or permeability to paracellular tracers) in some iPSC lines, suggesting that mutations in genes commonly associated with a neurological phenotype, may have an impact on non-neuronal cells, in particular at the BBB. Such observation aligned with our most recent study obtained from patients suffering from a familial form of Alzheimer's disease (FAD) associated with mutations with PSEN1 and PSEN2 genes [60]. The PSEN1 mutant iBMEC monolayers showed a remarkable impaired barrier phenotype compared to the control and PSEN2 mutant iBMECs. In conclusion, the use of iPSCs to model the BBB opens up the ability of researchers to identify and document a genotype–phenotype association between genes commonly not associated with a physiological role at the BBB and assess their possible contribution in a dysfunctional BBB.

7 Use of Induced Pluripotent Stem Cell-Based Model of the BBB in Drug Discovery and Toxicology

The ability to determine the penetration and diffusion of drug candidates across the blood–brain barrier (BBB) early in the drug discovery is a crucial step that as of today remains challenging. With exception of bovine- and porcine-based models, the use of in vitro models of the BBB based on current existing primary cell cultures and cell lines has been an important challenge for their use in drug discovery, due to their poor barrier properties and their rapid de-differentiation compared to their in vivo counterparts. Such important limitations, as well as the scalability issues associated with primary cultures, makes the use of such models limited in drug discovery and toxicology purposes.

However, the tight monolayer properties of the current iPSC-based BBB models maybe contribute to a progressive adoption of such models into academic and industrial settings.

Lippmann and colleagues initially documented a reduced in vitro–in vivo correlation (IVIC) in his 2012 study [8] and showed an acceptable correlation coefficient for a selected number of compounds differing by their physicochemical properties.

Our group later published a study investigating the permeability of alkylphosphocholine (APC) analogs and compared the permeability of the different compounds and their interaction with drug transporters [104].

8 Current Limitations and Challenges

In this chapter, we have extensively discussed the potential and benefits of the iPSC-based model of the BBB, by discussing the different advantages of such model.

However, such accolades should not blindside of the limitations and caveats of such models.

The first caveat and limitation of such a model is the lack of exhaustive study directly comparing iBMECs to actual human brain microvascular endothelial cells in terms of gene (transcriptomics) and protein (proteomics) expression, both qualitatively and quantitatively. There is evidence that such iBMECs are capable to form tight junction complexes with a tightness comparable to values reported in vivo. There is also evidence that such cells express several BBB markers, including drug transporters, similarly to those found at the human BBB. However, a piece of important information missing is a direct comparison of how comparable the expression of these transporters is between iBMECs and human brain endothelial cells. Therefore, future studies providing a direct benchmark of these iBMECs against their respective somatic cells will certainly increase the relevance (and highlight limitations) of iPSC-based models of the BBB.

Another limitation of the iPSC-based model of the BBB is the relative cell identity of the differentiated product. iBMECs express and display feature expected from brain microvascular endothelial cells; however, such feature appears temporary as these cells show signs of dedifferentiation over time, usually 3 days after purification [8]. Several iterations were made to the model to improve the stability of the model, including changes in the initial differentiation protocol into a more synthetic approach [49–51] and directed lineage differentiation through a mesoderm intermediate [48].

Yet, a recent pre-print study by Agaliu and colleagues suggests that the reprogramming of these iPSCs into iBMECs maybe not complete and could display some epithelial-like features as well, albeit Shusta and colleagues partially refuted such claim by providing evidence of the expression of VE-cadherin by these cells [105].

Finally, an issue of iPSC-derived BMECs is inherent to intra- and inter-individual variability that may lead to different outcomes in terms of barrier phenotype. A major limitation of patient-derived iPSCs is the availability of only one clone of iPSC colonies derived from the patient somatic cells. Due to the random insertion nature of transcription factors needed to derivate such somatic cells into iPSCs, we cannot exclude that such insertion may impair certain genes function solicited during the differentiation phase. This can have serious consequences when characterizing and phenotyping iBMECs from patients suffering from a certain genetic disease. For instance, Searson and colleagues showed evidence that two iPSC clones obtained from the same diseased patient (suffering from a familial form of Alzheimer's disease and presenting a single-point mutation in the PSEN1 gene) resulted in different outcomes in the barrier function, with one of these clones showing deficiencies in the barrier function.

9 Future Directions

iPSC-based models of the human blood–brain barrier (BBB) are about to celebrate their tenth anniversary since the original publication by Lippmann and colleagues [8]. Since the initial inception, this model has gained tremendous popularity with the community as a tool for drug discovery and disease modeling. Yet, there are still challenges to be overcome and needed to allow such models to mature.

An important issue that needs to be addressed to improve such a model is to have a consistent and comprehensive differentiation approach, which allows researchers to obtain iBMECs mimicking human brain microvascular endothelial cells isolated from patients.

We believe such differentiation can only be improved if we hold a better grasp of the developmental biology of the BBB. A better understanding of the signaling pathways involved in the developmental BBB will significantly help research provide a more directed differentiation of iPSCs into mature BMECs.

Another limitation currently encountered is the limited supply of patients' iPSC cell lines to model diseases, which often lack the availability of adequate isogenic control lines to distinguish traits inherent to the diseases. The use of established iPSC lines in concomitance with gene editing (e.g., CRISPR/Cas9 technology) could partially address such issues.

Finally, the recent advances in stem cell biology, as well as next-generation sequencing and systems biology, can significantly help improve our ability to compare and contrast existing iPSC models and refine them to improve their in vitro–in vivo correlation in the future.We believe that such models have their future in personalized medicine and maybe eventually be used as an "organ-on-a-chip" to provide a patient-specific approach to neurological diseases and refine predictability in terms of therapeutic outcome.

References

1. Joo F, Karnushina I (1973) A procedure for the isolation of capillaries from rat brain. Cytobios 8(29):41–48
2. Jackson S, Meeks C, Vezina A et al (2019) Model systems for studying the blood-brain barrier: applications and challenges. Biomaterials 214:119217. https://doi.org/10.1016/j.biomaterials.2019.05.028
3. Helms HC, Abbott NJ, Burek M et al (2016) In vitro models of the blood-brain barrier: an overview of commonly used brain endothelial cell culture models and guidelines for their use. J Cereb Blood Flow Metab 36(5):862–890. https://doi.org/10.1177/0271678X16630991
4. Lyck R, Ruderisch N, Moll AG et al (2009) Culture-induced changes in blood-brain barrier transcriptome: implications for amino-acid transporters in vivo. J Cereb Blood Flow Metab 29(9):1491–1502. https://doi.org/10.1038/jcbfm.2009.72
5. Yusof SR, Avdeef A, Abbott NJ (2014) In vitro porcine blood-brain barrier model for permeability studies: pCEL-X software pKa (FLUX) method for aqueous boundary layer correction and detailed data analysis. Eur J Pharm Sci 65:98–111. https://doi.org/10.1016/j.ejps.2014.09.009
6. Patabendige A, Skinner RA, Abbott NJ (2013) Establishment of a simplified in vitro porcine blood-brain barrier model with high transendothelial electrical resistance. Brain Res 1521:1–15. https://doi.org/10.1016/j.brainres.2012.06.057
7. Culot M, Lundquist S, Vanuxeem D et al (2008) An in vitro blood-brain barrier model for high throughput (HTS) toxicological screening. Toxicol In Vitro 22(3):799–811. https://doi.org/10.1016/j.tiv.2007.12.016
8. Lippmann ES, Azarin SM, Kay JE et al (2012) Derivation of blood-brain barrier endothelial cells from human pluripotent stem cells. Nat Biotechnol 30(8):783–791. https://doi.org/10.1038/nbt.2247
9. Murugan V (2009) Embryonic stem cell research: a decade of debate from Bush to Obama. Yale J Biol Med 82(3):101–103
10. Vazin T, Freed WJ (2010) Human embryonic stem cells: derivation, culture, and differentiation: a review. Restor Neurol Neurosci 28(4):589–603. https://doi.org/10.3233/RNN-2010-0543
11. Takahashi K, Yamanaka S (2006) Induction of pluripotent stem cells from mouse embryonic and adult fibroblast cultures by defined factors. Cell 126(4):663–676. https://doi.org/10.1016/j.cell.2006.07.024
12. Yagi M, Yamanaka S, Yamada Y (2017) Epigenetic foundations of pluripotent stem cells that recapitulate in vivo pluripotency. Lab Investig 97(10):1133–1141. https://doi.org/10.1038/labinvest.2017.87
13. Yu J, Vodyanik MA, Smuga-Otto K et al (2007) Induced pluripotent stem cell lines derived from human somatic cells. Science 318(5858):1917–1920. https://doi.org/10.1126/science.1151526
14. Ebben JD, Zorniak M, Clark PA et al (2011) Introduction to induced pluripotent stem cells: advancing the potential for personalized medicine. World Neurosurg 76(3–4):270–275. https://doi.org/10.1016/j.wneu.2010.12.055
15. Inoue H, Nagata N, Kurokawa H et al (2014) iPS cells: a game changer for future medicine. EMBO J 33(5):409–417
16. Soldner F, Jaenisch R (2017) In vitro modeling of complex neurological diseases. In: Jaenisch R, Zhang F, Gage F (eds) Genome editing in neurosciences, Cham, pp 1–19. https://doi.org/10.1007/978-3-319-60192-2_1
17. Wong PC, Cai H, Borchelt DR et al (2002) Genetically engineered mouse models of neurodegenerative diseases. Nat Neurosci 5(7):633–639. https://doi.org/10.1038/nn0702-633
18. Mehta D, Jackson R, Paul G et al (2017) Why do trials for Alzheimer's disease drugs keep failing? A discontinued drug perspective for 2010-2015. Expert Opin Investig Drugs 26(6):735–739
19. Zhang SC, Wernig M, Duncan ID et al (2001) In vitro differentiation of transplantable neural precursors from human embryonic stem cells. Nat Biotechnol 19(12):1129–1133. https://doi.org/10.1038/nbt1201-1129
20. Marchetto MC, Carromeu C, Acab A et al (2010) A model for neural development and treatment of Rett syndrome using human induced pluripotent stem cells. Cell 143(4):527–539. https://doi.org/10.1016/j.cell.2010.10.016
21. Mattis VB, Svendsen CN (2011) Induced pluripotent stem cells: a new revolution for clinical neurology? Lancet Neurol 10(4):383–394

22. Tiscornia G, Vivas EL, Belmonte JCI (2011) Diseases in a dish: modeling human genetic disorders using induced pluripotent cells. Nat Med 17(12):1570–1576. https://doi.org/10.1038/nm.2504
23. Zhang Y, Xie X, Hu J et al (2020) Prospects of directly reprogrammed adult human neurons for neurodegenerative disease modeling and drug discovery: iN vs. iPSCs models. Front Neurosci 14:546484. https://doi.org/10.3389/fnins.2020.546484
24. Ebert AD, Yu J, Rose FF Jr et al (2009) Induced pluripotent stem cells from a spinal muscular atrophy patient. Nature 457(7227): 277–280. https://doi.org/10.1038/nature07677
25. Chang T, Zheng W, Tsark W et al (2011) Brief report: phenotypic rescue of induced pluripotent stem cell-derived motoneurons of a spinal muscular atrophy patient. Stem Cells 29(12): 2090–2093. https://doi.org/10.1002/stem.749
26. Zhang X, Li Z, Liu Y et al (2021) Great expectations: induced pluripotent stem cell technologies in neurodevelopmental impairments. Int J Med Sci 18(2):459–473. https://doi.org/10.7150/ijms.51842
27. Cheffer A, Flitsch LJ, Krutenko T et al (2020) Human stem cell-based models for studying autism spectrum disorder-related neuronal dysfunction. Mol Autism 11(1):99. https://doi.org/10.1186/s13229-020-00383-w
28. Burnight ER, Bohrer LR, Giacalone JC et al (2018) CRISPR-Cas9-mediated correction of the 1.02 kb common deletion in CLN3 in induced pluripotent stem cells from patients with batten disease. CRISPR J 1:75–87. https://doi.org/10.1089/crispr.2017.0015
29. Lukovic D, Moreno-Manzano V, Rodriguez-Jimenez FJ et al (2017) hiPSC disease modeling of rare hereditary cerebellar ataxias: opportunities and future challenges. Neuroscientist 23(5):554–566. https://doi.org/10.1177/1073858416672652
30. Gough G, O'Brien NL, Alic I et al (2020) Modeling down syndrome in cells: from stem cells to organoids. Prog Brain Res 251: 55–90. https://doi.org/10.1016/bs.pbr.2019.10.003
31. Tidball AM, Parent JM (2016) Concise review: exciting cells: Modeling genetic epilepsies with patient-derived induced pluripotent stem cells. Stem Cells 34(1):27–33. https://doi.org/10.1002/stem.2203
32. Abu Diab M, Eiges R (2019) The contribution of pluripotent stem cell (PSC)-based models to the study of Fragile X Syndrome (FXS). Brain Sci 9(2). https://doi.org/10.3390/brainsci9020042
33. Gomathi M, Balachandar V (2017) Novel therapeutic approaches: Rett syndrome and human induced pluripotent stem cell technology. Stem Cell Investig 4:20. https://doi.org/10.21037/sci.2017.02.11
34. Devineni A, Tohme S, Kody MT et al (2016) Stepping back to move forward: a current review of iPSCs in the fight against Alzheimer's disease. Am J Stem Cells 5(3):99–106
35. Sances S, Bruijn LI, Chandran S et al (2016) Modeling ALS with motor neurons derived from human induced pluripotent stem cells. Nat Neurosci 19(4):542–553. https://doi.org/10.1038/nn.4273
36. Csobonyeiova M, Polak S, Danisovic L (2020) Recent overview of the use of iPSCs Huntington's disease modeling and therapy. Int J Mol Sci 21(6). https://doi.org/10.3390/ijms21062239
37. Kouroupi G, Antoniou N, Prodromidou K et al (2020) Patient-derived induced pluripotent stem cell-based models in Parkinson's disease for drug identification. Int J Mol Sci 21(19). https://doi.org/10.3390/ijms21197113
38. Weksler BB, Subileau EA, Perriere N et al (2005) Blood-brain barrier-specific properties of a human adult brain endothelial cell line. FASEB J 19(13):1872–1874. https://doi.org/10.1096/fj.04-3458fje
39. Daneman R, Agalliu D, Zhou L et al (2009) Wnt/beta-catenin signaling is required for CNS, but not non-CNS, angiogenesis. Proc Natl Acad Sci U S A 106(2):641–646. https://doi.org/10.1073/pnas.0805165106
40. Liebner S, Corada M, Bangsow T et al (2008) Wnt/beta-catenin signaling controls development of the blood-brain barrier. J Cell Biol 183(3):409–417. https://doi.org/10.1083/jcb.200806024
41. Crone C, Olesen SP (1982) Electrical resistance of brain microvascular endothelium. Brain Res 241(1):49–55. https://doi.org/10.1016/0006-8993(82)91227-6
42. Mizee MR, Wooldrik D, Lakeman KA et al (2013) Retinoic acid induces blood-brain barrier development. J Neurosci 33(4): 1660–1671. https://doi.org/10.1523/JNEUROSCI.1338-12.2013
43. Lippmann ES, Al-Ahmad A, Azarin SM et al (2014) A retinoic acid-enhanced, multicellular human blood-brain barrier model derived from stem cell sources. Sci Rep 4:4160. https://doi.org/10.1038/srep04160

44. Stebbins MJ, Lippmann ES, Faubion MG et al (2018) Activation of RARalpha, RARgamma, or RXRalpha increases barrier tightness in human induced pluripotent stem cell-derived brain endothelial cells. Biotechnol J 13(2). https://doi.org/10.1002/biot.201700093
45. Katt ME, Xu ZS, Gerecht S et al (2016) Human brain microvascular endothelial cells derived from the BC1 iPS cell line exhibit a blood-brain barrier phenotype. PLoS One 11(4):e0152105. https://doi.org/10.1371/journal.pone.0152105
46. Cecchelli R, Aday S, Sevin E et al (2014) A stable and reproducible human blood-brain barrier model derived from hematopoietic stem cells. PLoS One 9(6):e99733. https://doi.org/10.1371/journal.pone.0099733
47. Wilhelm I, Fazakas C, Krizbai IA (2011) In vitro models of the blood-brain barrier. Acta Neurobiol Exp (Wars) 71(1):113–128
48. Qian T, Maguire SE, Canfield SG et al (2017) Directed differentiation of human pluripotent stem cells to blood-brain barrier endothelial cells. Sci Adv 3(11):e1701679. https://doi.org/10.1126/sciadv.1701679
49. Patel R, Alahmad AJ (2016) Growth-factor reduced Matrigel source influences stem cell derived brain microvascular endothelial cell barrier properties. Fluids Barriers CNS 13(6):6. https://doi.org/10.1186/s12987-016-0030-5
50. Neal EH, Marinelli NA, Shi Y et al (2019) A simplified, fully defined differentiation scheme for producing blood-brain barrier endothelial cells from human iPSCs. Stem Cell Rep 12(6):1380–1388. https://doi.org/10.1016/j.stemcr.2019.05.008
51. Canfield SG, Stebbins MJ, Morales BS et al (2017) An isogenic blood-brain barrier model comprising brain endothelial cells, astrocytes, and neurons derived from human induced pluripotent stem cells. J Neurochem 140(6): 874–888. https://doi.org/10.1111/jnc.13923
52. Patel R, Page S, Al-Ahmad AJ (2017) Isogenic blood-brain barrier models based on patient-derived stem cells display inter-individual differences in cell maturation and functionality. J Neurochem 142(1):74–88. https://doi.org/10.1111/jnc.14040
53. Stebbins MJ, Gastfriend BD, Canfield SG et al (2019) Human pluripotent stem cell-derived brain pericyte-like cells induce blood-brain barrier properties. Sci Adv 5(3):eaau7375. https://doi.org/10.1126/sciadv.aau7375
54. Mantle JL, Min L, Lee KH (2016) Minimum transendothelial electrical resistance thresholds for the study of small and large molecule drug transport in a human in vitro blood-brain barrier model. Mol Pharm 13(12):4191–4198. https://doi.org/10.1021/acs.molpharmaceut.6b00818
55. Wilson HK, Canfield SG, Shusta EV et al (2014) Concise review: tissue-specific microvascular endothelial cells derived from human pluripotent stem cells. Stem Cells 32(12): 3037–3045. https://doi.org/10.1002/stem.1797
56. DeStefano JG, Xu ZS, Williams AJ et al (2017) Effect of shear stress on iPSC-derived human brain microvascular endothelial cells (dhBMECs). Fluids Barriers CNS 14(1):20. https://doi.org/10.1186/s12987-017-0068-z
57. Lim RG, Quan C, Reyes-Ortiz AM et al (2017) Huntington's disease iPSC-derived brain microvascular endothelial cells reveal WNT-mediated Angiogenic and blood-brain barrier deficits. Cell Rep 19(7):1365–1377. https://doi.org/10.1016/j.celrep.2017.04.021
58. Vatine GD, Al-Ahmad A, Barriga BK et al (2017) Modeling psychomotor retardation using iPSCs from MCT8-deficient patients indicates a prominent role for the blood-brain barrier. Cell Stem Cell 20(6):831–843.e835. https://doi.org/10.1016/j.stem.2017.04.002
59. Katt ME, Mayo LN, Ellis SE et al (2019) The role of mutations associated with familial neurodegenerative disorders on blood-brain barrier function in an iPSC model. Fluids Barriers CNS 16(1):20. https://doi.org/10.1186/s12987-019-0139-4
60. Raut S, Patel R, Al-Ahmad AJ (2021) Presence of a mutation in PSEN1 or PSEN2 gene is associated with an impaired brain endothelial cell phenotype in vitro. Fluids Barriers CNS 18(1):3. https://doi.org/10.1186/s12987-020-00235-y
61. Berndt P, Winkler L, Cording J et al (2019) Tight junction proteins at the blood-brain barrier: far more than claudin-5. Cell Mol Life Sci 76(10):1987–2002. https://doi.org/10.1007/s00018-019-03030-7
62. Al-Ahmad AJ (2017) Comparative study of expression and activity of glucose transporters between stem cell-derived brain microvascular endothelial cells and hCMEC/D3 cells. Am J Physiol Cell Physiol 313(4):C421–C429. https://doi.org/10.1152/ajpcell.00116.2017
63. Albekairi TH, Vaidya B, Patel R et al (2019) Brain delivery of a potent opioid receptor agonist, Biphalin during ischemic stroke: role

of Organic Anion Transporting Polypeptide (OATP). Pharmaceutics 11(9). https://doi.org/10.3390/pharmaceutics11090467

64. Nakagawa S, Deli MA, Kawaguchi H et al (2009) A new blood-brain barrier model using primary rat brain endothelial cells, pericytes and astrocytes. Neurochem Int 54(3–4): 253–263. https://doi.org/10.1016/j.neuint.2008.12.002
65. Al Ahmad A, Gassmann M, Ogunshola OO (2009) Maintaining blood-brain barrier integrity: pericytes perform better than astrocytes during prolonged oxygen deprivation. J Cell Physiol 218(3):612–622. https://doi.org/10.1002/jcp.21638
66. Weidenfeller C, Svendsen CN, Shusta EV (2007) Differentiating embryonic neural progenitor cells induce blood-brain barrier properties. J Neurochem 101(2):555–565. https://doi.org/10.1111/j.1471-4159.2006.04394.x
67. Demeuse P, Kerkhofs A, Struys-Ponsar C et al (2002) Compartmentalized coculture of rat brain endothelial cells and astrocytes: a syngenic model to study the blood-brain barrier. J Neurosci Methods 121(1):21–31. https://doi.org/10.1016/s0165-0270(02)00225-x
68. Cucullo L, McAllister MS, Kight K et al (2002) A new dynamic in vitro model for the multidimensional study of astrocyte-endothelial cell interactions at the blood-brain barrier. Brain Res 951(2):243–254. https://doi.org/10.1016/s0006-8993(02)03167-0
69. Dehouck MP, Meresse S, Delorme P et al (1990) An easier, reproducible, and mass-production method to study the blood-brain barrier in vitro. J Neurochem 54(5): 1798–1801. https://doi.org/10.1111/j.1471-4159.1990.tb01236.x
70. Janzer RC, Raff MC (1987) Astrocytes induce blood-brain barrier properties in endothelial cells. Nature 325(6101):253–257. https://doi.org/10.1038/325253a0
71. Arthur FE, Shivers RR, Bowman PD (1987) Astrocyte-mediated induction of tight junctions in brain capillary endothelium: an efficient in vitro model. Brain Res 433(1): 155–159. https://doi.org/10.1016/0165-3806(87)90075-7
72. Rhea EM, Logsdon AF, Hansen KM et al (2020) The S1 protein of SARS-CoV-2 crosses the blood-brain barrier in mice. Nat Neurosci. https://doi.org/10.1038/s41593-020-00771-8
73. Buzhdygan TP, DeOre BJ, Baldwin-Leclair A et al (2020) The SARS-CoV-2 spike protein alters barrier function in 2D static and 3D microfluidic in-vitro models of the human blood-brain barrier. Neurobiol Dis 146: 105131. https://doi.org/10.1016/j.nbd.2020.105131
74. Alimonti JB, Ribecco-Lutkiewicz M, Sodja C et al (2018) Zika virus crosses an in vitro human blood brain barrier model. Fluids Barriers CNS 15(1):15. https://doi.org/10.1186/s12987-018-0100-y
75. Kim BJ, Shusta EV, Doran KS (2019) Past and current perspectives in modeling bacteria and blood-brain barrier interactions. Front Microbiol 10:1336. https://doi.org/10.3389/fmicb.2019.01336
76. Kim BJ, Bee OB, McDonagh MA et al (2017) Modeling group B streptococcus and blood-brain barrier interaction by using induced pluripotent stem cell-derived brain endothelial cells. mSphere 2(6). https://doi.org/10.1128/mSphere.00398-17
77. Patel R, Hossain MA, German N et al (2018) Gliotoxin penetrates and impairs the integrity of the human blood-brain barrier in vitro. Mycotoxin Res 34(4):257–268. https://doi.org/10.1007/s12550-018-0320-7
78. Nishihara H, Gastfriend BD, Soldati S et al (2020) Advancing human induced pluripotent stem cell-derived blood-brain barrier models for studying immune cell interactions. FASEB J 34(12):16693–16715. https://doi.org/10.1096/fj.202001507RR
79. Martinez A, Al-Ahmad AJ (2019) Effects of glyphosate and aminomethylphosphonic acid on an isogeneic model of the human blood-brain barrier. Toxicol Lett 304:39–49. https://doi.org/10.1016/j.toxlet.2018.12.013
80. Park TE, Mustafaoglu N, Herland A et al (2019) Hypoxia-enhanced blood-brain barrier chip recapitulates human barrier function and shuttling of drugs and antibodies. Nat Commun 10(1):2621. https://doi.org/10.1038/s41467-019-10588-0
81. Haley MJ, Lawrence CB (2017) The blood-brain barrier after stroke: structural studies and the role of transcytotic vesicles. J Cereb Blood Flow Metab 37(2):456–470. https://doi.org/10.1177/0271678X16629976
82. Zhang Z, Yan J, Shi H (2016) Role of hypoxia inducible factor 1 in Hyperglycemia-exacerbated blood-brain barrier disruption in ischemic stroke. Neurobiol Dis 95:82–92. https://doi.org/10.1016/j.nbd.2016.07.012
83. Yan J, Zhou B, Taheri S et al (2011) Differential effects of HIF-1 inhibition by YC-1 on the

overall outcome and blood-brain barrier damage in a rat model of ischemic stroke. PLoS One 6(11):e27798. https://doi.org/10.1371/journal.pone.0027798

84. Yan J, Zhang Z, Shi H (2012) HIF-1 is involved in high glucose-induced paracellular permeability of brain endothelial cells. Cell Mol Life Sci 69(1):115–128. https://doi.org/10.1007/s00018-011-0731-5
85. Ogunshola OO, Al-Ahmad A (2012) HIF-1 at the blood-brain barrier: a mediator of permeability? High Alt Med Biol 13(3): 153–161. https://doi.org/10.1089/ham.2012.1052
86. Engelhardt S, Patkar S, Ogunshola OO (2014) Cell-specific blood-brain barrier regulation in health and disease: a focus on hypoxia. Br J Pharmacol 171(5):1210–1230. https://doi.org/10.1111/bph.12489
87. Engelhardt S, Al-Ahmad AJ, Gassmann M et al (2014) Hypoxia selectively disrupts brain microvascular endothelial tight junction complexes through a hypoxia-inducible factor-1 (HIF-1) dependent mechanism. J Cell Physiol 229(8):1096–1105. https://doi.org/10.1002/jcp.24544
88. Deli MA, Abraham CS, Kataoka Y et al (2005) Permeability studies on in vitro blood-brain barrier models: physiology, pathology, and pharmacology. Cell Mol Neurobiol 25(1): 59–127. https://doi.org/10.1007/s10571-004-1377-8
89. Page S, Raut S, Al-Ahmad A (2019) Oxygen-glucose deprivation/Reoxygenation-induced barrier disruption at the human blood-brain barrier is partially mediated through the HIF-1 pathway. NeuroMolecular Med 21(4):414–431. https://doi.org/10.1007/s12017-019-08531-z
90. Page S, Munsell A, Al-Ahmad AJ (2016) Cerebral hypoxia/ischemia selectively disrupts tight junctions complexes in stem cell-derived human brain microvascular endothelial cells. Fluids Barriers CNS 13(1):16. https://doi.org/10.1186/s12987-016-0042-1
91. Natah SS, Srinivasan S, Pittman Q et al (2009) Effects of acute hypoxia and hyperthermia on the permeability of the blood-brain barrier in adult rats. J Appl Physiol (1985) 107(4): 1348–1356. https://doi.org/10.1152/japplphysiol.91484.2008
92. Witt KA, Mark KS, Sandoval KE et al (2008) Reoxygenation stress on blood-brain barrier paracellular permeability and edema in the rat. Microvasc Res 75(1):91–96. https://doi.org/10.1016/j.mvr.2007.06.004
93. Yeh WL, Lu DY, Lin CJ et al (2007) Inhibition of hypoxia-induced increase of blood-brain barrier permeability by YC-1 through the antagonism of HIF-1alpha accumulation and VEGF expression. Mol Pharmacol 72(2): 440–449. https://doi.org/10.1124/mol.107.036418
94. Koto T, Takubo K, Ishida S et al (2007) Hypoxia disrupts the barrier function of neural blood vessels through changes in the expression of claudin-5 in endothelial cells. Am J Pathol 170(4):1389–1397. https://doi.org/10.2353/ajpath.2007.060693
95. Kaur C, Sivakumar V, Zhang Y et al (2006) Hypoxia-induced astrocytic reaction and increased vascular permeability in the rat cerebellum. Glia 54(8):826–839. https://doi.org/10.1002/glia.20420
96. Hayashi K, Nakao S, Nakaoke R et al (2004) Effects of hypoxia on endothelial/pericytic co-culture model of the blood-brain barrier. Regul Pept 123(1–3):77–83. https://doi.org/10.1016/j.regpep.2004.05.023
97. Fischer S, Wiesnet M, Marti HH et al (2004) Simultaneous activation of several second messengers in hypoxia-induced hyperpermeability of brain derived endothelial cells. J Cell Physiol 198(3):359–369. https://doi.org/10.1002/jcp.10417
98. Witt KA, Mark KS, Hom S et al (2003) Effects of hypoxia-reoxygenation on rat blood-brain barrier permeability and tight junctional protein expression. Am J Physiol Heart Circ Physiol 285(6):H2820–H2831. https://doi.org/10.1152/ajpheart.00589.2003
99. Schoch HJ, Fischer S, Marti HH (2002) Hypoxia-induced vascular endothelial growth factor expression causes vascular leakage in the brain. Brain 125(Pt 11):2549–2557. https://doi.org/10.1093/brain/awf257
100. Fischer S, Wobben M, Marti HH et al (2002) Hypoxia-induced hyperpermeability in brain microvessel endothelial cells involves VEGF-mediated changes in the expression of zonula occludens-1. Microvasc Res 63(1):70–80. https://doi.org/10.1006/mvre.2001.2367
101. Plateel M, Dehouck MP, Torpier G et al (1995) Hypoxia increases the susceptibility to oxidant stress and the permeability of the blood-brain barrier endothelial cell monolayer. J Neurochem 65(5):2138–2145. https://doi.org/10.1046/j.1471-4159.1995.65052138.x
102. Lee CAA, Seo HS, Armien AG et al (2018) Modeling and rescue of defective blood-brain barrier function of induced brain microvascular endothelial cells from childhood cerebral adrenoleukodystrophy patients. Fluids

Barriers CNS 15(1):9. https://doi.org/10.1186/s12987-018-0094-5

103. Kinarivala N, Morsy A, Patel R et al (2020) An iPSC-derived neuron model of CLN3 disease facilitates small molecule phenotypic screening. ACS Pharmacol Transl Sci 3(5): 931–947. https://doi.org/10.1021/acsptsci.0c00077

104. Clark PA, Al-Ahmad AJ, Qian T et al (2016) Analysis of cancer-targeting Alkylphosphocholine analogue permeability characteristics using a human induced pluripotent stem cell blood-brain barrier model. Mol Pharm 13(9):3341–3349. https://doi.org/10.1021/acs.molpharmaceut.6b00441

105. Lippmann ES, Azarin SM, Palecek SP et al (2020) Commentary on human pluripotent stem cell-based blood-brain barrier models. Fluids Barriers CNS 17(1):64. https://doi.org/10.1186/s12987-020-00222-3

Chapter 4

Induced Pluripotent Stem Cell (iPSC)-Derived Endothelial Cells to Study Bacterial–Brain Endothelial Cell Interactions

Eric R. Espinal, S. Jerod Sharp, and Brandon J. Kim

Abstract

Bacterial meningitis is a serious infection of the central nervous system (CNS) that occurs when blood-borne bacteria are able to exit the cerebral vasculature and cause inflammation. The blood–brain barrier (BBB) and the meningeal blood–CSF barrier (mBCSFB) are composed of highly specialized brain endothelial cells (BECs) that possess unique phenotypes when compared to their peripheral endothelial counterparts. To cause meningitis, bacterial pathogens must be able to interact and penetrate these specialized BECs to gain access to the CNS. In vitro models have been employed to study bacterial–BEC interactions; however, many lack BEC phenotypes. Induced pluripotent stem cell (iPSC) technologies have enabled the derivation of brain endothelial-like cells that phenocopy BECs in culture. Recently, these iPSC-BECs have been employed to examine the host–pathogen interaction at the endothelial brain barriers. Using two clinically relevant human meningeal pathogens, this chapter describes the use of iPSC-BECs to study various aspects of BEC–bacterial interaction.

Key words Induced pluripotent stem cells, Brain endothelial cells, Host–pathogen interaction, Bacterial meningitis, Meningitis

1 Introduction

The endothelial brain barriers, such as the blood–brain barrier (BBB) and the meningeal blood–CSF barrier (mBCSFB), are primarily composed of highly specialized brain endothelial cells (BECs) that serve to maintain proper central nervous system (CNS) homeostasis [1–4]. BECs are unique when compared to peripheral endothelial cells and possess attributes that contribute to their tightly restrictive properties [1, 2, 4, 5]. Namely, BECs express complex tight junctions, exhibit greatly restricted endocytosis, and express an array of nutrient influx and drug efflux transporters [1–5]. In vitro modeling of BECs has relied on culture of

Eric R. Espinal and S. Jerod Sharp contributed equally with all other contributors.

Nicole Stone (ed.), *The Blood-Brain Barrier: Methods and Protocols*,
Methods in Molecular Biology, vol. 2492, https://doi.org/10.1007/978-1-0716-2289-6_4,

primary or immortalized BECs and offer great utility as they can be highly scalable and easy to work with. However, primary and immortalized cells in culture can suffer from interspecies variations and the observation that once removed from the brain microenvironment; BECs tend to lose their unique characteristics [6, 7]. Induced pluripotent stem cell-derived BEC-like cells (iPSC-BECs) have been demonstrated to possess characteristic BEC phenotypes such as tight barrier properties, expression and localization of tight junction proteins, and expression of functional transporters [6–19]. Further development of iPSC technologies and models have enabled the derivation of astrocytes, neurons, and pericytes enabling the ability to model the entire neurovascular unit from a single stem cell source [20, 21]. iPSC-BEC and iPSC neurovascular unit models have been successfully employed to examine various CNS disorders such as Huntington's disease, Allan–Herndon–Dudley syndrome, Alzheimer's disease, amyotrophic lateral sclerosis (ALS), stroke, and others [22–31]. More recently, we and others have utilized iPSC-BECs to model host–pathogen interaction at the endothelial brain barriers [32–36]. The presented methodology provides an updated procedure to derive BEC-like cells from naïve iPSCs and infect them with two different clinically relevant meningeal bacterial pathogens *Neisseria meningitidis* and Group B *Streptococcus* and highlights standard assays to examine aspects of the host–pathogen interaction between bacteria and BECs [37, 38].

2 Materials

2.1 General

1. Micropipettes.
2. 1.5 mL microfuge tubes.

Preparation of materials for iPSC culture and BEC differentiation:

2.1.1 Aliquoting Matrigel

1. Matrigel matrix solution.
2. Biological safety cabinet.
3. Ice and ice bucket.

2.1.2 Matrigel Coating of Tissue Culture Plasticware for iPSC Culture and Differentiation

1. Aliquoted 2.5 mg Matrigel (*see* Subheadings 2.1.1 and 3.1.1).
2. Dulbecco's Modified Eagle Medium/F12 (DMEM/F12).
3. Pipette-aid.
4. 50 mL conical tubes.
5. Serological pipettes.
6. Micropipettes.
7. 6-well tissue culture plastic plate.

8. T75 and T25 tissue culture plastic flasks.
9. Biological safety cabinet.
10. CO_2 incubator.

2.1.3 Stem Cell Maintenance Media Preparation

1. Stem cell basal maintenance medium (e.g., StemFlex, mTeSR1, E8).
2. 50x supplement (supplied with the basal medium).
3. Biological safety cabinet.
4. Pipette-aid.
5. Serological pipettes.

2.1.4 Unconditioned Medium (UM) Preparation

1. DMEM/F12.
2. Knock-out serum replacement (KOSR).
3. Non-essential amino acids (NEAA).
4. GlutaMAX solution.
5. Beta-mercaptoethanol (B-ME).
6. 500 mL sterile vacuum filter.
7. Autoclaved 500 mL glass bottle.
8. Aspirator.
9. Pipette-aid.
10. Serological pipettes.
11. Biological safety cabinet.

2.1.5 Endothelial Cell (EC) Medium +/– bFGF Preparation

1. Human endothelial serum free medium (hESFM).
2. B27 supplement.
3. bFGF.
4. 200 mL vacuum filter unit.
5. 250 mL autoclaved glass bottle.
6. Pipette-aid.
7. Micropipettes.
8. Aspirator.
9. Biological safety cabinet.

2.2 Maintenance and Differentiation of Brain Endothelial-like Cells from Human iPSCs

2.2.1 Maintenance of iPSCs and Passaging

1. iPSC cell line in culture on 6-well plates.
2. Versene.
3. 6-well Matrigel-coated plastic plates.
4. StemFlex media (or equal).
5. CO_2 incubator.
6. Aspirator.

7. Pipette-aid.
8. Serological pipettes.
9. Biological safety cabinet.
10. Inverted light microscope.

2.2.2 Differentiation of Brain Endothelial-like Cells from Human iPSCs (Days −3 to 8)

1. iPSC cell line in culture on 6-well plates.
2. Accutase.
3. 15 mL conical tubes.
4. 50 mL conical tubes.
5. Centrifuge.
6. Hemocytometer.
7. 0.4% trypan blue in PBS.
8. T75 or T25 Matrigel-coated flasks.
9. ROCK inhibitor.
10. StemFlex media.
11. UM media.
12. EC media without bFGF.
13. EC media with bFGF.
14. Retinoic acid.
15. Inverted light microscope.
16. Aspirator.
17. Biological safety cabinet.

2.2.3 Coating of Trans-Wells and Cell Culture Plates for iPSC-BEC Purification

1. 12, 24, and 48-well plastic tissue culture plate.
2. Corning 3460 12-well trans-well insert.
3. Collagen IV.
4. Fibronectin.
5. Sterile tissue culture grade water.
6. CO_2 incubator.
7. Micropipettes.
8. Forceps.
9. 50 mL conical tubes.
10. Pipette-aid.
11. Serological pipettes.
12. Biological safety cabinet.

2.2.4 Purification of iPSC-BECs

1. Coated trans-well inserts and coated well plates.
2. 50 mL conical.
3. Serological pipettes.

4. Pipette-aid.
5. Micropipettes.
6. hESFM.
7. EC without bFGF.
8. EC with bFGF.
9. Retinoic acid.
10. Accutase.
11. Centrifuges.
12. Hemocytometer.
13. Inverted light microscope.
14. Aspirator.
15. Biological safety cabinet.

2.3 Validation of Transepithelial Electrical Resistance (TEER) and Staining for BEC Markers

2.3.1 TEER Measurements to Confirm Barrier Properties

1. Differentiated and purified iPSC-BECs on a trans-well insert.
2. EVOMII TEER meter.
3. EVOMII electrodes.
4. 70% ethanol.
5. Biological safety cabinet.

2.3.2 Immunostaining of BEC Markers

1. Aspirator.
2. Phosphate-buffered saline (PBS).
3. Fetal bovine serum (FBS).
4. Micropipettes.
5. Paraformaldehyde.
6. 100% methanol.
7. Primary antibodies.
8. Secondary antibodies.
9. Biological safety cabinet.
10. Fluorescence microscope.

2.4 Bacterial Preparation and Growth Conditions

2.4.1 Preparation and Growth of *Neisseria Meningitidis*

1. *Neisseria meningitidis* (Nm) serogroup B (strain MC58) in frozen stock.
2. GC agar plates.
3. Sterilized (autoclaved) proteose peptone media (15 g/L protease peptone, 5 g/L NaCl, 0.5 g/L starch, and 20 mL phosphate buffer solution [1.47 mM potassium phosphate monobasic, 0.287 μM potassium phosphate dibasic pH 7.25–7.5]).

4. 2 M $MgCl_2$ filter sterilized.
5. 2 M $NaHCO_3$ filter sterilized.
6. Filter sterilized Kellogg's supplement (400 g/L glucose, 10 g/L glutamine, 20 mg/L thiamine pyrophosphate, and 50 mg/L iron(III) nitrate).
7. Phosphate-buffered saline (PBS).
8. 50 mL conical tubes.
9. 37 °C incubator.
10. 37 °C shaking incubator.
11. Sterilized cotton swabs.
12. Micropipettes.
13. Pipette-aid.
14. Serological pipettes.
15. Spectrophotometer capable of reading at OD_{600}.
16. Sterile plastic cuvette tube.
17. Biological safety cabinet.

2.4.2 Preparation and Growth of Group B *Streptococcus*

1. Group B *Streptococcus* (GBS) serotype III multi-locus sequence type 17 (strain COH1) in frozen stock.
2. Todd Hewitt broth.
3. Phosphate-buffered saline (PBS).
4. 15 mL conical tube.
5. Sterile glass test tubes.
6. Micropipettes.
7. Pipette-aid.
8. Serological pipettes.
9. Spectrophotometer capable of reading at OD_{600}.
10. 37 °C incubator.
11. Biological safety cabinet.

2.5 Infection of iPSC-BECs with Bacteria and Representative Analysis

2.5.1 Initiation of Infection of iPSC-BECs with Bacteria and Incubation

1. Successfully differentiated iPSC-BECs on day 10 (*see* Subheading 2.2).
2. EC without bFGF.
3. PBS.
4. Bacterial resuspended at an OD of 0.4 in PBS (*see* Subheadings 2.4.1 and 2.4.2).
5. 15 mL conical tubes.
6. Micropipettes.
7. Pipette-aid.

8. Serological pipettes.
9. 37 °C + 5% CO_2 incubator.
10. Biological safety cabinet.

2.6 Measurement of Gene Expression by qPCR

2.6.1 Collection and Isolation of RNA from iPSC-BECs

1. RNA isolation kit.
2. On-column DNase from the respective RNA isolation kit manufacturer.
3. Microcentrifuge.
4. RNase free microfuge tubes.
5. Nuclease-free eater.
6. B-ME.
7. Micropipettes.

2.6.2 cDNA Synthesis and qPCR

1. cDNA synthesis kit.
2. Thermocycler.
3. qPCR thermocycler.
4. NanoDrop 2000 spectrophotometer.
5. SYBR green master mix.
6. Nuclease-free water.
7. Primers.
8. 96-well qPCR plate.
9. Micropipettes.

2.7 Western Blot Analysis for Proteins of Interest

2.7.1 Collection of Protein Lysates from Infected iPSC-BECs and Assessment of Protein Concentration

1. Radioimmunoprecipitation assay buffer (RIPA).
2. HALT protease/phosphatase inhibitor cocktail (or equal).
3. Sterile microfuge tubes.
4. BCA assay kit (or equal).
5. 96-well non-treated plates.
6. Colorimetric plate reader.
7. Micropipettes.

2.7.2 Western Blot Analysis

1. SDS-PAGE precast gels.
2. SDS-PAGE buffer.
3. Gel running apparatus.
4. Blotting transfer buffer.
5. Blotting transfer apparatus.
6. Nitrocellulose membranes.
7. Filter paper.

8. Tris-buffered saline (24 mM Tris-HCl, 150 mM NaCl, pH 7.6) + 0.05% Tween 20 (TBST).
9. 5% dried milk dissolved in TBST.
10. Primary antibodies.
11. Secondary antibodies.
12. ECL western blotting substrate.
13. Blotting imager.

2.8 Quantification of Adherent and Invasive Bacteria

2.8.1 Quantification of Adherent and Invasive *Neisseria Meningitidis*

1. GC agar plates.
2. 20% saponin in water.
3. PBS.
4. Gentamicin (10 mg/mL).
5. Sterile 96-well plastic plates.
6. 250 mL glass bottle to collect liquid waste.
7. Micropipettes.
8. Multichannel pipettes.
9. Biological safety cabinet.

2.8.2 Quantification of Adherent and Invasive Group B *Streptococcus*

1. Todd Hewitt agar plates.
2. 0.025% Triton X-100 in PBS.
3. 0.25% Trypsin-EDTA.
4. PBS.
5. Gentamicin.
6. Penicillin G.
7. Sterile 96-well plastic plates.
8. Aspirator.
9. Micropipettes.
10. Multichannel pipettes.
11. Biological safety cabinet.

3 Methods

3.1 Preparation of Materials for iPSC Culture and BEC Differentiation

This protocol describes the process of aliquoting Matrigel and coating plasticware with Matrigel for iPSC culture as previously described [37].

3.1.1 Aliquoting Matrigel

1. The day prior to aliquoting Matrigel, remove Matrigel from −20 °C storage and place at 4 °C to thaw overnight. Do not let the Matrigel warm above 4 °C (*see* **Note 1**).

2. The next day prepare the biological safety cabinet (BSC) and sterilize the work area. Place ice, micropipettes, microfuge tubes, and DMEM/F12 into the BSC.
3. Replace the chilled and thawed Matrigel solution back into the −20 °C freezer for at least 15 minutes to cool the solution (*see* **Note 2**).
4. Place the cooled Matrigel on ice inside the BSC. Using a 1 mL micropipette, aliquot 2.5 mg of the Matrigel solution into 1.5 mL microfuge tubes and immediately place on ice (*see* **Note 3**). Continue until the entire Matrigel solution is aliquoted. Store the Matrigel aliquots at −20 °C.

3.1.2 Matrigel Coating of Tissue Culture Plasticware

1. Prepare and sterilize the BSC and working area.
2. Place all items needed into the BSC such as DMEM/F12, tissue culture plates and flasks to be coated, micropipettes, etc.
3. Thawing of Matrigel and resuspension must happen quickly to prevent Matrigel from warming above 4 °C prior to resuspension (*see* **Note 1**).
4. Add 30 mL of DMEM/F12 to a 50 mL conical tube.
5. Obtain a 2.5 mg Matrigel aliquot from the −20 °C freezer and quickly add approximately 1 mL of the DMEM/F12 from the conical tube using a 1 mL micropipette. Pipette up and down until completely thawed and transfer immediately to the 50 mL conical tube containing the remaining medium.
6. Using a 10 mL serological pipette, gently mix the Matrigel solution by pipetting up and down.
7. Use the Matrigel-DMEM/F12 solution to coat tissue culture plastic using the following volumes as a guide.
   (a) 1 mL per well of a 6-well plate
   (b) 4 mL per T25 flask
   (c) 12 mL per T75 flask
8. Matrigel prepared plasticware must be incubated at least 4 h at 37 °C + 5% CO_2. Typically, Matrigel is prepared the day prior and allowed to coat at least overnight.
9. Matrigel-coated plasticware can be prepared up to 2 weeks before it is used. To prevent wells from drying out add an equal volume of DMEM/F12 as **step 7** periodically.

3.1.3 Stem Cell Maintenance Media Preparation

The following example used is for StemFlex media preparation; however, other maintenance media may be used such as mTeSR1 and E8. Similar steps are required for the preparation of mTeSR1 and E8.

1. Prepare the BSC and sterilize the working area.
2. Using good sterile technique, pipette the entire 50 mL 50× supplement to the 450 mL of StemFlex basal media (*see* **Note 4**).

3. Store the mixed medium at 4 °C for up to 2 weeks or until the medium is completely used.

3.1.4 Unconditioned Medium (UM) Preparation

The goal of this protocol is to generate UM media that is used for the first few days of differentiation of iPSCs into BECs.

1. Prepare and sterilize the working area and the BSC.
2. Use a sterilized (autoclaved) 500 mL glass bottle and tightly affix the 500 mL vacuum filter unit to the top of the glass bottle.
3. Using the pipette-aid and serological pipettes add the following:
   (a) 392.5 mL DMEM/F12
   (b) 100 mL KOSR
   (c) 5 mL NEAA
   (d) 2.5 mL GlutaMAX
4. Using a micropipette, add 3.5 μL of B-ME to the top of the filter reservoir and to the mixed media (*see* **Note 5**).
5. Attach the aspirator hose to the vacuum filter unit and filter the final UM media into the empty glass bottle.
6. Sterilely cap the bottle and store UM at 4 °C for up to 2 weeks.

3.1.5 Endothelial Cell (EC) Medium +/− bFGF Preparation

The goal of this protocol is to generate EC medium +/− bFGF. Both versions are required for successful differentiation of iPSC-BECs.

1. Prepare and sterilize the working area inside the BSC.
2. Use a sterilized (autoclaved) 250 mL glass bottle and tightly affix the 200 mL vacuum filter unit to the top of the glass bottle.
3. Combine 199 mL of hESFM with 1 mL B27 to the top of the filter unit.
   (a) For EC + bFGF, add 20 ng/mL final concentration of bFGF to the top of the filter unit.
4. Attach aspirator hose to the vacuum filter unit and filter the final EC +/− bFGF into the empty glass bottle.
5. Sterilely cap the bottle and store EC +/− bFGF at 4 °C for up to 2 weeks.

3.2 Maintenance and Differentiation of Brain Endothelial-Like Cells from Human iPSCs

3.2.1 Maintenance of iPSCs and Passaging

This protocol utilizes the IMR90-4 cell line as an example, but other iPSC lines such as CC3, CD10, CD12, DF19-9-11 T, 83iCTR, 00iCTR, and CS03iCTRn2 have been successfully utilized for differentiation into BECs [7, 15, 17, 19–21, 33, 34, 39]. The following describes how to maintain iPSCs in culture and how to passage iPSCs to maintain culture (Fig. 1). iPSCs are kept at 37 °C + 5% CO_2 on 6-well plates with 2 mL of StemFlex per well.

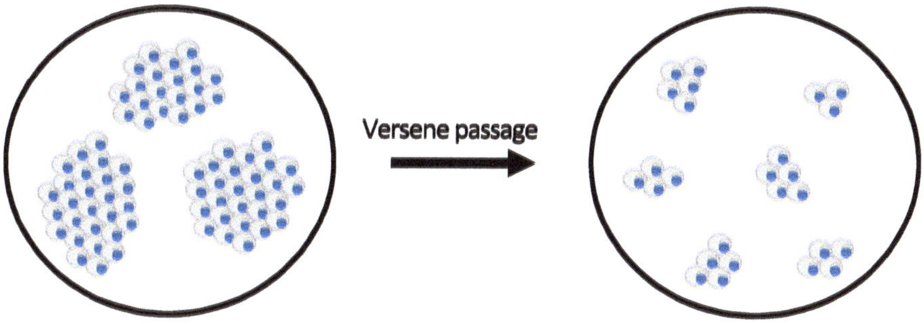

Fig. 1 Schematic of iPSC splitting. When iPSC colonies are not confluent, split using Versene and the next passage should have smaller colonies of stem cells on the plate

1. Each day that iPSCs are in culture, prepare and sterilize the biological safety cabinet and working area.
2. Daily maintain the iPSC culture by aspirating spent StemFlex media and with a serological pipette gently add 2 mL of fresh StemFlex media per well. Replace the maintenance plate back into the 37 °C + 5% CO_2 incubator (*see* **Note 6**).
3. Passage iPSCs before confluence and before the iPSC colonies touch each other.
4. For passage select a single well for passage that is the best iPSC well (*see* **Note 7**).
5. Aspirate the culture medium and add 1 mL of Versene and incubate at 37 °C for 7 minutes.
6. While incubation is taking place, obtain a pre-coated Matrigel 6-well plate previously prepared (*see* Subheading 3.1.2) and replace the Matrigel solution with 2 mL of fresh StemFlex medium per well.
7. After incubating in Versene for 7 minutes, gently aspirate the Versene taking care to not aspirate cells attached to the plate.
8. With a serological pipette, add 6 mL of fresh StemFlex medium and gently rinse the well bottom a few times until all of the cells are completely detached (*see* **Note 8**).
9. From this resuspension, use a serological pipette and seed the new 6-well plate with varying densities. Typically 1:6 or 1:12 for normal maintenance is sufficient (*see* **Note 9**).
10. Move the plate to the 37 °C + 5% CO_2 incubator and distribute the seeded cells equally across the wells by gently moving the plate back and forth, then left and right, pausing between alternating shaking motions to prevent any circulating flow.

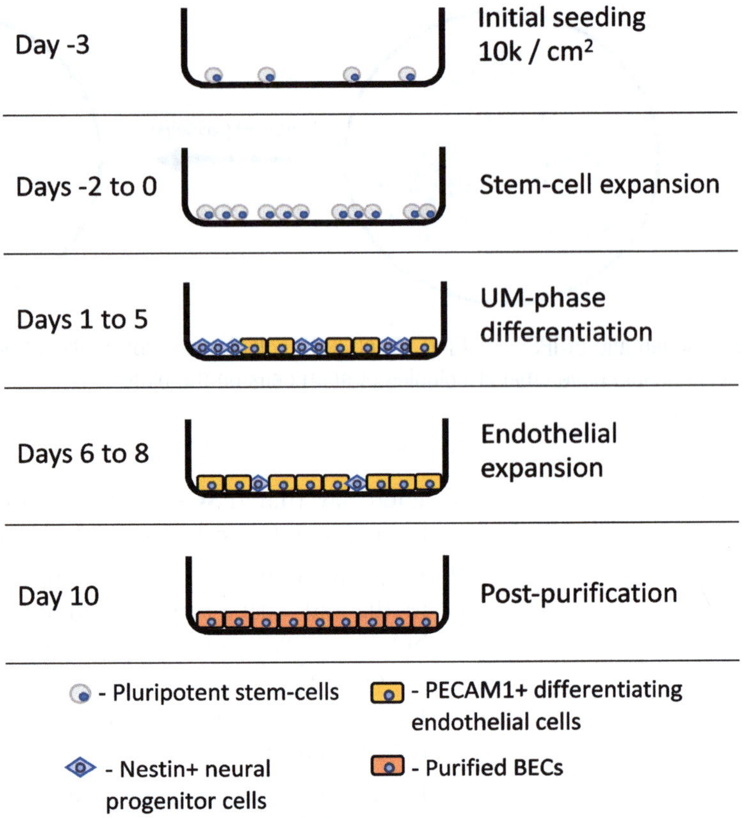

Fig. 2 Schematic of iPSC-BEC differentiation. Seeding of iPSCs for differentiation require initial expansion, followed by differentiation. Finally, after purification of iPSC-BECs, yield a pure population of brain-like endothelial cells

3.2.2 Differentiation of Brain Endothelial-Like Cells from Human iPSCs (Days −3 to 8)

This protocol covers the methods used in the differentiation of brain endothelial-like cells from human iPSCs adapted from previously described methods (Fig. 2) [15, 19, 37, 38]. All of the steps must be conducted inside a properly prepared and sterilized biological safety cabinet.

1. Prior to seeding for differentiation, split the best looking well for maintenance (*see* Subheading 3.2.1). Depending on scale and need, select up to four other wells to be split for initiation of differentiation on day −3.
2. Aspirate StemFlex media and add 1 mL of Accutase per well using a serological pipette, and incubate at 37 °C for 7 minutes.
3. After the incubation cells will be lifted off of the plate. Resuspend the cells in the Accutase using a 1 mL pipette ensuring the cells are all washed off of the plate and in single cell suspension.

4. Transfer the resuspended cells from each well and combine into a 15 mL conical tube with at least 2 mL of fresh StemFlex media per 1 mL of resuspended cells.
5. Spin down the cell suspension at 1500 × g for 5 minutes at room temperature.
6. Resuspend the cell pellet in 1 mL of StemFlex media per well of iPSCs used (e.g., if 4 wells were Accutased and resuspended, resuspend the pellet in 4 mL of fresh StemFlex) and count the cells using a hemocytometer (*see* **Note 10**).
7. Calculate the density of resuspended cells and determine what is required to reach a final seeding density of 10,000 cells per cm^2.
8. For beginning a differentiation in a T75 flask, add 7.5×10^5 cells to 12 mL of StemFlex along with ROCK inhibitor (Y27632 dihydrochloride) at a final concentration of 10 μM (*see* **Note 11**).
9. Distribute the cells equally by shaking the flask back and forth and left to right pausing between steps. Incubate in the 37 °C + 5% CO_2 incubator.
10. The next day (−2) aspirate the spent StemFlex media + ROCK inhibitor and replace with fresh 12 mL of StemFlex (*see* **Note 12**).
11. On day −1 replace the spent StemFlex media with 12 mL of fresh StemFlex.
12. The next day (day 0), initiate the differentiation by aspirating the spent StemFlex media with 12 mL of UM media. Replace UM daily until day 5 of the differentiation (*see* **Note 13**).
13. On day 6, selectively expand the endothelial cell population by aspirating the spent UM media and replacing with EC media + bFGF + 10 μM RA (*see* **Note 14**), and incubate in this EC + bFGF + RA for 2 days without replacing the media on day 7.

3.2.3 Coating of Trans-Wells and Cell Culture Plates for iPSC-BEC Purification

This protocol explains the process of coating of cell culture plates and trans-well inserts with a mixture of collagen IV and fibronectin for the purification of BECs on day 8 [37]. All of the steps must be conducted inside a properly prepared and sterilized biological safety cabinet.

1. Coating of collagen IV and fibronectin occurs on day 7 for purification on day 8.
2. Calculate how much volume is required for trans-wells and cell culture plates. Use the following estimations as a guide for each well.

(a) 12-well trans-well inserts—200 μL

(b) 24-well plate—250 μL

(c) 48-well plate—125 μL

3. Transfer the necessary trans-wells carefully and sterilely with forceps into a fresh sterile 12-well plate.
4. For the coating of trans-wells, combine collagen IV, fibronectin, and sterile tissue culture grade water at a ratio of 4:1:5 in a 50 mL conical tube.
5. Using a micropipette, transfer the collagen IV, fibronectin, and water to the top of the trans-well inserts.
6. For the coating of cell culture plates, combine collagen IV, fibronectin, and sterile tissue culture grade water at a ratio of 4:1:45 in a 50 mL conical tube.
7. Using a micropipette, transfer the collagen IV, fibronectin, and water to the wells to be coated.
8. Incubate the trans-wells and the cell culture plates at least 4 h or incubate at 37 °C + 5% CO_2 overnight.

3.2.4 Purification of iPSC-BECs

This portion of the protocol explains the process of purifying iPSC-BECs by subculturing the differentiated cells onto the collagen IV- and fibronectin-coated plates or trans-well inserts [37, 38]. All of these steps must be conducted inside a properly prepared and sterilized biological safety cabinet.

1. Aspirate EC medium and add Accutase (12 mL per T75 flask or 4 mL per T25 flask).
2. Incubate at 37 °C + 5% CO_2 until at least 90% of cells have detached from the flask (*see* **Note 15**).
3. During the incubation time, remove the collagen IV/fibronectin coating solution from previously prepared plates and trans-well inserts and allow them to dry in the BSC. This process takes approximately 20 minutes for the trans-well inserts to dry.
4. Once cells have detached, use a 10 mL serological pipette to vigorously wash the attached cells off the flask and pipette up and down to achieve a single cell suspension (*see* **Note 16**).
5. Transfer the resuspended cells to a fresh 50 mL conical tube and dilute with at least an equal volume of fresh hESFM. Mix the cell suspension by pipetting up and down with a 10 mL serological pipette and count cells using a hemocytometer. Calculate the total cells in the entire 50 mL conical tube (e.g., 1 million cells/mL in 24 mL would yield 24 million cells).
6. Pellet the cells at 1500 × g for 10 minutes.

7. Resuspend the cells in an appropriate volume of freshly prepared EC + bFGF + RA to achieve a final cell suspension of 2×10^6 cells/mL for seeding on membrane inserts. For example, if there are 24 million cells in total, resuspend in 12 mL of EC + bFGF + RA.
8. To seed one million cells onto each coated 12-well trans-well insert, add 500 μL on the top of a 12-well insert and 1.5 mL of EC + bFGF + RA medium on the bottom (*see* **Note 17**).
9. For seeding on 24- and 48-well plates, dilute the cell suspension 1:2 with EC + bFGF + RA to a final concentration of 1×10^6 cells/ mL. Seed 500,000 cells onto coated 24-well plates by adding 500 μL of this suspension to each well and 250,000 cells onto coated 48-well plates by adding 250 μL to each well.
10. Distribute the cells equally by shaking the plates back and forth and left to right pausing between steps. Incubate in the 37 °C + 5% CO_2 incubator overnight.
11. The following day, carefully aspirate the media from the well plates and trans-wells both top and bottom (*see* **Note 18**).
12. Replace the same volume with EC without bFGF and without RA and incubate overnight at 37 °C + 5% CO_2.
13. The final day (10) the iPSC-BECs are ready for infection experiments (*see* **Note 19**).

3.3 Validation of TEER and Staining for BEC Markers

3.3.1 TEER Measurements to Confirm Barrier Properties

This protocol describes how to measure trans-endothelial electrical resistance across the layer of purified iPSC-BECs on a trans-well using a voltmeter. Typically TEER is read on day 9 after differentiation and day 10 [37, 38]. Measurements should be conducted prior to changing media and with sterilized electrodes in a biological safety cabinet (BSC).

1. Prepare the BSC and sterilize the working area.
2. Spray off the EVOMII and electrode with 70% ethanol and place inside the BSC.
3. Disinfect the electrode by submerging the "chop sticks" in 70% ethanol for at least 5 minutes (*see* **Note 20**).
4. Remove electrode from ethanol and allow to dry completely while not allowing it to touch anything else in the hood.
5. Retrieve the iPSC-BECs on trans-well inserts from the incubator and measure TEER by ensuring the short side of the electrode is submerged on the top of the trans-well, while the long side of the electrode is submerged in the bottom compartment of the trans-well (*see* **Note 21**).

Marker	Antibody	Source and Clone or Product Number	Fixative	Dilution	Block Buffer	Secondary Antibody
PECAM-1 (CD31)	Rabbit polyclonal Anti-CD31	Thermo Fisher PA5-16301	Ice cold Methanol	1:25	10% Serum	Anti-Rabbit Alexa Fluor 488
Glut-1	Mouse IgG2a Anti-Glut1	Thermo Fisher MA5-11315 (SPM-498)	Ice cold Methanol	1:100	10% Serum	Anti-Mouse Alexa Fluor 488
VE-Cadherin	Mouse IgG1 Anti-VE-Cadherin	Santa Cruz sc-52752 (BV9)	Ice cold Methanol	1:25	10% Serum	Anti-Mouse Alexa Fluor 488
Occludin	Mouse IgG1 Anti-Occludin	Thermo Fisher 33-1500 (OC-3F10)	Ice cold Methanol	1:200	10% Serum	Anti-Mouse Alexa Fluor 488
ZO-1	Mouse IgG1 Anti-ZO1	Thermo Fisher 33-9100 (ZO1-1A12)	Ice cold Methanol	1:100	10% Serum	Anti-Mouse Alexa Fluor 488
Claudin-5	Mouse IgG1 Anti-Claudin-5	Thermo Fisher 35-2500 (4C3C2)	Ice cold Methanol	1:50	10% Serum	Anti-Mouse Alexa Fluor 488
P-gp	Mouse IgG1 Anti-Pgp	Thermo Fisher MA5-13854 (F4)	Ice cold Methanol	1:25	10% Serum	Anti-Mouse Alexa Fluor 488
BCRP	Mouse IgG2b Anti-BCRP	Sigma-Millipore MAB4155 (5D3)	4% PFA	1:50	10% Serum	Anti-Mouse Alexa Fluor 488
MRP1	Mouse IgG1 Anti-MRP1	Sigma-Millipore MAB4100 (QCRL-1)	Ice cold Methanol	1:25	10% Serum	Anti-Mouse Alexa Fluor 488

Fig. 3 Antibodies and staining conditions for the validation of iPSC-BECs

3.3.2 Immunostaining of BEC Markers

This protocol aims to validate the quality of the fully differentiated and purified iPSC-BECs. Monolayers are stained for characteristic markers of BECs on day 10 of the process as previously described [37]. The example provided assumes that iPSC-BECs were purified onto a coated 48-well plate; however, this protocol can be scaled to whichever plate is used.

1. Work within a BSC until after the cells are fixed, and then it can be conducted at the bench top in a non-sterile environment.
2. Remove plate with purified iPSC-BECs and place inside of the BSC.
3. Aspirate medium and wash 1× with an equal volume of PBS (250 μL on a 48-well plate).
4. Fix cells in accordance with the particular marker being stained either 4% paraformaldehyde or ice-cold 100% methanol for 15 minutes at room temperature. See Fig. 3 for fixing conditions.
5. Aspirate fixing reagent and wash 3× with PBS and block with 10% FBS in PBS at room temperature for at least 1 h.
6. Aspirate blocking reagent and incubate with primary antibodies diluted in blocking solution and incubate overnight at 4 °C. See Fig. 3 for primary antibody dilutions.
7. The following day, aspirate the primary antibodies and wash 3× with PBS before adding the secondary antibodies diluted in the blocking solution at room temperature for at least 1 h protected from light. See Fig. 3 for secondary antibody dilutions [37].
8. Aspirate the secondary antibodies and wash 2× in PBS. Then add DAPI at a 1:5000 dilution to PBS for the final wash and incubate for 15 minutes at room temperature.
9. Image markers on a fluorescence microscope.

The protocol outlines how to grow and prepare Nm for use during infection experiments on iPSC-BECs modified from Fig. 4 (*see* **Note 22**) [40].

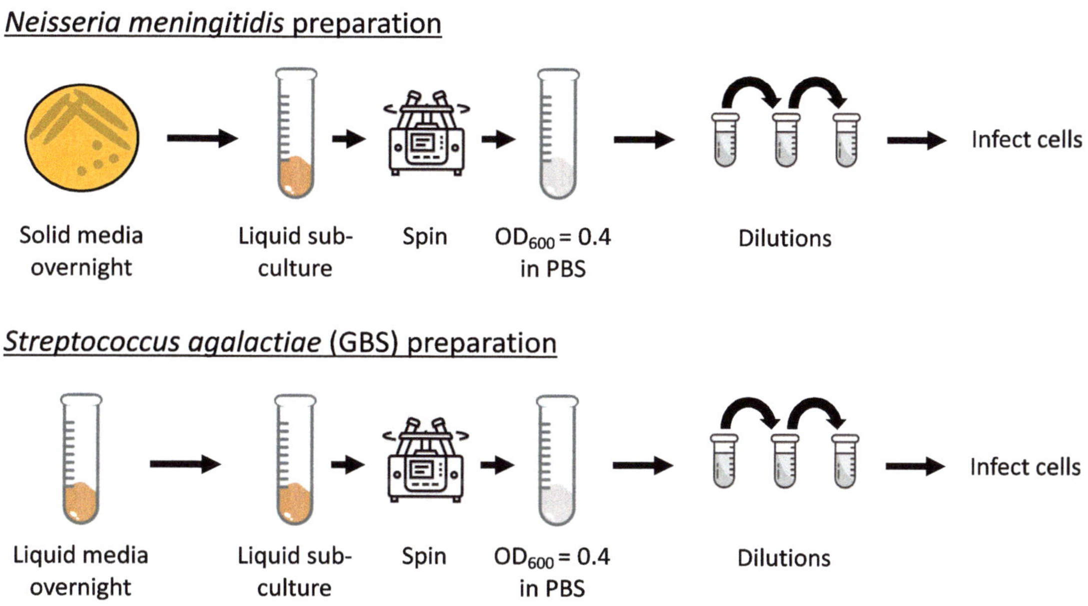

Fig. 4 Schematic of bacterial preparation. Starting *Neisseria meningitidis* on solid media to limit liquid waste, bacteria are grown and prepared for infection (top). Starting GBS in liquid media, bacteria are grown and prepared for infection (bottom)

3.4 Bacterial Preparation and Growth Conditions

3.4.1 Preparation and Growth of Neisseria Meningitidis

1. The day prior to the experiment, obtain the frozen stock of Nm and using a sterile pipette tip streak out on to a GC agar plate. Incubate this plate at 37 °C + 5% CO_2 overnight.
2. The following day make PPM + supplements (PPM+) by combining 10 mL PPM with 50 μL of 2 M $MgCl_2$, 50 μL of 2 M $NaHCO_3$, and 100 μL Kellogg's supplement in a 50 mL conical tube.
3. Using a sterile cotton swab, inoculate the 10 mL of PPM+ in the 50 mL conical tube (*see* **Note 23**).
4. With the lid tightly sealed, incubate shaking at 200 RPM at 37 °C for at least 1.5 hs.
5. After incubation, spin down the culture in the 50 mL conical at 4000 × g for 10 minutes.
6. After pelleting and working inside the BSC, use a 1 mL micropipette and carefully aspirate into a 250 mL glass bottle. Work carefully ensuring to not generate aerosols.
7. Working in the BSC, resuspend the pellet in 250 μL of sterile PBS.
8. In the sterile plastic capped cuvette, add 3 mL of fresh PBS and with a micropipette carefully add the bacterial suspension until an OD_{600} of 0.4 is achieved. When the OD_{600} = 0.4, the suspension is approximately 1×10^8 CFU/mL and can be used for calculating the exact MOI for infection experiments.

3.4.2 Preparation and Growth of Group B Streptococcus

This protocol describes the method to prepare GBS for infection experiments on iPSC-BECs adapted from well-established protocols (Fig. 4) [41–44].

1. On the day prior to infection experiments and working sterilely, start an overnight culture by inoculating from a frozen stock into 10 mL of Todd Hewitt broth (THB). Incubate overnight at 37 °C in static culture.
2. The next day, add 5 mL of THB sterilely into a sterilized glass test tube with a cap and subculture the overnight culture by adding between 150 and 200 μL of the overnight culture into the 5 mL fresh THB. Incubate at 37 °C for 2–2.5 h or until the culture reaches an OD_{600} of 0.4 to 0.6.
3. After incubation, transfer the culture to a fresh 15 mL conical tube and pellet the culture by spinning at 4000 × g for 10 minutes.
4. Aspirate the media careful not to disturb the pellet and resuspend the pellet with 250 μL of sterile PBS.
5. In a fresh glass test tube, add 4 mL of fresh sterile PBS and with a micropipette carefully add the bacterial suspension until an OD_{600} of 0.4 is achieved. When $OD_{600} = 0.4$, the suspension is approximately 1×10^8 CFU/mL and can be used for calculating the exact MOI for infection experiments.

3.5 Infection of iPSC-BECs with Bacteria

3.5.1 Initiation of Infection of iPSC-BECs with Bacteria and Incubation

This protocol describes the general steps to infect iPSC-BECs with either Nm or GBS for experimentation. Following these steps will allow for the infection of iPSC-BECs, and all downstream applications can be divergent with different incubation times or sample collection. All steps should be conducted inside of a BSC, apart from the incubation steps.

1. While bacteria are incubating (*see* Subheadings 3.4.1 and 3.4.2) retrieve day 10 differentiated iPSC-BECs and aspirate the spent media taking care to not disrupt the monolayer.
2. Replace the aspirated media with 400 μL of fresh EC without bFGF and incubate in the "dirty" infection 37 °C + 5% CO_2 incubator.
3. On 24-well plates, iPSC-BECs are estimated at about 1×10^5 cells per well. Calculate the multiplicity of infection (MOI) that is required for the experiment. This protocol will use an MOI of 10 as an example.
4. Prepare bacteria per the protocols described in Subheadings 3.4.1 and 3.4.2.
5. In a sterile 15 mL conical tube, dilute the OD_{600} 0.4 (1×10^8 CFU/mL) 1:10 in EC without bFGF to achieve a concentration of 1×10^7 CFU/mL.

6. Retrieve the 24-well plate to be infected, and carefully pipette 100 µL of the diluted culture into each well to be infected. This will place 1×10^6 CFU/well achieving an MOI of 10.
7. Place the infected plate into the infection 37 °C + 5% CO_2 incubator for the appropriate time required for the specific experiment.

3.6 Measurement of Gene Expression by qPCR

The following procedure outlines the steps to perform quantitative PCR on samples isolated from infected iPSC-BECs. The process will go through the steps for RNA extraction, generation of cDNA, and qPCR analysis. These techniques are common and gene expression analysis can be conducted on any gene of interest with properly designed primer sets.

3.6.1 Collection and Isolation of RNA from iPSC-BECs

1. After infecting iPSC-BECs for the time necessary, isolate RNA by aspirating the media into the liquid waste container and add the first buffer from the RNA extraction kit (*see* **Note 24**). Lyse cells by pipetting up and down with the 1 mL micropipette. Store samples at −80 °C prior to extraction.
2. Follow instructions for RNA isolation to extract RNA from cell lysates. To avoid any potential genomic DNA contamination, conduct the on-column DNase treatment.
3. Elute the purified extracted RNA with nuclease-free water into a sterile microfuge tube. Keep RNA samples on ice to prevent any RNase activity. These can be stored at −80 °C.

3.6.2 cDNA Synthesis and qPCR

1. Thaw RNA on ice.
2. Estimate RNA concentrations using a NanoDrop 2000 spectrophotometer.
3. Calculate the volume needed to conduct cDNA synthesis with at least 200 ng of RNA, nuclease-free water, and enzyme mix.
4. With the appropriate cDNA synthesis kit, follow instructions per the manual and run the program on a standard thermocycler.
5. After synthesis, dilute the cDNA in nuclease-free water 1:10 and store cDNA at −20 °C for up to 1 year.
6. Prepare validated qPCR primers to a working concentration of 10 mM in nuclease-free water (*see* **Note 25**).
7. To a 96-well qPCR plate, add 1 µL of diluted cDNA sample to each well along with 10.5 µL of sterile water.
8. For each well to be run, prepare a master mix of primers and SYBR green (0.5 µL of forward and 0.5 µL of reverse primers per well along with 12.5 µL of SYBR green). For example, if you are running 6 wells with a given primer set, add 3 µL of forward primer, 3 µL of reverse primer, and 75 µL of SYBR green master mix (*see* **Note 26**).

9. Add 13.5 μL of the primers/SYBR green master mix to each well.
10. Perform the following program on a qPCR machine.
    (a) 95 °C for 15 minutes.
    (b) 95 °C for 15 seconds.
    (c) 60 °C for 1 minute.
    (d) Read the plate and GO TO step B 45×.
    (e) Optional melt curve: 30–99 °C in 1 °C increments (*see* **Note 27**).
11. Use the ΔΔCT calculation to compare gene expression levels to that of a housekeeping gene such as GAPDH.

3.7 Western Blot Analysis for Proteins of Interest

The following describes the process to generate protein lysates from infected iPSC-BECs and estimate relative protein abundance using western blotting.

3.7.1 Collection of Protein Lysates from Infected iPSC-BECs and Assessment of Protein Concentration

1. Prepare lysis buffer by combining RIPA buffer with the 100× protease/phosphatase inhibitor cocktail at a ratio of 100:1. Be sure to prepare enough RIPA buffer + inhibitors to lyse all of the wells necessary.
2. To collect protein lysates and after incubating the infected iPSC-BECs for the appropriate time, move the infected plate inside of the BSC.
3. Carefully aspirate the infected media into a liquid waste collection bottle.
4. For each 24-well infected, add 200 μL of the RIPA + inhibitor mix to each well.
5. Using a pipette tip, scrape the plate to physically remove all of the cells attached to the bottom of the well.
6. Transfer the entire volume to a labeled microfuge tube. These samples can be stored at −20 °C for future analysis if necessary.
7. Estimate protein concentration in each sample by following the steps in the BCA kit of choice and reading on a colorimetric plate reader. Be sure to include a standard BSA curve beginning with 2000 μg/mL and performing seven 1:2 fold dilutions.
8. Analyze by performing linear regression on the standard curve. With the resulting regression, estimate the protein concentration for each sample being tested.

3.7.2 Western Blot Analysis

1. Onto a polyacrylamide gel, load uniform amounts of protein for all samples along with a lane with protein standards, and run the gel at 100–150 V until the loading dye nears the bottom of the gel (*see* **Note 28**).

Marker	Antibody	Source and Clone or Product Number
p38	Rabbit polyclonal Anti-p38	Sigma-Merck M0800
phospho p38	Rabbit polyclonal Anti-p38 pTyr322	Sigma-Merck SAB4301534
JNK	Rabbit polyclonal Anti-JNK	Sigma-Merck J4500
phospho JNK	Mouse Anti-JNK pThr183 pTyr185	Sigma-Merck ZRB1173 (3F7)
p44/42 ERK1/2	Mouse Anti-ERK1/2	Cell Signaling Technologies 4696 (L34F12)
phospho p44/42 ERK1/2	Rabbit polyclonal Anti-ERK1/2 pThr202 pTyr204	Cell Signaling Technologies 9101
FAK	Rabbit polyclonal Anti-FAK	Sigma-Merck 06-543
phospho FAK	Rabbit polyclonal Anti-FAK pTyr397	Sigma-Merck SAB4504181

Fig. 5 Antibodies for representative proteins of interest for western blotting

2. Transfer the gel to a nitro-cellulose membrane utilizing a blotting transfer apparatus. Assemble the transfer sandwich as follows: sponges, filter paper, gel, membrane, filter paper, sponges. Transfer at 350 mA for 1–1.5 h.
3. After the transfer, block membrane in 5% milk in TBST at room temperature for at least 1 h (*see* **Note 29**).
4. Dilute primary antibodies in blocking buffer either 1:500 or 1:1000 and proceed to incubate overnight at 4 °C on a rocker. Primary antibody examples for some selected proteins of interest are outlined in Fig. 5 (*see* **Note 30**).
5. On the next day, remove the primary antibody dilution and wash 3× for 10 minutes each at room temperature in TBST.
6. In the same blocking buffer, dilute secondary antibodies 1:2000 and incubate at room temperature rocking for at least 1 h.
7. Remove the secondary antibody dilution and wash 3× for 10 minutes each at room temperature in TBST.
8. Mix ECL substrate 1:1 as outlined in the kit and wash over the membrane for at least 30 seconds.
9. Image the blot within 30 minutes of ECL incubation using an imager.

3.8 Quantification of Adherent and Invasive Bacteria

The following protocol requires the generation of iPSC-BECs and infection with bacteria as outlined in Subheading 3.5. The protocol provides a method to quantify the number of bacteria either adherent or invaded to a monolayer of iPSC-BECs and is modified from previous protocols [33, 36]. All steps outlined must be performed in a properly sterilized biological safety cabinet.

3.8.1 Quantification of Adherent and Invasive Neisseria Meningitidis

Quantification of Adherent *Neisseria Meningitidis*

1. Incubate bacteria for the appropriate time for the specific experiment. Typically, this can be 30 min at 37 °C + 5% CO_2 as outlined in Subheading 3.5.
2. After the time point has been reached, hand aspirate the media using a 1 mL micropipette into a liquid waste container and wash 1× with 500 μL of PBS. When handling Nm, DO NOT use the vacuum aspirator (*see* **Note 31**).

3. Aspirate the PBS by hand aspiration with a 1 mL micropipette into a liquid waste container.
4. Add 300 μL of 1% saponin in PBS and mix by pipetting up and down at least 20×.
5. On a 96-well plate, perform 5× 1:10 serial dilutions in sterile PBS.
6. Using a multichannel pipette, plate out 20 μL samples of each dilution onto GC agar plates and incubate overnight at 37 °C + 5% CO_2.
7. The following day count the colonies and record on which dilution countable colonies can be enumerated. It is important to count between 20 and 200 colonies to obtain reliable results.
8. Data can be presented as CFU/initial inoculum.

Quantification of Invasive *Neisseria Meningitidis*

1. Incubate bacteria for the appropriate time for the specific experiment. Typically, this can be between 2 and 8 hs at 37 °C + 5% CO_2 as outlined in Subheading 3.5.
2. After the time point has been reached, hand aspirate the media using a 1 mL micropipette into a liquid waste container and wash 1× with 500 μL of PBS. When handling Nm, DO NOT use the vacuum aspirator (*see* **Note 31**).
3. Aspirate the PBS by hand aspiration with a 1 mL micropipette into a liquid waste container.
4. Add 1 mL of EC without bFGF + 200 μg/mL gentamicin per well and incubate at 37 °C + 5% CO_2 for 2 h.
5. After incubation in antibiotic media, hand aspirate the media using a 1 mL micropipette into a liquid waste container and wash 1× with 500 μL of PBS. When handling Nm, DO NOT use the vacuum aspirator (*see* **Note 31**).
6. Add 300 μL of 1% saponin in PBS and mix by pipetting up and down at least 20×.
7. On a 96-well plate, perform 5× 1:10 serial dilutions in sterile PBS.
8. Using a multichannel pipette, plate out 20–μL samples of each dilution onto GC agar plates and incubate overnight at 37 °C + 5% CO_2.
9. The following day count the colonies and record on which dilution countable colonies can be enumerated. It is important to count between 20 and 200 colonies to obtain reliable results.
10. Data can be presented as CFU/initial inoculum.

3.8.2 Quantification of Adherent and Invasive Group B Streptococcus

Quantification of Adherent Group B *Streptococcus*

1. Incubate bacteria for the appropriate time for the specific experiment. Typically, this can be for 30 minutes at 37 °C + 5% CO_2 as outlined in Subheading 3.5.
2. After the time point has been reached, aspirate the media using the aspirator.
3. Wash the monolayer 5× with 500 μl PBS aspirating the washing PBS after each step.
4. Remove cells off of the plate with 100 μl of Trypsin-EDTA solution and incubate at 37 °C + 5% CO_2 for 5–10 minutes.
5. Selectively lyse the human cells by adding 400 μl 0.025% Triton X-100 in PBS and pipetting up and down 30×.
6. On a 96-well plate, perform 5× 1:10 serial dilutions in sterile PBS.
7. Using a multichannel pipette, plate out 20 μl samples of each dilution onto Todd Hewitt agar plates and incubate overnight at 37 °C.
8. The following day, count the colonies and record on which dilution countable colonies can be enumerated. It is important to count between 20 and 200 colonies to obtain reliable results.
9. Data can be presented as CFU/initial inoculum.

Quantification of Invasive Group B *Streptococcus*

1. Incubate bacteria for the appropriate time for the specific experiment. Typically, this can be between 2 and 8 hs at 37 °C + 5% CO_2 as outlined in Subheading 3.5.
2. After the time point has been reached, aspirate the media using the aspirator.
3. Wash the monolayer 3× with 500 μl PBS aspirating the washing PBS after each step.
4. Add 1 mL of EC without bFGF + 100 μg/mL gentamicin +5 μg/mL penicillin G per well. Incubate the plate at 37 °C + 5% CO_2 for 2 h.
5. After incubation with antibiotic media, wash the monolayer 3× with 500 μl PBS aspirating the washing PBS after each step.
6. Remove cells off the plate with 100 μl of Trypsin-EDTA solution and incubate at 37 °C + 5% CO_2 for 5–10 minutes.
7. Selectively lyse the human cells by adding 400 μl 0.025% Triton X-100 in PBS and pipetting up and down 30×.
8. On a 96-well plate, perform 5× 1:10 serial dilutions in sterile PBS.
9. Using a multichannel pipette, plate out 20 μl samples of each dilution onto Todd Hewitt agar plates and incubate overnight at 37 °C.

10. The following day count the colonies and record on which dilution countable colonies can be enumerated. It is important to count between 20 and 200 colonies to obtain reliable results.
11. Data can be presented as CFU/initial inoculum.

4 Notes

1. Work quickly and smoothly when handling the Matrigel and always work with Matrigel and aliquots on ice. Matrigel forms a gel above 4 °C and cannot be aliquoted or used once it has solidified.
2. Replacing the Matrigel into the −20 °C for a few minutes prior to aliquoting chills the Matrigel without freezing. This allows for aliquoting to take place while never letting the Matrigel's temperature to rise above 4 °C.
3. Concentrations of the specific lots of Matrigel can vary so be sure to check the manufacturers Certificate of Analysis for the lot being used to determine the volume required for aliquots. Usually Matrigel is around 10 mg/mL.
4. Never pour the 50× supplement into the basal medium. Pipette using a serological pipette as this is the most sterile way to mix these components.
5. Be sure to add the B-ME last when making UM. Adding B-ME first will damage the filter unit and the sterilization process will fail. Also be sure to note the volume of B-ME, as it is common to accidentally add 3.5 mL instead of 3.5 μL. Adding too much B-ME will kill all cells that this UM is used on.
6. Stem cell maintenance media generally contains bFGF that is relatively unstable; however, it is required for the maintenance of stem cells in a pluripotent state. Replacing the media daily ensures that the concentrations of bFGF remain sufficient to keep the pluripotent potential of the cells. Current media such as StemFlex use a stabilized recombinant bFGF and may be changed less often.
7. Selecting the best and most healthy looking well for passage to maintain culture will ensure the longevity of the maintenance culture. Look for a well that possesses well-formed colonies, spaces between colonies, and little to no aberrant colonies. The edges of the iPSC colonies should be sharp and clearly defined.
8. When passaging iPSCs for maintenance, take care not to pipette too fast. iPSCs in culture are healthier when they remain clumped and pipetting too quickly could break them apart reducing the attachment and viability.

9. Using the ratios of 1:6 and 1:12, iPSCs will generally be ready for split approximately twice per week. iPSCs needed to be maintained at a density sufficient to ensure growth. Alternate seeding may be required to keep the maintenance culture going. Using 1:2 or 1:3 may be used; however, these densities may be needed to be split more often.
10. It is recommended to dilute 1:1 with 0.4% trypan blue to distinguish between live and dead cells when counting. Depending on the initial density of iPSCs, one well of a 6-well plate will typically yield about 1–2 × 10^6 cells.
11. It is critical to add ROCK inhibitor at this step to enhance survival of the dissociated single stem cells.
12. On day −2, the cells should appear evenly spread across the flask in singlets exhibiting a spread, mesenchymal-like morphology due to the ROCK inhibitor treatment. The cells should return to their normal epithelial phenotype after the replacement of fresh StemFlex media.
13. The cells typically reach confluence after 2–3 days in UM which can be observed with the naked eye or through an inverted bright field microscope. As the differentiation progresses, nestin positive "neural tracts" become visible with PECAM1 positive cells in between as previously observed and described [7].
14. Prior to adding the EC + bFGF, add RA to a final concentration of 10 μM fresh on day 6. Successful differentiation can also be achieved without supplement of RA at days 6 and 8; however, omission of RA will yield BECs with significantly reduced TEER.
15. Accutase of differentiated iPSC-BECs can take up to an hour incubation. The addition of RA to the process extends this time. RA absent differentiations typically take about 20–30 minutes.
16. Successfully achieving a single cell suspension is critical for accurately and reliably enumerating the cells to achieve BEC monolayers.
17. Seeding of trans-well inserts is scalable to the size of the inserts. The protocol is set up for 12-well inserts; however if 24-well inserts are utilized, 500 k cells can be seeded with 250 μl of the cell suspension.
18. Aspirating the EC + bFGF + RA from the purified trans-wells can be tricky. Take care not to physically disrupt the monolayer of cells or puncturing the trans-well insert. Additionally, some media is often trapped underneath the trans-well insert itself in the bottom compartment. Using the aspirator carefully nudge the trans-well on the outside to fully aspirate this media.

19. Successfully differentiated and purified BECs typically peak TEER at day 10 and express characteristic markers of brain endothelial cells such as PECAM1 (CD31), VE-cadherin, and Glut-1; efflux transporters such as P-glycoprotein, MRP1, and BCRP; and tight junction proteins such as ZO-1, occludin, and claudin-5. Refer to Lippmann et al. and Stebbins et al. and others for details and images of cell types, morphologies, and expression of cell type specific markers during the differentiation process [7, 37].
20. Disinfecting the electrode is critical to ensuring that no contamination of iPSC-BECs occurs. If necessary, longer incubation in 70% ethanol or decontamination of the electrode using 5% sodium hypochlorite solution is possible.
21. It is critical to read TEER quickly after removal from the incubator as temperature change may impact TEER measurement. Ensuring that both electrodes are covered by liquid is important. If needed, tilt the well to achieve this before setting the plate down again prior to measuring.
22. *Neisseria meningitidis* can be transmitted by aerosols. It is critical to keep the volume of liquid culture to a minimum. Nm should only be grown in liquid culture inside of sealed capped tubes and pipetted inside of a BSC.
23. While inoculating PPM+ using the cotton swab, gather as much of the culture off the solid medium and inoculate by rolling the cotton swab in the PPM+ media.
24. Some RNA isolation kits require that B-ME is added to the first buffer. Follow all the kit instructions exactly.
25. Prior to qPCR experiments, validate primers by conducting a dilution series and calculating the primer efficiency. Using primers between 90% and 110% is recommended. DNA gel electrophoresis can be conducted on the products to visualize if there are multiple products. Good primer design is necessary for quality qPCR results.
26. A master mix is advised to reduce the variability in qPCR. To account for potential loss during pipetting, make a half-well more volume than is needed to ensure each experimental well has the same amount of master mix. For example, if running 12 wells, make enough master mix for 12.5 wells.
27. A melt curve can detect if there are potentially secondary products from the qPCR reaction. A single product would produce a single peak in an ideal situation. However, a single peak can result from two products with similar melting temperatures as well as two peaks can result from a single product that is incompletely melted. Visualizing products on a gel

could help confirm these results, but the melt curve is a useful tool for initial examination of products.

28. A visible ladder such as SeeBlue2 plus is helpful to estimate where the protein of interest may be in real time.
29. While TBST is recommended, PBS + Tween 20 can also be used. However, if blotting for phosphorylated proteins, TBST will yield less potential background due to the buffering agent.
30. Primary antibodies diluted in 5% milk and TBST can be stored with 0.02% (w/v) sodium azide for at least 3 months at 4 °C. Replace the antibody dilution in blocking buffer into a 15 mL conical tube and store at 4 °C to be used at a later time.
31. As *Neisseria meningitidis* can be transmitted by aerosols, it is critical that all aspiration steps do not use the vacuum aspirator.

Acknowledgements

B.J.K. is supported by laboratory startup funding provided by the University of Alabama. Eric R. Espinal and S. Jerod Sharp contributed equally to this work.

References

1. Abbott NJ, Patabendige AAK, Dolman DEM et al (2010) Structure and function of the blood-brain barrier. Neurobiol Dis 37:13–25
2. Engelhardt B, Vajkoczy P, Weller RO (2017) The movers and shapers in immune privilege of the CNS. Nat Immunol 18(2):123–131
3. Le Guennec L, Coureuil M, Nassif X, Bourdoulous S (2020) Strategies used by bacterial pathogens to cross the blood–brain barrier. Cell Microbiol 22(1):e13132
4. Rua R, McGavern DB (2018) Advances in meningeal immunity. Trends Mol Med 24(6): 542–559
5. Kim BJ, Shusta EV, Doran KS (2019) Past and current perspectives in modeling bacteria and blood-brain barrier interactions. Front Microbiol 10:1336
6. Helms HC, Abbott NJ, Burek M et al (2015) In vitro models of the blood-brain barrier: an overview of commonly used brain endothelial cell culture models and guidelines for their use. J Cereb Blood Flow Metab 36(5):862–890
7. Lippmann ES, Azarin SM, Kay JE et al (2012) Derivation of blood-brain barrier endothelial cells from human pluripotent stem cells. Nat Biotechnol 30:783–791. https://doi.org/10.1038/nbt.2247
8. Nishihara H, Gastfriend BD, Soldati S et al (2020) Advancing human induced pluripotent stem cell-derived blood-brain barrier models for studying immune cell interactions. FASEB J. https://doi.org/10.1096/fj.202001507RR
9. Park TE, Mustafaoglu N, Herland A et al (2019) Hypoxia-enhanced blood-brain barrier chip recapitulates human barrier function and shuttling of drugs and antibodies. Nat Commun. https://doi.org/10.1038/s41467-019-10588-0
10. Praça C, Rosa SC, Sevin E et al (2019) Derivation of brain capillary-like endothelial cells from human pluripotent stem cell-derived endothelial progenitor cells. Stem Cell Rep. https://doi.org/10.1016/j.stemcr.2019.08.002
11. Qian T, Maguire SE, Canfield SG et al (2017) Directed differentiation of human pluripotent stem cells to blood-brain barrier endothelial cells. Sci Adv. https://doi.org/10.1126/sciadv.1701679
12. Appelt-Menzel A, Cubukova A, Günther K et al (2017) Establishment of a human blood-

brain barrier co-culture model mimicking the neurovascular unit using induced Pluri- and multipotent stem cells. Stem Cell Rep. https://doi.org/10.1016/j.stemcr.2017.02.021

13. Goldeman C, Andersen M, Al-Robai A et al (2021) Human induced pluripotent stem cells (BIONi010-C) generate tight cell monolayers with blood-brain barrier traits and functional expression of large neutral amino acid transporter 1 (SLC7A5). Eur J Pharm Sci. https://doi.org/10.1016/j.ejps.2020.105577
14. Grifno GN, Farrell AM, Linville RM et al (2019) Tissue-engineered blood-brain barrier models via directed differentiation of human induced pluripotent stem cells. Sci Rep. https://doi.org/10.1038/s41598-019-50193-1
15. Hollmann EK, Bailey AK, Potharazu AV et al (2017) Accelerated differentiation of human induced pluripotent stem cells to blood–brain barrier endothelial cells. Fluids Barriers CNS 14:1–13. https://doi.org/10.1186/s12987-017-0059-0
16. Jamieson JJ, Linville RM, Ding YY et al (2019) Role of iPSC-derived pericytes on barrier function of iPSC-derived brain microvascular endothelial cells in 2D and 3D. Fluids Barriers CNS. https://doi.org/10.1186/s12987-019-0136-7
17. Lippmann ES, Al-Ahmad A, Azarin SM et al (2014) A retinoic acid-enhanced, multicellular human blood-brain barrier model derived from stem cell sources. Sci Rep 4:4160. https://doi.org/10.1038/srep04160
18. Lippmann ES, Al-Ahmad A, Palecek SP, Shusta EV (2013) Modeling the blood-brain barrier using stem cell sources. Fluids Barriers CNS. https://doi.org/10.1186/2045-8118-10-2
19. Neal EH, Marinelli NA, Shi Y et al (2019) A simplified, fully defined differentiation scheme for producing blood-brain barrier endothelial cells from human iPSCs. Stem Cell Rep 12:1380–1388. https://doi.org/10.1016/j.stemcr.2019.05.008
20. Canfield SG, Stebbins MJ, Faubion MG et al (2019) An isogenic neurovascular unit model comprised of human induced pluripotent stem cell-derived brain microvascular endothelial cells, pericytes, astrocytes, and neurons. Fluids Barriers CNS. https://doi.org/10.1186/s12987-019-0145-6
21. Canfield SG, Stebbins MJ, Morales BS et al (2017) An isogenic blood–brain barrier model comprising brain endothelial cells, astrocytes, and neurons derived from human induced pluripotent stem cells. J Neurochem. https://doi.org/10.1111/jnc.13923
22. Workman MJ, Svendsen CN (2020) Recent advances in human iPSC-derived models of the blood-brain barrier. Fluids Barriers CNS. 17(1):30
23. Vatine GD, Barrile R, Workman MJ et al (2019) Human iPSC-derived blood-brain barrier chips enable disease modeling and personalized medicine applications. Cell Stem Cell. https://doi.org/10.1016/j.stem.2019.05.011
24. Vatine GD, Al-Ahmad A, Barriga BK et al (2017) Modeling psychomotor retardation using iPSCs from MCT8-deficient patients indicates a prominent role for the blood-brain barrier. Cell Stem Cell. https://doi.org/10.1016/j.stem.2017.04.002
25. Shin Y, Choi SH, Kim E et al (2019) Blood–brain barrier dysfunction in a 3D in vitro model of Alzheimer's disease. Adv Sci. https://doi.org/10.1002/advs.201900962
26. Qosa H, Lichter J, Sarlo M et al (2016) Astrocytes drive upregulation of the multidrug resistance transporter ABCB1 (P-glycoprotein) in endothelial cells of the blood–brain barrier in mutant superoxide dismutase 1-linked amyotrophic lateral sclerosis. Glia. https://doi.org/10.1002/glia.23003
27. Page S, Raut S, Al-Ahmad A (2019) Oxygen-glucose deprivation/Reoxygenation-induced barrier disruption at the human blood–brain barrier is partially mediated through the HIF-1 pathway. NeuroMolecular Med. https://doi.org/10.1007/s12017-019-08531-z
28. Mohamed LA, Markandaiah SS, Bonanno S et al (2019) Excess glutamate secreted from astrocytes drives upregulation of P-glycoprotein in endothelial cells in amyotrophic lateral sclerosis. Exp Neurol. https://doi.org/10.1016/j.expneurol.2019.04.002
29. Lim RG, Quan C, Reyes-Ortiz AM et al (2017) Huntington's disease iPSC-derived brain microvascular endothelial cells reveal WNT-mediated Angiogenic and blood-brain barrier deficits. Cell Rep. https://doi.org/10.1016/j.celrep.2017.04.021
30. Lee CAA, Seo HS, Armien AG et al (2018) Modeling and rescue of defective blood-brain barrier function of induced brain microvascular endothelial cells from childhood cerebral adrenoleukodystrophy patients. Fluids Barriers CNS. https://doi.org/10.1186/s12987-018-0094-5
31. Katt ME, Mayo LN, Ellis SE et al (2019) The role of mutations associated with familial neurodegenerative disorders on blood-brain barrier function in an iPSC model. Fluids

Barriers CNS. https://doi.org/10.1186/s12987-019-0139-4

32. Alimonti JB, Ribecco-Lutkiewicz M, Sodja C et al (2018) Zika virus crosses an in vitro human blood brain barrier model. Fluids Barriers CNS. https://doi.org/10.1186/s12987-018-0100-y
33. Kim BJ, Bee OB, McDonagh MA et al (2017) Modeling group B *streptococcus* and blood-brain barrier interaction by using induced pluripotent stem cell-derived brain endothelial cells. mSphere. https://doi.org/10.1128/mSphere.00398-17
34. Kim BJ, McDonagh MA, Deng L et al (2019) Streptococcus agalactiae disrupts P-glycoprotein function in brain endothelial cells. Fluids Barriers CNS 16:26. https://doi.org/10.1186/s12987-019-0146-5
35. Patel R, Hossain MA, German N, Al-Ahmad AJ (2018) Gliotoxin penetrates and impairs the integrity of the human blood-brain barrier in vitro. Mycotoxin Res. https://doi.org/10.1007/s12550-018-0320-7
36. Martins Gomes SF, Westermann AJ, Sauerwein T et al (2019) Induced pluripotent stem cell-derived brain endothelial cells as a cellular model to study Neisseria meningitidis infection. Front Microbiol. https://doi.org/10.3389/fmicb.2019.01181
37. Stebbins MJ, Wilson HK, Canfield SG et al (2016) Differentiation and characterization of human pluripotent stem cell-derived brain microvascular endothelial cells. Methods 101: 93–102. https://doi.org/10.1016/j.ymeth.2015.10.016
38. Endres LM, Schubert-Unkmeir A, Kim BJ (2020) Neisseria meningitidis infection of induced pluripotent stem-cell derived brain endothelial cells. J Vis Exp. https://doi.org/10.3791/61400
39. Sances S, Ho R, Vatine G et al (2018) Human iPSC-derived endothelial cells and microengineered organ-chip enhance neuronal development. Stem Cell Rep. https://doi.org/10.1016/j.stemcr.2018.02.012
40. Kim BJ, Schubert-Unkmeir A (2019) In vitro models for studying the interaction of neisseria meningitidis with human brain endothelial cells. Methods Mol Biol 1969:135–148
41. Doran KS, Liu GY, Nizet V (2003) Group B streptococcal β-hemolysin/cytolysin activates neutrophil signaling pathways in brain endothelium and contributes to development of meningitis. J Clin Invest 112:736–744. https://doi.org/10.1172/JCI200317335
42. Kim BJ, Hancock BM, Bermudez A et al (2015) Bacterial induction of Snail1 contributes to blood-brain barrier disruption. J Clin Invest 125:2473–2483. https://doi.org/10.1172/JCI74159
43. Banerjee A, Kim BJ, Carmona EM et al (2011) Bacterial pili exploit integrin machinery to promote immune activation and efficient blood-brain barrier penetration. Nat Commun 2: 462. https://doi.org/10.1038/ncomms1474
44. Nizet V, Kim KS, Stins M et al (1997) Invasion of brain microvascular endothelial cells by group B streptococci. Infect Immun 65: 5074–5081

Chapter 5

An hiPSC-Derived In Vitro Model of the Blood–Brain Barrier

Mary Goodwin-Trotman, Krushangi Patel, and Alessandra Granata

Abstract

Human induced pluripotent stem cells (hiPSC) offer a tractable system to model the blood–brain barrier (BBB). Here we detail the assembly of a triple co-culture hiPSC-BBB model, using hiPSC-derived brain microvascular endothelial cells (BMEC), astrocytes, and mural cells (MC). Transendothelial electrical resistance (TEER) and sodium fluorescein (NaFl) permeability can be used to test the barrier properties. The model has applications in studying BBB-related pathology and for drug screening.

Key words Transwell®, Blood–brain barrier, hiPSC, Co-culture, TEER

1 Introduction

hiPSCs, derived from human dermal fibroblasts, were first reported in 2007 by Yamanaka and his team [1]. hiPSCs are relatively inexpensive and can be used for rapid and scalable assays more easily than in vivo animal models [2]. hiPSCs circumvent the technical challenge of isolating primary human cells, which is complicated and time-consuming [2]. hiPSCs can also be used to generate intermediate populations, such as neural stem cells [3] or neural crest [4], which are highly proliferative, are reproducible, and can be cryopreserved, shortening the differentiation time while maintaining equal performance.

hiPSCs also overcome inter-species differences [5], making the results more relevant for pharmacological testing [2]. They remove not only inter-species differences but also inter-individual differences in genetic background, since an hiPSC line maintains the background of the individual of origin [1]. Importantly, all cell types of the BBB, including BMEC, astrocytes, and MC, can be generated from the same hiPSC line, therefore creating an isogenic model [6].

To date, BBB disease models derived from patient hiPSCs have been successfully implemented in disease modeling for vascular

Nicole Stone (ed.), *The Blood-Brain Barrier: Methods and Protocols*,
Methods in Molecular Biology, vol. 2492, https://doi.org/10.1007/978-1-0716-2289-6_5,

disorders such as a form of monogenic small vessel disease, cerebral autosomal dominant arteriopathy with subcortical infarcts and leukoencephalopathy (CADASIL) [7, 8], and the neurological disorder, Huntingdon's disease [9]. Recently, hiPSC-BBB models have been able to acquire in vivo-like properties owing to their realistic drug transport capabilities [10], highlighting possibilities for their use in drug screening.

Once designed, BBB models must be thoroughly tested to demonstrate their relevance to the in vivo BBB. Functional assays are employed as key determinants of equating artificial in vitro performance to in vivo function [11]. The most widely used functional test of BMEC function is transepithelial electrical resistance (TEER). In vitro, this is recorded with an epithelial voltohmmeter which uses two pairs of electrodes [12], one applying current and one measuring the output, on each side of a monolayer, for example, a Transwell® membrane [13]. Permeability assays with fluorescent compounds can also be used as cost-effective measures of barrier properties [14]. Here, we use a combination of TEER reading and NaFl, a 376.27 Da fluorescent tracer permeability assay to thoroughly assess BBB integrity.

Thus, we have demonstrated that an isogenic hiPSC-BBB model can be readily established using a Transwell® membrane, an approach that has also been adopted by other investigators [15, 16].

2 Materials

Prepare all media in advance, filter using a 0.22 μm filter unit and store at 4 °C. Refer to the previously published protocols to generate hiPSC-BMEC, astrocytes, and MC. hiPSCs were cultured in E8 media.

2.1 hiPSC

E8 media: 250 mL DMEM/F-12 with 5 mL Insulin Transferrin Selenium, 1.8 mL (7.5%) sodium bicarbonate, 2.5 mL (6.4 mg/ml) L-ascorbic acid, 1.56 mL (4 μg/ml) FGF-2, and 250 μL (1.74 ng/ml) TGFβ.

2.2 hiPSC-BMEC

1. 0.5% Matrigel®: dilute on ice in DMEM/F12 to 0.5% [17].
2. EndoSFM+B-27: 49.5 mL Human Endothelial Serum-Free Medium with 250 μL B-27 [17].
3. BMEC media: 49.5 mL Human Endothelial Serum-Free Medium with 250 μL B-27, 250 μl (4 μg/mL) FGF-2, and 25 μL (20 mM) retinoic acid.

4. Fibronectin/collagen IV: 1 mg/mL collagen IV (from human placenta, Bornstein and Traub Type IV) and 1 mg/ mL fibronectin (from bovine plasma) and 1× phosphate-buffered saline (PBS) according to a 4:1:5 ratio.
5. Sodium fluorescein (NaFl): Dilute NaFl in EndoSFM+B27 at 1:100 dilution. Avoid direct light.

2.3 hiPSC-MC

1. CDM-PVA media: prepare 500 ml (250 ml Ham's F12 Nutrient Mix, 250 ml IMDM, 5 ml concentrated lipids, 20 ul 1-thio glycerol, 250 ul transferrin, 350 ul insulin, 5 ml polyvinyl alcohol (PVA)) [18].
2. FSB media: CDM-PVA supplemented with 12 ng/mL FGF-2 and 10-uM SB431542 [18].
3. MC media: CDM-PVA supplemented with 10 ng/mL PDGF-BB and 2 ng/mL TGF-b1 [18].
4. MC mature media: DMEM/F12 medium supplemented with 10% fetal bovine serum (FBS) [18].
5. Mouse embryonic fibroblast (MEF) media: 250 mL advanced DMEM/F-12 supplemented with 25 mL (10%) FBS, 2.5 mL (2 mM) L-glutamine, and 1.75 μL β mercaptoethanol.
6. 0.1% Gelatin: gelatin (from porcine skin) diluted in water and kept overnight in a water bath at 60 °C to dissolve.

2.4 hiPSC-Astrocytes

1. Neural induction media (NIM) [3]: 47.5 mL TeSR™-E6, 2.5 mL TeSR™-E6 supplement (20×), 25 μl (10 μM) SB431542, 5 μl (1 μM) LDN193189, and 10 μl (2 μM) XAV939.
2. Neural maintenance media (NMM) [3]: 125 mL neurobasal, 125 mL DMEM:F12+GlutaMAX, 1.25 mL N-2 supplement, 2.5 mL B-27 supplement, 1.25 mL GlutaMAX, 1.25 mL non-essential amino acids (NEAA), 1.25 mL (100 mM) sodium pyruvate, 250 μL (50 mM) 2 mercaptoethanol, and 62.5 μL (10 mg/mL) insulin.
3. Astrocyte media [19]: To 50 mL NMM, add 5 μL (5 mg/mL) CNTF, 5 μL (5 mg/mL) BMP-2, and 10 μl (4 μg/mL) FGF-2.
4. VTN-N: VTN-N diluted 1:100 in 1× PBS.
5. 0.01% poly-L-ornithine (PLO): used as supplied by the manufacturer, no preparation is required.
6. Laminin: Dilute Laminin L2020 1:300 in 1× PBS.

2.5 Cell Culture

1. StemPro™ Accutase™ Cell Dissociation Reagent.
2. TrypLE express.
3. 1× PBS.
4. ROCK inhibitor.

5. TeSR™-E6.
6. ReLeSR™.
7. TrypLE express.
8. 24-well plate Transwells®.
9. 24-well cell culture plates.
10. 0.22 μm media filter unit.
11. Black 96-well plate.

2.6 Equipment

1. Cell culture centrifuge.
2. Microbiological safety cabinet/laminar flow hood.
3. Hemocytometer.
4. EVOM2 voltohmmeter and STX2 electrodes (World Precision Instruments).
5. Incubator set to 37 °C and 5% CO_2.
6. Fluorescent 96-well plate reader.

3 Methods

All procedures should be carried out in a microbiological safety cabinet using aseptic technique.

3.1 hiPSC-BMEC Culture

hiPSC are differentiated according to a previous protocol [17], described below.

1. 24 h before seeding, coat 6-well plates with 0.5% Matrigel® in DMEM/F12 and store at 37 °C overnight.
2. The next day, wash 70–80% confluent hiPSC with 1× PBS and incubate with StemPro™ Accutase™ Cell Dissociation Reagent at 37 °C for 4 min.
3. Collect cells, dilute 1:4 with E8 and centrifuge at 300 g for 3 min.
4. Resuspend in E8, count with a hemocytometer, and plate at a density of $1.56 \times 10^4/\text{cm}^2$ with ROCK inhibitor.
5. 24 h later, change media to TeSR™-E6 and refresh daily for 4 days.
6. On day 5, replace media with BMEC media for 2 days.
7. One day prior to subculture (on day 6), coat fibronectin/collagen IV 24-well Transwells® overnight at 37 °C (*see* **Note 1**).
8. Following 2 days of BMEC media treatment (on day 7), aspirate the spent media and wash once with 1× PBS.

9. Apply 1 mL of Accutase per 6-well plate and leave at 37 °C for 45 min up to 1 h to dissociate to single cells (*see* **Note 2**).
10. Collect the disassociated single cells and centrifuge at 300 g for 3 min.
11. Count the cells and seed onto fibronectin/collagen IV coated 24-well plate Transwells® at a density of 2.5 × 10^5/well in BMEC media as previously reported [20], *see* **Note 3**.
12. Record the first TEER measurement (*see* Part 3.4) and then refresh media to BMEC media without FGF-2 or RA. No further media changes are required [17, 21].

3.2 hiPSC-MC Culture

hiPSC are differentiated into neural crest (NC) and collected after four passages in FSB media to mature from neural crest cells to MC, as previously reported [18] and described below.

1. 24 h before seeing, coat wells with 0.1% gelatin for 20 min at room temperature. Aspirate the gelatin and incubate the plate overnight at 37 °C in MEF medium.
2. The next day, wash 70–80% confluent hiPSC with 1× PBS and incubate with ReLeSR™ for 4 min at room temperature.
3. Collect cells and plate onto 0.1% Gelatin in CDM-PVA.
4. 24 h after plating, wash cells once with 1× PBS and culture in FSB media.
5. Refresh FSB daily for 5 days.
6. On day 5, passage with TrypLE express by applying 0.5 mL of TrypLE per 6-well plate for 4 min and thereafter passage when confluent, until passage 4.
7. After four passages in FSB media, collect cells with TrypLE express.
8. Seed cells in FSB media onto 0.1% gelatin coated plates at 50% confluence.
9. After 24 h, replace FSB media with MC media.
10. Feed hiPSC-MC with MC media every 2 days and passage every 3 days or when they reach confluence with TrypLE at a 1:2 or 1:3 split ratio as appropriate.
11. hiPSC-MC can be maintained for up to 3 months or ten passages in DMEM:F12 with 10% FBS serum (*see* **Note 4**).
12. When hiPSC-MC are required for co-culture, passage and seed into 24-well plates at a 1:2 or 1:3 split ratio as appropriate.

3.3 hiPSC-Astrocyte Culture

hiPSCs were first differentiated into neural stem cells (NSCs) using an available protocol [3], but modified so the initial stage begins with TeSR-E6 media. Following this, NSC were differentiated to astrocytes with a previously reported protocol [19], described below.

3.3.1 To Differentiate hiPSC to NSCs

1. 24 h before seeding, coat 6-well plates with 1:100 VTN-N for 1 h at room temperature, aspirate and replace with 1× PBS, and store at 37 °C overnight.
2. The next day, wash 70–80% confluent hiPSC with 1× PBS and incubate with StemPro™ Accutase™ Cell Dissociation Reagent at 37 °C for 4 min.
3. Collect cells, dilute 1:4 with E8 and centrifuge at 300 g for 3 min.
4. Resuspend cells in E8, count and plate in E8 medium with ROCK inhibitor at a density of $2.0 \times 10^5/cm^2$ on VTN-N.
5. 24 h later, change media to NIM and refresh daily for 10–12 days.
6. Passage with StemPro™ Accutase™ Cell Dissociation Reagent at a 1:1 dilution onto VTN-N into NMM.
7. 24 h later, coat 6 well plates with Laminin L2020 diluted 1:100 in 1× PBS overnight at 37 °C.
8. After 3–5 days (when 100% confluent), passage cells with StemPro™ Accutase™ Cell Dissociation Reagent, count and plate at a density of $5.0 \times 10^4/cm^2$ onto laminin.

3.3.2 To Differentiate NSCs to Astrocytes

1. One day prior to plating, coat 24-well culture plates with PLO (0.01%) for 1 h at room temperature before washing with 1× PBS and coating overnight at 37 °C with Laminin L2020 diluted 1:300 in 1× PBS.
2. Seed NSCs at 50% density into astrocyte media, on 24-well plates.
3. To passage, apply StemPro™ Accutase™ Cell Dissociation Reagent per 24-well plate and leave for 4–6 min.
4. Collect the single cells and centrifuge at 300 g for 3 min.
5. Aspirate the spent media and resuspend in astrocyte media before plating at 50% confluency (*see* **Note 5**).

3.4 TEER

3.4.1 Collection of Readings

Begin taking TEER readings 1 day after subculture onto fibronectin/collagen IV Transwells®. Perform three replicates and ensure to include a single blank Transwell® with the same matrix coating and media but no cells seeded as a background reading [21].

TEER should be performed in three parallel Transwells® from the same BMEC differentiation. Ensure that the temperature is controlled for (*see* **Note 6**).

1. Clean the EVOM2 voltohmmeter with 70% ethanol and set up in the laminar flow hood. Set function to "Ohms" to measure resistance.

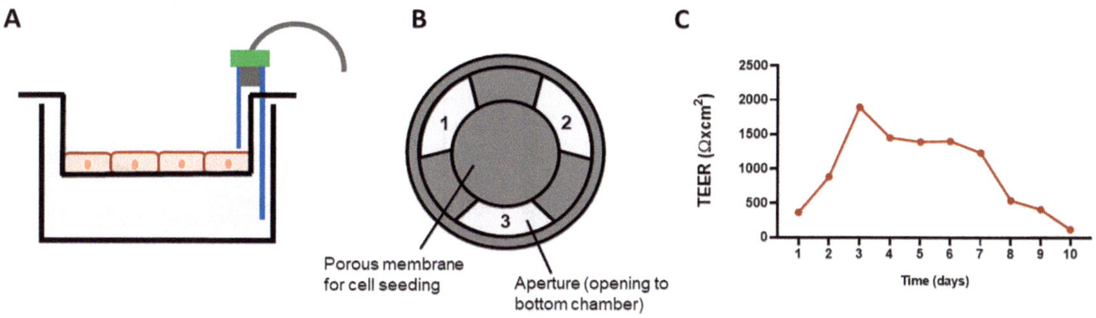

Fig. 1 (**a**) STX2 electrode placement for hiPSC-BMEC monoculture and (**b**) the location of the three apertures of the Transwell®. (**c**) TEER obtained from wild-type hiPSC-BMEC for 10 days post plating onto Transwells®

2. Clean the STX2 electrodes with 70% ethanol and insert into the EVOM2. Keep 70% ethanol in the laminar flow hood to clean the electrodes before and after each reading.
3. Retrieve the Transwell® plate from the incubator and leave at room temperature to acclimatize before taking readings (*see* **Note 7**).
4. Begin by taking three TEER readings of the blank, one from each aperture of the Transwell (Fig. 1a, b; *see* **Note 8**). Clean the electrodes in 70% ethanol before taking the next reading (*see* **Note 9**).
5. Proceed to take readings from the remaining BMEC seeded Transwell®, ensuring to take three readings from each Transwell® (*see* **Note 10**).
6. Repeat **steps 1–5** every day thereafter.

3.4.2 Calculating the Result

Using Ohm's law, the ratio of voltage to current is computed, before subtraction of a blank measurement and multiplication by the surface area [13], to arrive at $\Omega \times cm^2$ and take the mean average of the replicate values (Fig. 1c).

1. Calculate the average TEER for each Transwell® (three readings each).
2. Subtract the average TEER of the blank Transwell® from each of the experimental Transwell® averages.
3. Multiple these values with the area of the Transwell®.

$$\Omega \times cm^2 = (\text{Mean of 3 cell readings}) - (\text{mean of 3 blank readings}) \times \text{surface area}$$

Example wild-type (WT) data plotted as a time-course (Fig. 1c) demonstrate that TEER peaks around day 2/3 of subculture onto Transwells® [21].

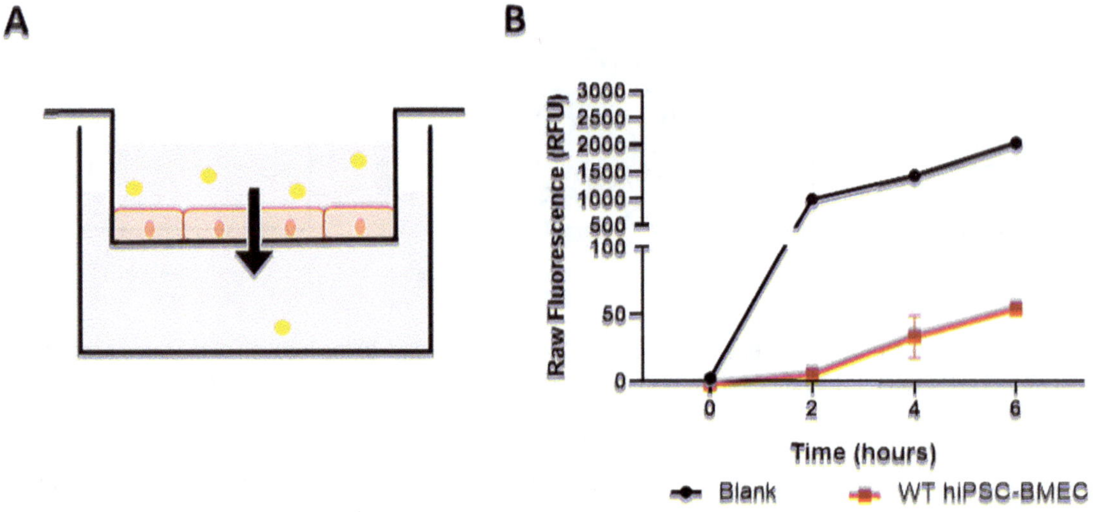

Fig. 2 (**a**) NaFl permeability strategy and (**b**) NaFl obtained from wild-type (WT) hiPSC-BMEC 2 days post-seeding onto Transwells®

3.5 NaFl Permeability

Perform the NaFl permeability assay when TEER is maximal (generally 2 days after subculture onto fibronectin/collagen IV Transwells®). As with TEER readings, ensure to include a blank, coated Transwell® as a background reading, and perform three replicates.

3.5.1 Collection of Readings

1. Remove media from underneath the Transwell® and replace with fresh EndoSFM+B27 (1 mL for 12 well, 0.5 mL for 24 well, Fig. 2a).
2. Remove media from the Transwell® and replace with diluted NaFl (1:100) in EndoSFM+B27 (1 mL for 12 well, 0.5 mL for 24 well).
3. Immediately collect a 100 μL sample from underneath the Transwell® and collect in a black 96-well plate (0 h reading). Cover the plate in tin foil and store at 4 °C until the next reading. Repeat for each replicate.
4. Repeat 100 μL sampling from under each Transwell® every 2 h for 8 h (15 samples in total).
5. Use a plate reader to analyze the samples and average the three Transwells® (*see* **Note 11**).

3.5.2 Calculating the Result

Plot the raw fluorescence of the hiPSC-BMEC monolayer and the blank (Fig. 2b). Data can also be reported as a percentage of the blank:

$$\%\text{of Blank} = \text{raw fluorescence of hiPSC-BMEC monolayer} - \text{raw fluorescence of blank} \times 100$$

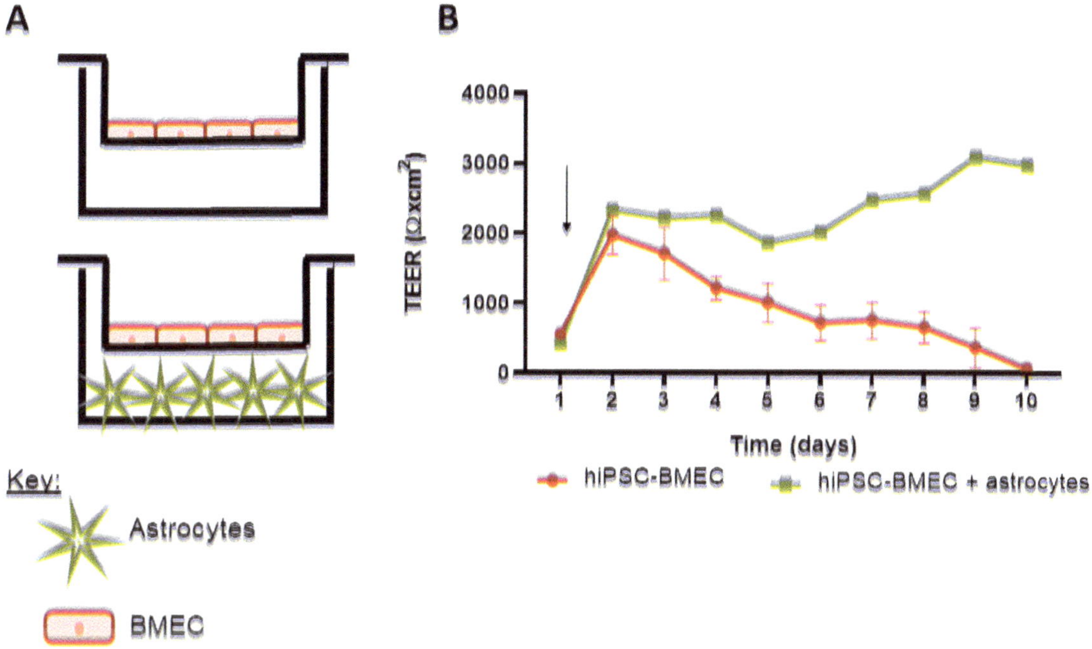

Fig. 3 (**a**) hiPSC-BMEC and astrocyte Transwell® strategy (**b**) WT hiPSC-BMEC and BMEC/astrocyte co-culture over 10 days. The arrow indicates the initiation of the co-culture

3.6 Co-culture hiPSC-BBB Model

Take TEER readings of hiPSC-BMEC 1 day post-subculture onto Transwells® before initiating any co-culture experiments. As before, ensure to have a blank well prepared.

3.6.1 hiPSC-BMEC/Astrocyte Co-culture

1. 3–4 days prior to initiation of co-culture, seed hiPSC-astrocytes into wells of 24-well plates (as described above), allow for enough wells for technical replicates, e.g., seed 3 wells.
2. 1 day prior to co-culture, seed hiPSC-BMEC on Transwells® in replicate for both co-culture and monoculture (control), e.g., seed 6 Transwells®.
3. On the day of co-culture, take TEER readings from the hiPSC-BMEC and media change to EndoSFM+B27. No further media changes are required.
4. Transfer half of the Transwell® to wells of a 24-well plates containing hiPSC-astrocytes and return to the incubator.
5. 24 h later and every day thereafter, take TEER readings and plot as a time-course (Fig. 3).

3.6.2 hiPSC-BMEC/MC Co-culture

1. 1–2 days prior to the initiation of co-culture, seed hiPSC-MC into 24-well plates (as described above), allow for enough wells for technical replicates, e.g., seed 3 wells.

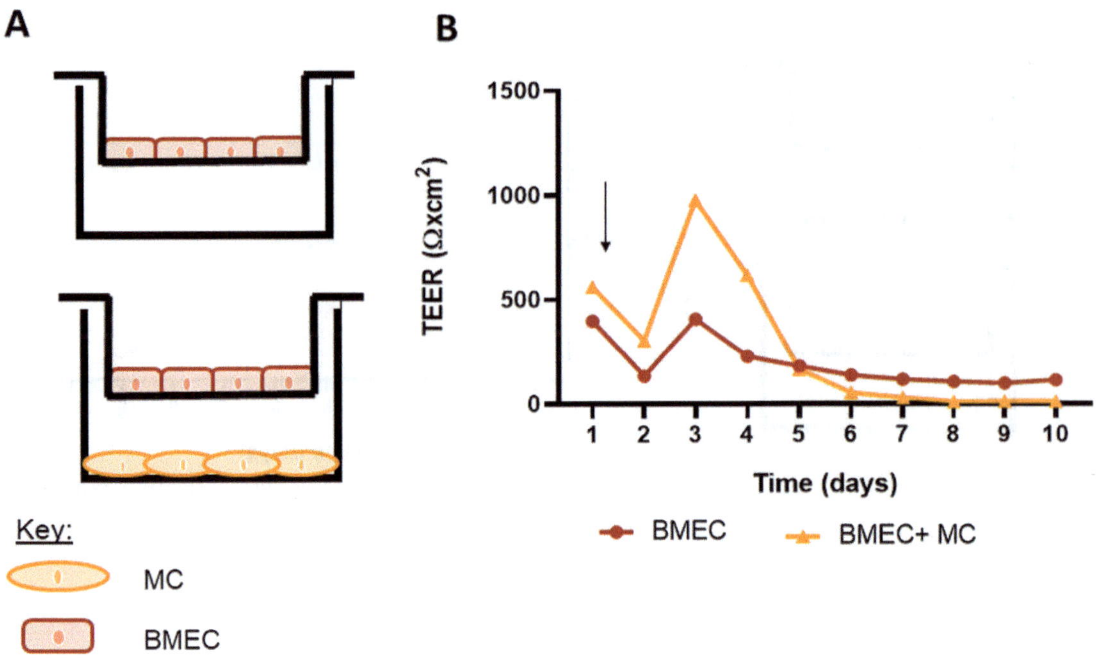

Fig. 4 (**a**) hiPSC-BMEC and MC Transwell® strategy. (**b**) WT hiPSC-BMEC and BMEC/MC co-culture over 10 days. The arrow indicates the initiation of the co-culture

2. 1 day prior to co-culture, seed hiPSC-BMEC on Transwells® in replicate for both co-culture and monoculture (control), e.g., seed 6 Transwells®.
3. On the day of co-culture, take TEER readings from the hiPSC-BMEC and media change to EndoSFM+B27. No further media changes are required.
4. Transfer half of the Transwells® to wells of a 24-well plates containing hiPSC-MC and return to the incubator overnight.
5. 24 h later and every day thereafter, take TEER readings and plot as a time-course (Fig. 4).

3.6.3 hiPSC-BMEC/Astrocyte/MC Triple Co-culture

1. 3–4 days prior to initiation of co-culture, seed hiPSC-astrocytes into 24-well plates (as described above), allow for enough wells for technical replicates for each co-culture arrangement, e.g., seed 6 wells.
2. On the day of co-culture (*see* **Note 12**), place half of the Transwells® upside down onto the lid of the culture plate (Fig. 5) and seed hiPSC-MC (the other half will be used as the control with BMEC alone).
3. Incubate in the laminar hood for 45 min to allow the cells to attach.

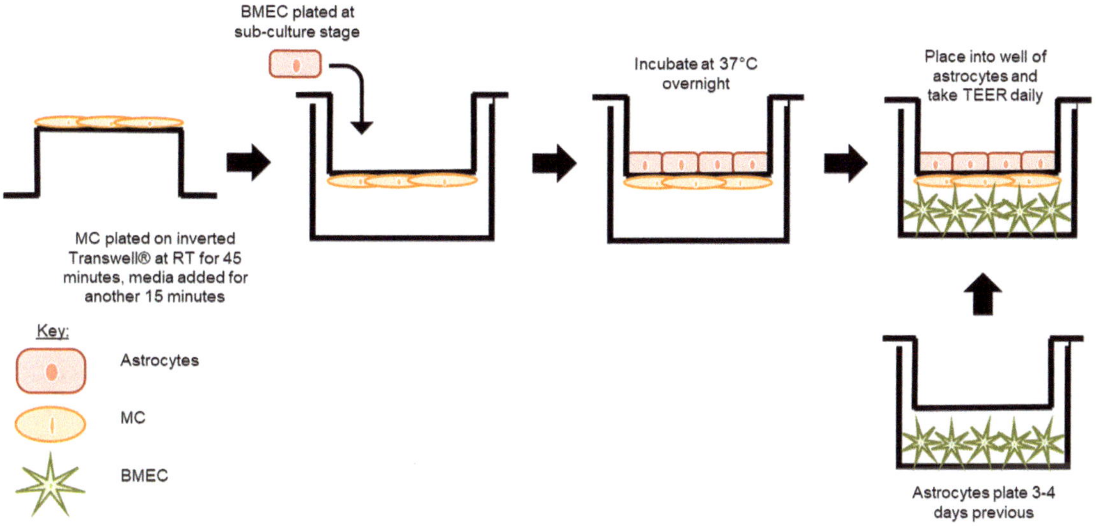

Fig. 5 Plating strategy for the hiPSC-BMEC, astrocyte, and MC triple co-culture Transwell® system

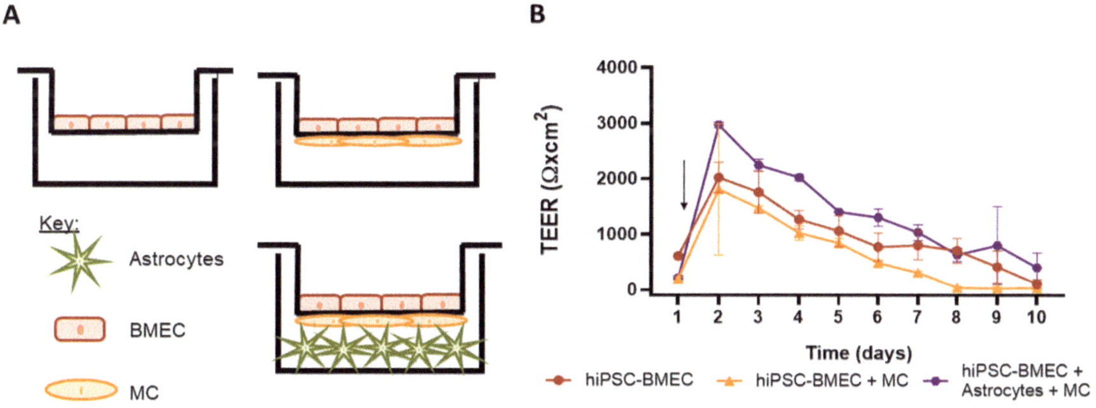

Fig. 6 (**a**) hiPSC-BMEC, astrocyte, and MC triple co-culture Transwell® strategy. (**b**) WT hiPSC-BMEC, BMEC/MC, and BMEC/astrocyte co-culture over 10 days. The arrow indicates the initiation of the co-culture

4. During this period, apply StemPro™ Accutase™ Cell Dissociation Reagent to the differentiating hiPSC-BMEC (as described above).
5. After 45 min, add 200 μL of media and incubate for a further 15 min.
6. After 15 min, return the Transwell® to the upright position and place it into the culture plate. Collect the hiPSC-BMEC and centrifuge at 300 g for 3 min.
7. Seed hiPSC-BMEC as described above and return the plate to the incubator.
8. The next day, take the first TEER reading and refresh the media to EndoSFM+B27. No further media changes are required.

9. Transfer half of the Transwells® to wells containing hiPSC-astrocytes.
10. Continue to take TEER readings every 24 h and plot as a time-course (Fig. 6).

4 Notes

1. Make sure to prepare an additional Transwell to use as a blank for TEER and NaFl.
2. Ensure that the StemPro™ Accutase™ Cell Dissociation Reagent treatment results in a suspension of single cells on gently tapping the plate, StemPro™ Accutase™ Cell Dissociation Reagent can be left for up to 1 h to achieve this [20, 21].
3. Plate the total number of cells required directly onto the Transwell® and shake to distribute, before topping up with media.
4. Mature SMC can survive at confluence without media change for up to 1 week. When necessary, split with TrypLE at an appropriate split ratio, ensuring cell to cell contact upon seeding. When changing media, aspirate half the volume of media and top with fresh media to allow cell-derived maturation factors to remain in the culture.
5. Astrocytes can be maintained for up to 6–7 passages and cryopreserved for later use [19].
6. The temperature must be consistent between the blank and all readings with a BMEC monolayer. To achieve this, ensure that each plate of Transwells® is removed from the incubator for the same duration of time.
7. Before taking readings, allow the Transwells® to acclimatize at room temperature as temperature can affect readings. To reduce variability in readings, be consistent in acclimatization time whenever measuring TEER.
8. TEER readings may fluctuate so ensure the reading is constant for some time before recording it.
9. Clean electrodes with 70% electrode between different cell lines/samples. For replicates of the same sample, wash with EndoSFM+B27 to prevent overexposure to 70% ethanol.
10. Take TEER readings daily until the resistance has plateaued.
11. To analyze the samples, set up the plate reader with an excitation wavelength of 460 nm, emission of 515 nm with a gain of 50, 25 flashes, and z-position of 20,000.
12. Coordinate the timing of the hiPSC-BMEC differentiation so co-culture can be initiated at the point of subculture, e.g., day 7.

Acknowledgements

This work was supported by the Rosetrees Trust, British Heart Foundation, Alzheimer's Association and Stroke Association.

References

1. Takahashi K, Tanabe K, Ohnuki M et al (2007) Induction of pluripotent stem cells from adult human fibroblasts by defined factors. Cell 131(5):861–872
2. Naik P, Cucullo L (2012) In vitro blood-brain barrier models: Current and perspective technologies [Internet]. J Pharm Sci. [cited 2017 Oct 23] 101:1337–1354. Available from: https://www.ncbi.nlm.nih.gov/pmc/articles/PMC3288147/pdf/nihms352244.pdf.
3. Shi Y, Kirwan P, Livesey FJ (2012) Directed differentiation of human pluripotent stem cells to cerebral cortex neurons and neural networks. Nat Protoc [Internet]. [cited 2017 Nov 21] 7(10):1836–1846. Available from: http://www.nature.com/doifinder/10.1038/nprot.2012.116
4. Serrano F, Bernard WG, Granata A et al (2019) A novel human pluripotent stem cell-derived neural crest model of Treacher Collins syndrome shows defects in cell death and migration. Stem Cells Dev [Internet]. [cited 2020 Mar 2] 28(2). Available from: www.liebertpub.com/scd
5. Oberheim NA, Takano T, Han X et al (2009) Uniquely hominid features of adult human astrocytes. J Neurosci 29(10):3276–3287
6. Chamberlain SJ (2016) Disease modelling using human iPSCs Introduction to iPSC Technology. Hum Mol Genet [Internet]. [cited 2020 May 22] 25(R2). Available from: https://academic.oup.com/hmg/article-abstract/25/R2/R173/2198192
7. Kelleher J, Dickinson A, Cain S et al (2019) Patient-specific iPSC model of a genetic vascular dementia syndrome reveals failure of mural cells to stabilize capillary structures. Stem Cell Rep 13(5):817–831
8. Ling C, Liu Z, Song M et al (2019) Modeling CADASIL vascular pathologies with patient-derived induced pluripotent stem cells. Protein Cell [Internet]. [cited 2019 Mar 6]. Available from: http://www.ncbi.nlm.nih.gov/pubmed/30778920
9. Lim RG, Quan C, Reyes-Ortiz AM et al (2017) Huntington's disease iPSC-derived brain microvascular endothelial cells reveal WNT-mediated angiogenic and blood-brain barrier deficits. Cell Rep [Internet]. [cited 2018 Apr 12] 19(7):1365–1377. Available from: http://linkinghub.elsevier.com/retrieve/pii/S2211124717304965
10. Le Roux G, Jarray R, Guyot A-C et al (2019) Proof-of-concept study of drug brain permeability between in vivo human brain and an in vitro iPSCs-human blood-brain barrier model. Sci Rep [Internet]. [cited 2019 Nov 12] 9. Available from: https://doi.org/10.1038/s41598-019-52213-6
11. Bhalerao A, Sivandzade F, Rahman Archie S et al (2020) In vitro modeling of the neurovascular unit: advances in the field. Fluids Barriers CNS [Internet]. [cited 2020 May 11] 17:22. Available from: https://doi.org/10.1186/s12987-020-00183-7
12. Benson K, Cramer S, Galla HJ (2013) Impedance-based cell monitoring: barrier properties and beyond. Fluids Barriers CNS 10:1–11
13. Srinivasan B, Kolli AR, Esch MB et al (2015) TEER measurement techniques for in vitro barrier model systems [Internet]. J Lab Autom. [cited 2018 Feb 16] 20:107–126. Available from: http://www.ncbi.nlm.nih.gov/pubmed/25586998
14. Helms HC, Abbott NJ, Burek M, et al (2016) In vitro models of the blood–brain barrier: an overview of commonly used brain endothelial cell culture models and guidelines for their use. J Cereb Blood Flow Metab [Internet]. [cited 2018 Apr 9] 36(5):862–890. Available from: http://www.ncbi.nlm.nih.gov/pubmed/26868179
15. Canfield SG, Stebbins MJ, Morales BS et al (2017) An isogenic blood-brain barrier model comprising brain endothelial cells, astrocytes, and neurons derived from human induced pluripotent stem cells. J Neurochem [Internet]. [cited 2017 Nov 3] 140(6):874–888. Available from: http://doi.wiley.com/10.1111/jnc.13923
16. Canfield SG, Stebbins MJ, Faubion MG et al (2019) An isogenic neurovascular unit model comprised of human induced pluripotent stem cell-derived brain microvascular endothelial cells, pericytes, astrocytes, and neurons. Fluids Barriers CNS [Internet]. [cited 2020 Nov 17] 16(1):25. Available from: https://

fluidsbarrierscns.biomedcentral.com/articles/10.1186/s12987-019-0145-6

17. Neal EH, Marinelli NA, Shi Y et al (2019) A simplified, fully defined differentiation scheme for producing blood-brain barrier endothelial cells from human iPSCs. [cited 2019 Nov 23]. Available from: https://doi.org/10.1016/j.stemcr.2019.05.008
18. Cheung C, Bernardo AS, Trotter MWB et al (2012) Generation of human vascular smooth muscle subtypes provides insight into embryological origin-dependent disease susceptibility. Nat Biotechnol 30(2):165–173
19. Shaltouki A, Peng J, Liu Q et al (2013) Efficient generation of astrocytes from human pluripotent stem cells in defined conditions. Stem Cells [Internet]. [cited 2017 Oct 24] 31(5): 941–952. Available from: http://onlinelibrary.wiley.com/doi/10.1002/stem.1334/abstract
20. Stebbins MJ, Wilson HK, Canfield SG et al (2016) Differentiation and characterization of human pluripotent stem cell-derived brain microvascular endothelial cells. Methods 101: 93–102
21. Hollmann EK, Bailey AK, Potharazu AV et al (2017) Accelerated differentiation of human induced pluripotent stem cells to blood–brain barrier endothelial cells. Fluids Barriers CNS [Internet] 14(1):9. Available from: http://fluidsbarrierscns.biomedcentral.com/articles/10.1186/s12987-017-0059-0

Chapter 6

A Three-Dimensional Brain-on-a-Chip Using Human iPSC-Derived GABAergic Neurons and Astrocytes

Lumei Liu, Youngmi Koo, Teal Russell, and Yeoheung Yun

Abstract

Brain-on-a-chip is a miniaturized engineering platform to mimic the structural and functional aspects of brain tissue. We describe a method to construct a three-dimensional (3D) brain-on-a-chip in this chapter. We firstly portray the method of a brain-on-a-chip model with cocultured mice neurons, microglia, and astrocytes to mimic brain tissue and membrane-free perfusion with endothelial cells, in which we successfully build the blood–brain barrier to screen neurotoxicity. Then we describe a method to construct a brain-on-a-chip with human induced pluripotent stem cell (iPSC)-derived neurons and astrocytes to simulate human brain behavior. This platform consists of neuronal tissue with extracellular matrix (ECM)-embedded GABAergic neurons and astrocytes and a perfusion channel with dynamic flow. We also include the broader applicability test of this model using an organophosphate (OP), malathion, to induce acute and chronic neurotoxicity, and then using butyrylcholinesterase (BuChE) as an exogenous bioscavenger of OP. Following the methods listed in this chapter, we are able to measure the neurotoxic effects on construct integrity, viability, and total AChE and BuChE activity.

Key words Brain-on-a-chip model, Three-dimensional cell culture, Organophosphate (OP), Induced pluripotent stem cell (iPSC), Neurotoxicity, Bioscavenger, ChE activity, Viability

1 Introduction

With the emerging of organ-on-a-chip technologies, the brain-on-a-chip platform has been explored to mimic the physiology of human brain tissue on an engineered microfluidic system [1–3]. A brain-on-a-chip model exhibits advantages over current well-accepted platforms [4, 5]. It can replicate the complex, multifunctional, and integrated neuro-gliovascular-immune system, which cannot be recapitulated by other traditional in vitro platforms such as two-dimensional (2D) cell culture and transwell models [6, 7]. The current challenges to the brain-on-a-chip include being (1) unable to reproduce the complexity of human brain; (2) hard to model the cerebrovascular interface; and (3) limited investigation of structural connections and cell-cell interactions [1, 8–

Nicole Stone (ed.), *The Blood-Brain Barrier: Methods and Protocols*,
Methods in Molecular Biology, vol. 2492, https://doi.org/10.1007/978-1-0716-2289-6_6,

10]. Targeting these challenges, there is a clear need to develop a dynamic three-dimensional (3D) brain-on-a-chip model that can provide clinically relevant information. The brain tissue models must (i) include different cell types, such as neurons and astrocytes with host extracellular matrix (ECM), and (ii) be able to replicate the spatiotemporal context of original brain tissue.

In this chapter, we first describe the method of a brain-on-a-chip model with cocultured murine-derived neurons, microglia, and astrocytes to mimic brain's neurovasculature tissue and membrane-free perfusion with endothelial cells, in which we successfully build a blood–brain barrier to screen neurotoxicity [7, 11]. Then we designate the method to construct a 3D brain tissue using human induced pluripotent stem cells (iPSCs)-derived GABAergic neurons and astrocytes to provide further results for human-relevant brain modeling for toxicity screening [12]. This iPSCs cocultured brain-on-a-chip model facilitates to address the technical challenge to provide high-throughput screening [8]. We validated this model by testing the organophosphate-induced neurotoxicity and the potential cure for neurotoxicity with a bioscavenger butyrylcholinesterase (BuChE). The human iPSCs combined brain-on-a-chip platform can be used for personalized medicine, customized drug screening, and disease modeling in the neuroscience field [13, 14]. These brain-on-a-chip models provide alternatives for animal testing to screen the new drugs, to treat potential neurotoxicity, and to understand the biochemical mechanisms underlying neurological disorders. This chapter provides a method to construct 3D brain-on-a-chip using commercial cell lines and human iPSCs which recapitulate complex brain function.

2 Materials

2.1 Solution Preparation

1. Culture medium for murine cell lines: Dulbecco's modified eagle's medium (DMEM), 10% fetal bovine serum (FBS), and 1% penicillin/streptomycin (P/S).
2. Neural Complete Maintenance Medium (NCMM) for human induced pluripotent stem cells (iPSCs): iCell Neural Base Medium 1, 2% Neural Supplement A, 1% P/S.
3. Organophosphate-induced neurotoxicity solution: 10^{-1}, 10^{-3}, and 10^{-5} M malathion (MT). 10^{-1} M MT: add 16.5 mg MT in 25 μL dimethyl sulfoxide (DMSO), then mix thoroughly, then add 475 μL iCell Neural Complete Maintenance Medium (NCMM). 10^{-3} M MT: 5 μL 10^{-1} M MT, 495 μL NCMM; 10^{-5} M MT: 5 μL 10^{-3} M MT, 495 μL NCMM.
4. Bioscavenger solution: 50 μM butyrylcholinesterase (BuChE) in NCMM.

5. 1X phosphate-buffered saline (PBS): 137 mM NaCl, 2.7 mM KCl, 8 mM Na_2HPO_4, and 2 mM KH_2PO_4 in deionized water (dH_2O).
6. 2.5% Trypsin-EDTA solution.

2.2 Murine Cell Lines and Human Induced Pluripotent Stem Cell (iPSC)

1. Murine cell lines include mouse brain neuroblastoma (N2a), microglia (BV-2), astrocyte (C8D1A), and immortalized murine brain endothelial cells (bEnd3). Store cell lines in liquid nitrogen before recovery.
2. Human iPSC-derived GABAergic neurons and astrocytes (iCell® GABANeurons and Astrocytes, FUJIFILM Cellular Dynamics, Inc.). Store GABAergic neurons and astrocytes in liquid nitrogen and thaw cell in 37 °C water bath right before seeding.

2.3 Hydrogel Preparation

1. Collagen matrix: 250 μL 4 mg/mL collagen I, pH 7.4. Add 25 μL PBS, 98 μL 10.21 mg/mL collagen (Rat tail, Conring), 126 μL DMEM. Adjust pH to 7.4 with 1.25 μL 0.5 M NaOH (*see* **Note 1**). Collagen preparation is performed on ice.
2. Matrigel: 5 mg/mL Matrigel™ Growth Factor Reduced Membrane Matrix (Corning). Add 10 mg/mL Matrigel to same amount of iCell Neural Complete Maintenance Medium (NCMM) on ice. Adjust pH to 7.4 with 0.5 M NaOH (*see* **Note 1**).

2.4 MIMETAS Culture Ware and Repeating Pipette

1. The 2-lane OrganoPlate® (MIMETAS, Netherlands) is used to construct the 3D microfluidic brain-on-a-chip model (*see* Fig. 1a) [7, 11, 12].
2. A repeating pipette with 0.5 ~ 5 μL capacity is necessary for consistent gel–cell seeding.
3. The combitips are 0.1 mL in sterile individual packages.

2.5 3D Cell Coculture Conditions

1. An incubator with 37 °C and 5% CO_2.
2. Use an interval rocker (MIMETAS) with a 7° angle to horizontal level and set it to perfusion mode to allow medium perfusion with bi-directional flow. Use the rocker for cell culture, treatment, and immunofluorescence.

2.6 Characterization Kit

1. LIVE/DEAD® Viability/Cytotoxicity Kit, live/dead staining solution: 2 μM calcein AM, 4 μM ethidium homodimer-1, and 2 μg/mL Hoechst in PBS.
2. Molecular Probes™ Amplex™ Acetylcholine/Acetylcholinesterase (AChE) Assay Kit, stock solution preparation:
   (a) 20 mM stock solution: warm Amplex Red reagent (A), DMSO (B) at room temperature, then dissolve 1 mg

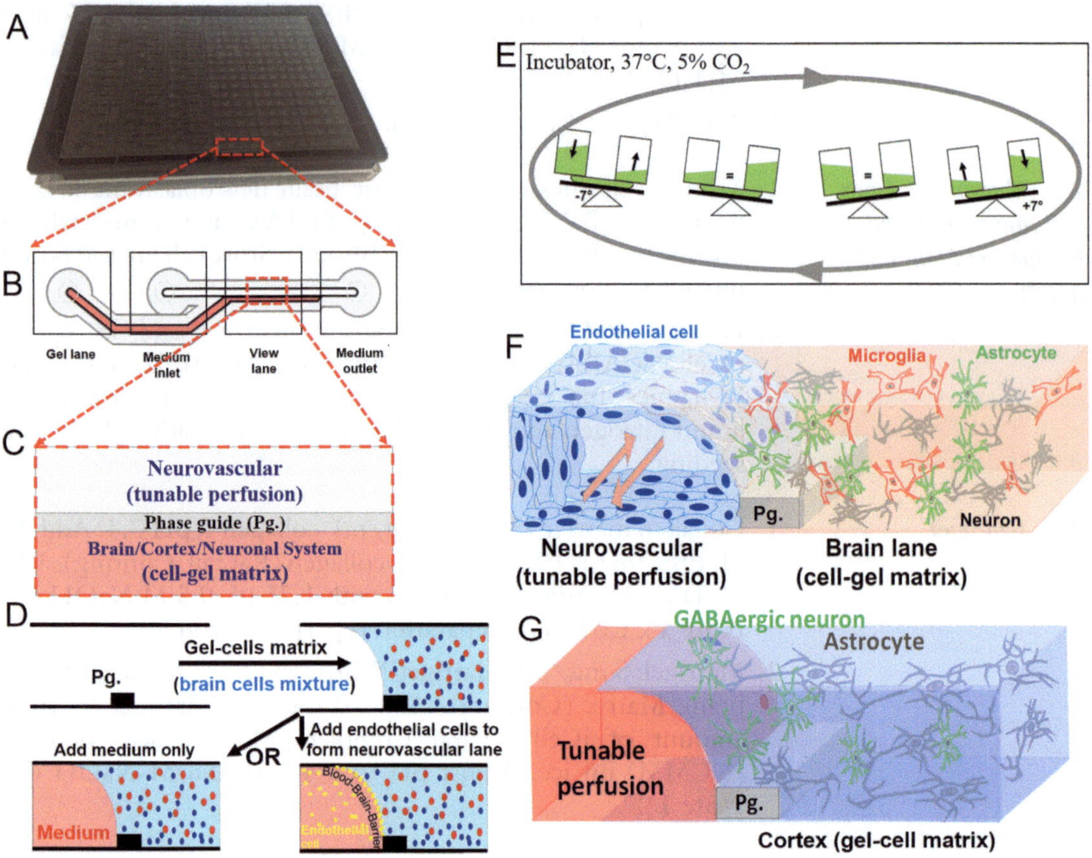

Fig. 1 Brain-on-a-chip platform design [7, 11, 12]. (**a**) MIMETAS OrganoPlate consisting of 384 wells. (**b**) The schematic structure of a two-lane microfluidic chamber consisting of 4 wells. (**c**) The schematic design (top view) of the observation column. (**d**) Cell seeding procedures. (**e**) The incubation of OrganoPlate in a dynamic perfusion condition. (**f**) Expecting cell coculture with murine cell lines after incubation. (**g**) Expecting cell coculture with human iPSC-derived GABAergic neurons and astrocytes

reagent A in 200 μL DMSO. Then store it at −20 °C and protect it from light.

(b) 1 X working solution of reaction buffer: Add 5 mL of 5 X reaction buffer stock solution (E) to 20 mL dH_2O. The 25 mL working solution is for 100 assays of 200 μL each.

(c) 20 U/mL stock solution of horseradish peroxidase (HRP): Dissolve the HRP (C) in 1 mL of 1 X reaction buffer. After use, aliquot solution into 1.5 mL tubes and store at −20 °C.

(d) 20 mM H_2O_2 working solution: Add 23 μL 3% H_2O_2 stock solution (D) into 977 μL dH_2O.

(e) 20 U/ml stock solution of choline oxidase: Add the content of the choline oxidase (F) in 600 μL of 1 X reaction buffer.

(f) 100 mM solution of acetylcholine: Dissolve 5 mg of acetylcholine chloride (G) in 275 μL dH_2O right before the experiment. Store the rest acetylcholine solution desiccated at −20 °C.

(g) 100 U/mL stock solution of acetylcholinesterase: Dissolve the content of acetylcholinesterase (H) in 600 μL 1 X reaction buffer. Aliquot the solution into 1.5 mL tubes and store at −20 °C.

2.7 Immunofluorescence

1. Murine brain neuroblastoma (N2a): Anti-MAP2 (1:100, Invitrogen PA110005).
2. Murine brain astrocyte (C8D1A): Goat GFAP polyclonal antibody (1:100, Invitrogen PA518598).
3. Murine immortalized brain endothelial cells (bEnd3): Claudin-5 (1:50, Invitrogen, 341,600).
4. Human iPSC GABA neurons: anti-β-tubulin III Alexa Fluor 488 (1:20, BD Biosciences, 560,381).
5. Human astrocytes: anti-GFAP (1:100, Invitrogen, PA5-18598).
6. Human cell synapses: anti-synaptophysin (1:50, Invitrogen, MA5-14532).
7. Murine brain neuroblastoma (N2a): Goat anti-Chicken Alexa Fluor 647 (1:100, Life Tech, A21449).
8. Murine brain astrocyte (C8D1A): Donkey anti-goad Alexa Fluor 488 (1:100. Invitrogen, A11055).
9. Murine immortalized brain endothelial cells (bEnd3): Goat anti-Rabbit Alexa Fluor 568 (1:50, Invitrogen A11036).
10. Human astrocytes: donkey anti-goat Alexa Fluor 647 (1:100, EMD Millipore, AP180SA6).
11. Human cell synapses: donkey anti-rabbit Alexa Fluor 568 (1: 50, Invitrogen, A10042).
12. Nuclei staining: Hoechst (1:2000, Invitrogen, H3570).
13. Permeabilization buffer: 0.1% Triton X-100 in PBS.
14. Blocking buffer: 10% normal donor horse serum in PBS.
15. 4% paraformaldehyde (PFA).

3 Methods

3.1 Murine Cell Seeding to 3D Brain-on-a-Chip

1. Culture N2a, BV-2, C8D1A, and bEnd3 separately in T75 flasks. When N2a, BV-2, and C8D1A become confluency, harvest the three cell lines first by detaching them using 2.5% Trypsin-EDTA. Then spin down at 1000 g for 5 min. (*see* **Note 2**).

2. Add 1 mL DMEM to resuspend cells then count with Biorad TC20 cell counter.
3. Calculate and add DMEM to cell suspension to 10^6 cells/mL N2a, 10^6 cells/mL BV-2, and 5×10^5 cells/mL C8D1A.
4. Aliquot cell suspension to 1.5 mL tubes, 1 mL/tube. Then spin down at 1000 g for 5 min.
5. Resuspend and mix N2a, C8D1A, and BV-2 cells in a total 320 μL collagen hydrogel to 3.12×10^6 cells/mL, 3.12×10^6 cells/mL, and 1.56×10^6 cells/mL, respectively. The cell–collagen mix stays on ice all the time.
6. Inject 2 μL cell–collagen mixture to the gel lane of a 2-lane OrganoPlate® using an Eppendorf repeater pipette (*see* **Note 3**), then incubate the plate at 37 °C, 5% CO_2 for 1 h for gel polymerization. Then add 20 μL DMEM into the gel lane inlet.
7. After polymerization, harvest bEnd.3 cells by detaching with 2.5% Trypsin-EDTA and spin down at 1000 g for 5 min. Then discard the supernatant.
8. Resuspend 1×10^7 cells/mL bEnd.3 in DMEM.
9. Then dispense 2 μL bEnd.3 in DMEM to neurovascular lane through medium inlet (*see* Fig. 1b).
10. Incubate the plate against the side of the incubator at a 75° angle for 4 h to allow bEnd3 cells to settle against the cell–gel matrix.
11. Then add 50 μL medium to the medium inlet and outlet, and place the plate on an interval rocker (MIMETAS) for medium perfusion inside the incubator (*see* Fig. 1e). The rocker (switching between +7° and − 7° inclination every 8 min) created a bi-directional flow with a mean flow rate of 2.02 μL/min and a mean shear rate of 0.13 Pa.
12. Refresh DMEM every other day (50 μL each in the inlet and outlet).

3.2 IPSC-Derived GABAergic Neurons and Astrocytes Seeding in the 3D Brain-on-a-Chip

1. Count cell number of GABAergic neurons and astrocytes with hemocytometer right after thawed in 37 °C water bath (*see* **Note 2**).
2. Mix GABAergic neurons and astrocytes in ratios of 4:1 and 1:4 (A1/N4 and A4/N1, respectively) for total amount 7.8×10^6 cell/mL.
3. Spin down cells at 1000 rpm for 5 min.
4. Discard supernatant and add the same amount of Matrigel. (The total amount of cell and Matrigel needs to be calculated before experiment according to specific experimental design.)

5. Pipette 2 μL of cell–gel matrix into each gel lane of a 2-lane OrganoPlate® using an Eppendorf repeater pipette (*see* **Note 3**).
6. Incubate the plate at 37 °C and 5% CO_2 for 1 h.
7. Pipette 20 μL medium into the gel lane, 50 μL medium into the inlet and outlet of the medium lane (100 μL total) of each well. The plate was incubated at 37 °C and 5% CO_2 on an interval rocker to allow medium perfusion with bi-directional flow.
8. Refresh medium every other day.

3.3 Brain-on-a-Chip Model Construct Examination Using Immunofluorescence

Perform all the following steps at room temperature.

1. Add 50 μL 4% PFA to both inlet and outlet of perfusion lane and incubate on the rocker for 15 min. (50 μL/50 μL to medium inlet/outlet, rocker).
2. Wash cells twice with phosphate-buffered saline (PBS) for 5 min. Repeat this step for once (50 μL/50 μL to medium inlet/outlet, rocker).
3. Permeabilize cells with Triton X-100 (0.1% in PBS) for 5 min. (50 μL/50 μL to medium inlet/outlet, rocker).
4. Wash cell with PBS for 1 min. (50 μL/50 μL to medium inlet/outlet, rocker).
5. Block with 10% normal donor horse serum in PBS for 1 h (50 μL/50 μL to medium inlet/outlet, rocker).

   Meanwhile, prepare primary antibodies in 1X PBS on ice.

6. Incubate cells with primary antibodies for 1.5 h (30 μL/10 μL to medium inlet/outlet, rocker).

Meanwhile, prepare for the secondary antibodies and Hoechst on ice.

7. Wash 3 times with PBS for 3 min (50 μL/50 μL to medium inlet/outlet, rocker).
8. Incubate cells with secondary antibodies for 1 h (30 μL/10 μL to medium inlet/outlet, rocker).
9. Wash 3 times with PBS for 3 min (50 μL/50 μL to medium inlet/outlet, rocker).
10. Use a two-photon confocal microscope and ZEN software to acquire fluorescent images. Set z-stack and tiling to capture an entire brain-on-a-chip. Representative images of brain-on-a-chip models are displayed in Figs. 2 and 3d tera-culture brain-on-a-chip model using murine cell lines; Fig. 3, brain-on-a-chip model using human iPSC-derived GABAergic neurons and astrocytes.

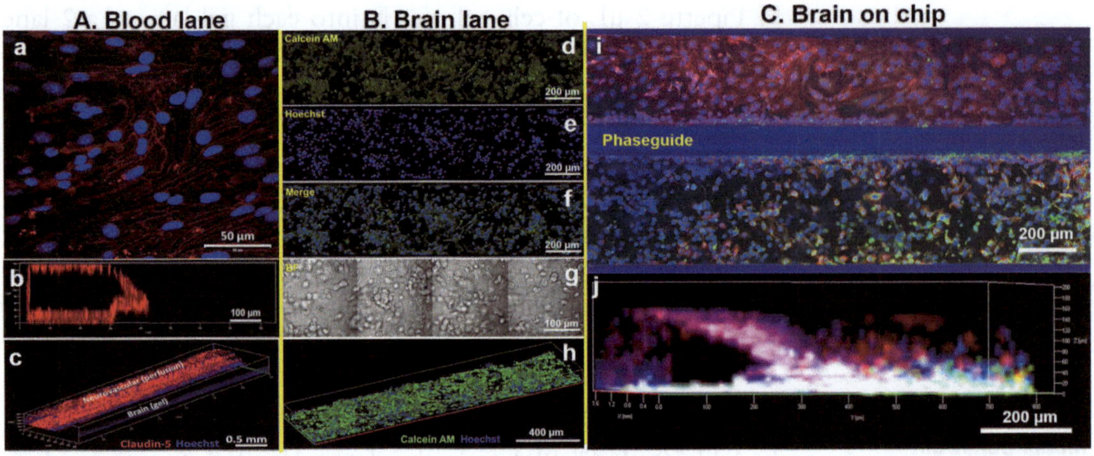

Fig. 2 Construction confocal imaging of blood lane and brain lane [11]. (**a**) Blood lane only; (a) 2D stained image of tight junction (claudin-5, red) and nucleus (Hoechst, blue) for neurovascular, (b) cross-sectional 3D image of neurovascular, (c) 3D image of neurovascular, (**b**) brain lane only; (d) live (calcein AM, green), (e) nucleus (Hoechst, blue), (f) merged image of live and nucleus images, (g) bright field image, (h) 3D image of merged construct of gel-3 different type cells matrix, (**c**) brain on chip by the tetra-culture (claudin-5, red; calcein AM, green, and Hoechst, blue); (i) 2D stained image for blood and brain lanes, (j) 3D images of brain construct with endothelial, neuroblastoma, astrocyte, and microglia cells

11. Store plate by sealing edges with Parafilm, wrap in aluminum foil, store at 4 °C for up to 2 weeks.

3.4 MT Exposure and BuChE Treatment

1. On the fifth day of 3D cell coculture of human iPSC-derived GABAergic neurons and astrocytes, remove the medium from each well.
2. Add 50 μL 10^{-1}, 10^{-3}, and 10^{-5} M MT solution to both inlet and outlet of perfusion lane (100 μL total) at each MT concentration (N = 3/MT concentration). This is one set of cells coculture.
3. After 20 min of MT exposure, add 10 μL (5 μL in/outlet) of 50 μM butyrylcholinesterase in medium to one set of MT-exposed perfusion lane. Leave another set of MT exposed cell coculture as control.
4. After 24 h, pipette medium out from each well and store medium at −20 °C for analysis of ChE activity.

3.5 Cell Viability Evaluation

1. Remove culture medium and rinse twice with PBS (50 μL/50 μL to medium inlet/outlet).
2. Incubate cell coculture with live/dead staining solution for 30 min at room temperature (50 μL/10 μL to medium inlet/outlet, rocker).
3. Wash cells with PBS for 1 min (50 μL/50 μL to medium inlet/outlet).

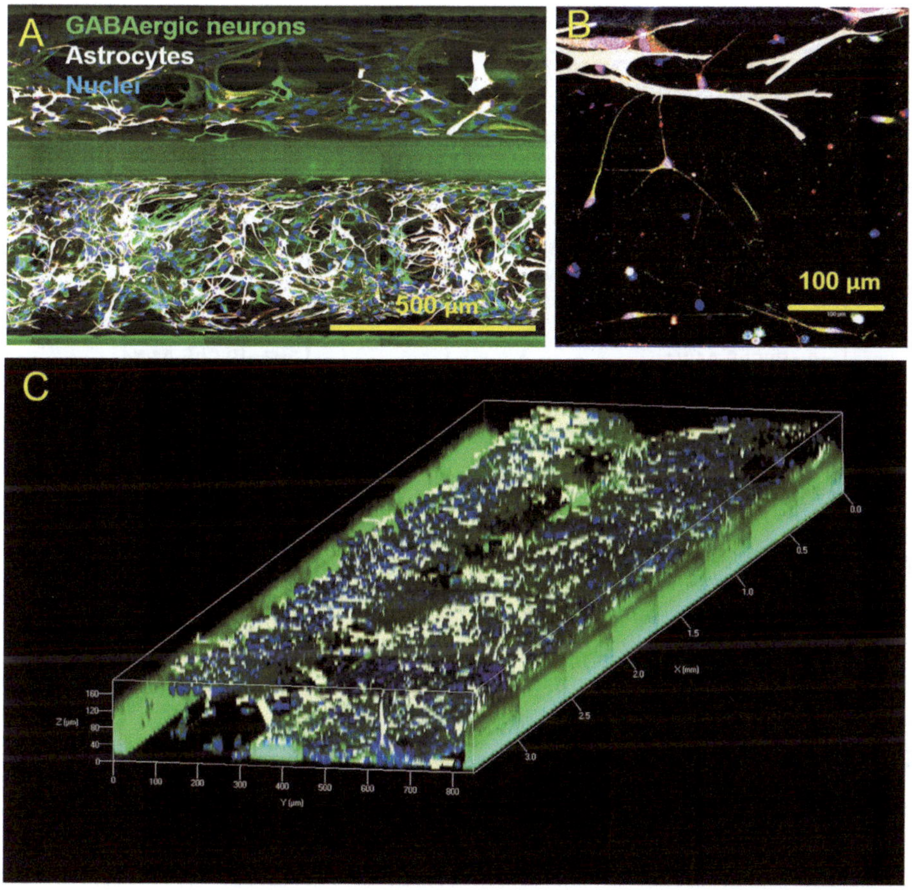

Fig. 3 Representative immunofluorescence images of brain-on-a-chip model using human iPSC-derived GABAergic neurons and astrocytes [12]. (**a**) Top view of cell coculture in the brain-on-a-chip model. (**b**) Synapse (red dots) formation on GABA neurons (green). (**c**) A representative 3D image of a brain-on-a-chip channel

4. Add 50 μL PBS to both medium inlet and outlet, and capture images under a fluorescent microscope.
5. Quantify the live (green) and dead (red) cells with ImageJ software, and calculate viability by dividing the number of live cells by the total number of cells and multiplying by 100% (Fig. 4a).

3.6 Acetylcholinesterase and Butyrylcholinesterase Activity Assay

1. Unknown samples preparation: Add 16 μL supernatant (gathered in **step 2.3**) to 384 μL 1X reaction buffer. Each reaction needs 100 μL for assay.
2. Prepare a positive control by diluting the 2 μL of 100 U/mL acetylcholinesterase stock solution (H) into 998 μL 1 X reaction buffer to produce a 0.2 U/mL acetylcholinesterase solution. Then add 500 μL 0.2 U/ML to 500 μL 1X reaction buffer. Use the same way to prepare 0.05 U/mL, 0.025 U/

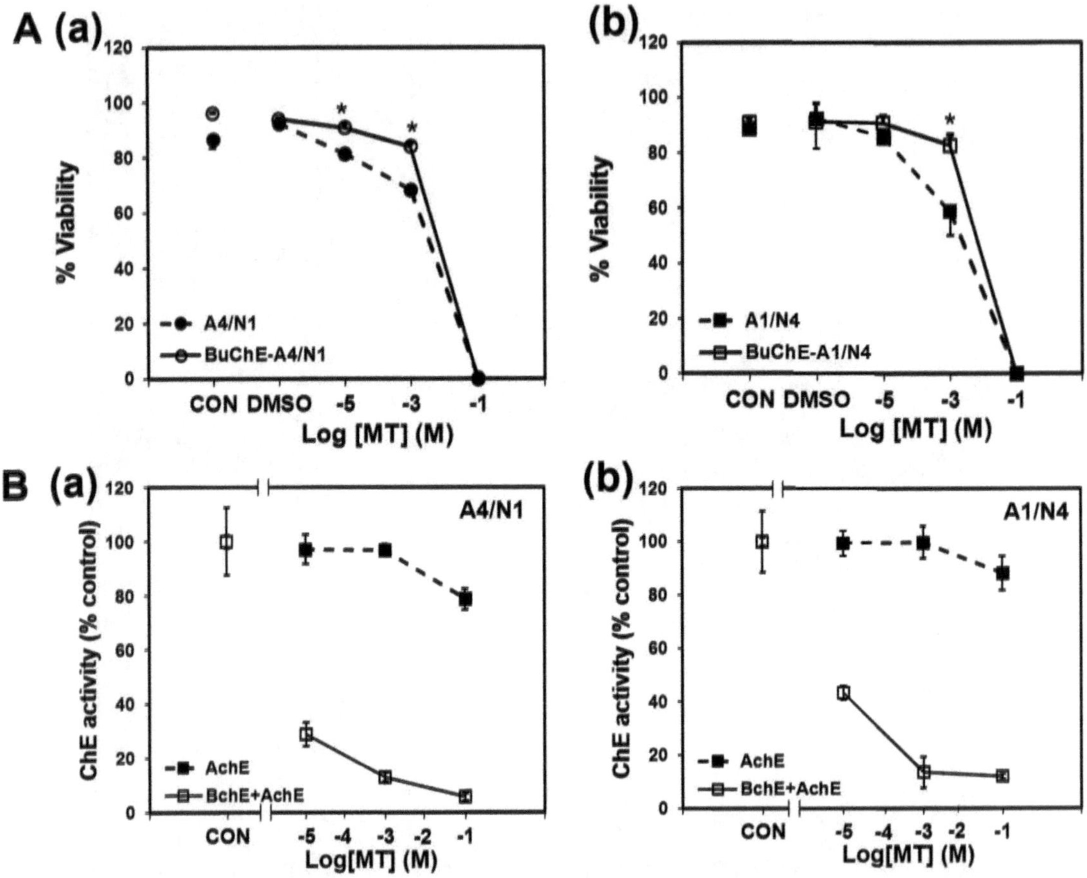

Fig. 4 The results of live/dead assay (**a**) and ChE assay (**b**) were analyzed with Origin® 2018 software [12]. The coculture of astrocytes and GABA neurons, (a) A4/N1 and (b) A1/N4, showed viability decrease when exposed to increased MT concentration, and the BuChE showed bioscavenger effect to MT

mL, 0.0125 U/mL, 0.00625 U/mL, 0.003125 U/mL, 0.0015625 U/mL, and 0.00078125 U/mL positive controls. Use the 1X reaction buffer without acetylcholinesterase as a negative control.

3. Prepare a second positive control by diluting the 20 mM H_2O_2 working solution to 10 μM with 1X reaction buffer.
4. Pipette 100 μL positive and negative controls, and diluted supernatant into each well of a 96 well plate.
5. Working solution of 400 μM Amplex Red reagent (2 U/mL HRP, 0.2 U/mL choline oxidase, and 100 μM acetylcholine): Add 200 μL Amplex Red reagent stock solution, 100 μL HRP stock solution, 100 μL choline oxidase stock solution, and 10 μL acetylcholine stock solution to 9.59 mL 1 X reaction buffer.

6. Add 100 μL previous working solution to samples and controls and incubate for 30 min at room temperature.
7. Read fluorescence using a CLARIOstar microplate reader at 590-nm at room temperature. The acetylcholinesterase and butyrylcholinesterase activity can be presented as Fig. 4b shows.

4 Notes

1. Add 0.25 μL 0.5 M NaOH each time to adjust PH.
2. In **step 3.1**, the murine cell lines need to be recovered in flasks to make cell–collagen mixture before seeding into the microfluidic chips, while the human iPSCs (*see* **step 3.2**) need to be counted and mixed with Matrigel right after thawing in a 37 °C water bath. There is no need of cell recovery before seeding into the microfluidic chip.
3. The MIMETAS microfluidic chip has a phase guide that is used to create a capillary pressure barrier to separate the brain lane from the medium lane. The small dimension of the wells allows high-throughput experiments. These features thus require specific 3D cell seeding skills. There are some highlights to make sure the successful seeding in **steps 3.1** and **3.2**:
   (a) The collagen and Matrigel needs to be on ice to maintain its viscosity and prevent gelation.
   (b) Cool repeating pipette tips before dispensing gel–cell mixture and keep it cool between dispensing.
   (c) Resuspend the gel–cell mixture on ice occasionally.
   (d) Hold the repeating pipette straight while dispensing the gel-cell mixture.
   (e) Observe the reflection of the channels: the well cannot be used if the gel–cell mixture flow over to the perfusion lane.
   (f) The medium in pipet needs to be dispensed against the wall to prevent bubbles.
   (g) Transport plates gently to avoid fierce agitation.
   (h) Put well plates symmetrically on the perfusion shaker.
   (i) Use murine cell lines with collagen (*see* **step 3.1**) to practice first before use Matrigel and iPSC (*see* **step 3.2**). There will be an over 90% successful seeding rate after practice.

Acknowledgments

This work was supported by the US Department of Defense contact # D01 W911SR-14–2–0001–0010 awarded through the MSRDC Consortium (DTRA and ECBC) at North Carolina A & T State University, NIH NIGMS grant (ISC3GM113728, YY is the recipient), NSF EAGER Grant (1649243, YY is the recipient), and Army Research Office (ARO74386, YY is the recipient).

References

1. Jahromi MAM, Abdoli A, Rahmanian M et al (2019) Microfluidic brain-on-a-chip: perspectives for mimicking neural system disorders. Mol Neurobiol 56(12):8489–8512
2. Ahadian S, Civitarese R, Bannerman D et al (2018) Organ-on-a-chip platforms: a convergence of advanced materials, cells, and microscale technologies. Adv Healthc Mater 7(2): 1700506
3. Nikolakopoulou P, Rauti R, Voulgaris D et al (2020) Recent progress in translational engineered in vitro models of the central nervous system. Brain 143(11):3181–3213
4. Ma C, Peng Y, Li H et al (2020) Organ-on-a-chip: a new paradigm for drug development. Trends Pharmacol Sci 42(2):119–133
5. Oddo A, Peng B, Tong Z et al (2019) Advances in microfluidic blood–brain barrier (BBB) models. Trends Biotechnol 37(12): 1295–1314
6. Bang S, Jeong S, Choi N et al (2019) Brain-on-a-chip: a history of development and future perspective. Biomicrofluidics 13(5):051301
7. Liu L, Koo Y, Akwitti C et al (2019) Three-dimensional (3D) brain microphysiological system for organophosphates and neurochemical agent toxicity screening. PLoS One 14(11): e0224657
8. Mungenast AE, Aron R, White JD et al (2019) Neural tissue microphysiological systems in the era of patient-derived pluripotent stem cells. In: Microfluidic cell culture systems. Elsevier, Amsterdam, Netherlands, pp, 249–296
9. Miccoli B, Braeken D, Li Y-CE (2018) Brain-on-a-chip devices for drug screening and disease modeling applications. Curr Pharm Des 24(45):5419–5436
10. Jorfi M, D'avanzo C, Kim DY et al (2018) Three-dimensional models of the human brain development and diseases. Adv Healthc Mater 7(1):1700723
11. Koo Y, Hawkins BT, Yun Y (2018) Three-dimensional (3D) tetra-culture brain on chip platform for organophosphate toxicity screening. Sci Rep 8(1):1–7
12. Liu L, Koo Y, Russell T et al (2020) Three-dimensional brain-on-chip model using human iPSC-derived GABAergic neurons and astrocytes: Butyrylcholinesterase post-treatment for acute malathion exposure. PLoS One 15(3): e0230335
13. Mahla RS (2016) Stem cells applications in regenerative medicine and disease therapeutics. Int J Cell Biol 2016:6940283
14. Papapetrou EP (2016) Induced pluripotent stem cells, past and future. Science 353(6303):991–992

Part III

Co-culture, Permeability, and Transwell Models of the BBB: Methods for Studying Shear Stress, Barrier Integrity and Breakdown

Chapter 7

An Improved In Vitro Porcine Blood–Brain Barrier Model for Permeability Screening and Functional Studies

Adjanie Patabendige

Abstract

The availability of good in vitro blood–brain barrier (BBB) models that closely mimic in vivo BBB features are essential for central nervous system (CNS) drug permeability screening and BBB functionality studies. Of the currently available monoculture primary BBB models, porcine brain endothelial cell models have the best barrier properties, which make them highly suitable for CNS drug permeability screening. In addition, they retain major BBB features such as BBB transporters, receptors, and enzymes and express BBB tight junctions. Therefore, porcine BBB models are also suitable for BBB functionality studies. This paper describes a procedure for extraction of brain microvessels from fresh porcine brains and the culture of pure primary porcine brain endothelial cells. In addition, techniques to improve culture purity and quality, and increase barrier tightness without using co-cultures are given. Using this method, a robust and reproducible in vitro BBB model can be established for CNS permeability screening and studying BBB functionality.

Key words Blood–brain barrier, Porcine brain endothelial cells, In vitro model, Microvessel extraction, Transendothelial electrical resistance, Permeability screening

1 Introduction

The blood–brain barrier (BBB) is a complex and highly specialized physiological barrier that protects the brain from blood-borne constituents to maintain homeostasis of the brain microenvironment. Complete replication of the structure and function of this barrier in vitro is not an easy task. However, several reliable and reproducible in vitro models have been developed over the years that mimic the major features of the in vivo BBB. In vitro BBB models are preferable to animal models for initial drug permeability screening as animal models are associated with high experimental cost and complexity. Of the available models, monoculture BBB models derived from porcine brain microvessels tend to give higher transendothelial electrical resistance (TEER, a measurement of tightness of the BBB) compared with models derived from rodent

Nicole Stone (ed.), *The Blood-Brain Barrier: Methods and Protocols*,
Methods in Molecular Biology, vol. 2492, https://doi.org/10.1007/978-1-0716-2289-6_7,

or human brains [1, 2]. Furthermore, monoculture models are preferred to co-culture/multi-culture or flow-based models, as simplicity, reproducibility, and reliability are important for central nervous system (CNS) drug permeability studies when screening a large number of compounds [3].

Patabendige et al. [4, 5] have established such a model system using primary porcine brain endothelial cells (PBEC). This chapter will describe the monoculture porcine BBB model in detail to support the adoption of the model by researchers interested in using it for BBB permeability screening and functionality studies. Isolating microvessels from porcine brains can be time-consuming and require expertise to obtain pure cultures with reproducible results. However, once isolated, the microvessels can be aliquoted and stored in liquid nitrogen for future use, similar to immortalized cell lines. Researchers who lack experience may find that this simplified protocol with technical notes to support troubleshooting will be especially helpful to establish the porcine brain endothelial cell model without significant difficulties.

2 Materials

Porcine brains can be obtained from a local abattoir, and extraction of the brains from the skull can be performed by the abattoir worker (this step does not require sterile technique). Ideally, the brains must be extracted within 1–2 hours of death and transported to the laboratory within 3–4 hours of death in ice-cold transport medium.

2.1 Transport and Initial Dissection

1. Six fresh porcine brains.
2. Transport medium: 1% penicillin (100 U/mL)/streptomycin (100 μg/mL) in 3 L L-15 medium. Store at 4 °C.
3. Wash solution: 1% penicillin/streptomycin in 1 L phosphate-buffered saline (PBS) without Ca^{2+}, Mg^{2+}. Store at 4 °C.
4. 3 x 1 L containers.
5. Ice.
6. Large polystyrene box for transport.
7. 70% ethanol.
8. Coplin jar.
9. Glass beakers.
10. Waste beaker.
11. Scalpels (no. 10) and blades.
12. Fine curved forceps.
13. Coarse curved forceps.

14. Petri dishes (150 mm diameter).
15. Gauze.
16. Paper towels.
17. Isolation medium: 10% fetal calf serum (FCS) and 1% penicillin/streptomycin in 2 L HEPES-buffered minimum essential medium. Store at 4 °C. Need 1600 mL.
18. 50 mL syringe
19. 40 mL glass homogenizer with tight and loose pestles.

2.2 Size-Selective Filtration and Enzymatic Digestion

1. Reusable filter holder unit for 50 mm diameter membranes with 500 mL receiver.
2. 5 × each of 60 μm and 150 μm pore size nylon meshes, cut into 50 mm diameter circles.
3. 250 mL isolation medium.
4. Fine forceps.
5. Petri dish (150 mm diameter).
6. Digest mix: 223 U/mg collagenase type III, 211 U/mg trypsin, and 2108 U/mg DNase I in 178 mL M199 medium. Syringe filter, using a 0.2 μm filter. Add 20 mL FCS and 2 mL penicillin/streptomycin. Aliquot into 50 mL tubes (40 mL each). Store at −20 °C for up to 6 months. Need 160 mL.
7. Humidified CO_2 incubator at 37 °C.
8. Rocker or shaker.

2.3 Centrifugation

1. 1000 μL pipette and tips.
2. Fine forceps.
3. Pipetting aid and 25 mL serological pipettes.
4. 150 mL of isolation medium.
5. 50 mL centrifuge tubes.
6. Refrigerated centrifuge.

2.4 Freezing

1. Freezing mix: 10% dimethyl sulfoxide (DMSO) in 25 mL FCS. Use immediately.
2. 24 × 2 mL cryogenic vials.
3. 1000 μL pipette and tips.
4. Freezing container.
5. −80 °C freezer.
6. Liquid nitrogen storage tank.

2.5 Coating

1. 2 × T75 flasks.
2. Rat tail collagen, type I solution: 300 μg/mL in sterile distilled water. Prepare immediately before use. Need 4 mL.
3. PBS.
4. Human fibronectin solution: 7.5 μg/mL in sterile distilled water. Prepare immediately before use. Need 4 mL.

2.6 Thawing and Growth

1. 37 °C water bath.
2. 1 mL plastic Pasteur pipette.
3. Pipetting aid and 10 mL serological pipettes.
4. 2 × coated T75 flasks.
5. PBEC growth medium: 10% bovine plasma-derived serum (BPDS), 1% glutamine (2 mM), 1% penicillin/streptomycin, 125 μg/mL heparin, and 4 μg/mL puromycin in 20 mL Dulbecco's modified eagle medium (DMEM). Syringe filter, using a 0.2 μm filter. Prepare immediately before use.
6. 1 × frozen vial of PBEC.

2.7 Passaging and Seeding onto Transwell Inserts

1. Pipetting aid and 10 mL serological pipettes.
2. Inverted microscope.
3. Hemocytometer.
4. 1% trypan blue solution.
5. Centrifuge.
6. 10 mL centrifuge tubes.
7. PBS without Ca^{2+}, Mg^{2+}.
8. 2 mL Trypsin-EDTA
9. 24 × Transwell "clear" inserts (12 mm diameter, 0.4 μm pore size) in 12-well plates.
10. PBEC culture medium: 10% BPDS, 1% glutamine (2 mM), 1% penicillin/streptomycin, and 125 μg/mL heparin in 50 mL DMEM. Syringe filter, using a 0.2 μm filter. Prepare immediately before use.

2.8 Differentiation

1. Serum-free PBEC switch medium: 1% glutamine (2 mM), 1% penicillin/streptomycin, and 125 μg/mL heparin and 50 μg/mL hydrocortisone in 25 mL DMEM. Syringe filter, using a 0.2 μm filter. Prepare immediately before use.
2. 500 μL pCPT-cAMP. Dissolve 100 mg in 8 mL sterile distilled water. Store at −20 °C.
3. 25 μL RO-20-1724. Dissolve 19.5 mg in 2 ml DMSO. Store at −20 °C.
4. 1.2 mL BPDS.

3 Methods

Once the porcine brains are transported to the laboratory, they must be handled inside a class II biological safety cabinet. Practice sterile technique. Autoclave or use 70% ethanol as necessary to maintain sterility and to decontaminate surfaces. All solutions and media must be kept ice-cold during the transport and initial dissection of porcine brains, and then at 37 °C during the cell growth phase. The cells must be maintained in a humidified CO_2 incubator at 37 °C.

3.1 Preparations (1 Day Before the Procedure)

1. Sterilize by autoclaving dissecting instruments (fine forceps, coarse forceps, and scalpel), homogenizer and pestles, filter holder unit, 5 × each of 60 μm and 150 μm nylon meshes cut into 50 mm diameter circles, gauze and 1 L transport containers.
2. Prepare 3 L of sterile transport medium. Store at 4 °C.
3. Prepare 1 L of wash solution. Store at 4 °C.
4. Prepare 2 L of isolation medium. Store at 4 °C.
5. Prepare 200 mL of digest mix. Store at −20 °C.

3.2 Set Up on the Day of the Procedure

1. Switch on the class II biological safety cabinet and sterilize the surfaces.
2. Ensure the following are at hand: sterilized dissecting instruments, sterile scalpel blades, glass beakers, waste beaker, petri dishes, gauze, paper towels, nylon meshes, centrifuge tubes, 70% ethanol in a Coplin jar, and 70% ethanol in a spray bottle.
3. On ice: transport medium, wash solution and isolation medium.
4. Switch on the centrifuge (4 °C).
5. Switch on the water bath (37 °C).
6. Leave the digest mix and FCS out to thaw.

3.3 Acquire and Transport Porcine Brains

1. Half-fill each 1 L container with ice-cold transport medium. Place in a large polystyrene container with ice.
2. Collect six porcine brains from the abattoir within 1–2 hours of death. Wash each brain briefly with ice-cold transport medium and place in the containers (two brains per container). Fill up the containers with remaining transport medium (*see* **Note 1**).
3. Pack the containers in a polystyrene box containing ice to keep the brains cold during transport to the laboratory (ideally within 3–4 hours of death).

3.4 Initial Dissection

1. Add ice-cold isolation medium into a beaker. Remove one brain from the transport container, and place into the beaker to wash the brain. Transfer to another beaker with fresh ice-cold isolation medium.
2. Place a sterile piece of gauze on a Petri dish. Remove the brain from the beaker, and place it on your hand directly above the Petri dish (Fig. 1A). Using fine curved forceps, carefully remove the meninges (Fig. 1B), making sure to remove the meninges inside the sulci (*see* **Note 2**).
3. Place another piece of sterile gauze on a second Petri dish. Roll the brain on the gauze to remove surface leptomeningeal cells. Keeping the brain cold will keep it intact and make this step easier to perform. Transfer to a new beaker containing ice-cold isolation medium.
4. Clean hands using paper towels and spray with 70% ethanol before moving on to the next brain. Repeat **steps 1–3** for each brain.
5. Place a cleaned brain on your hand. Using coarse curved forceps, remove the white matter (Fig. 1C) (*see* **Note 3**).
6. Transfer the remaining brain matter to a new beaker containing ice-cold isolation medium. Repeat **steps 5** and **6** for all brains.
7. Using a scalpel, chop the brain matter into about 1 cm^3 pieces (Fig. 1D).
8. Add 100 mL of isolation medium into a beaker. Using forceps, transfer the brain matter into a 50 mL syringe (without needle). Replace the plunger and extrude into the beaker containing isolation medium. Repeat this process for the remaining brain matter.
9. Half-fill the homogenizer with brain matter. Top up with isolation medium and gently homogenize for 15 strokes with the loose pestle and then with the tight pestle. Pour the homogenate into a sterile beaker. Repeat the homogenization process until all brain matter is homogenized evenly (Fig. 1E) (*see* **Note 4**).

3.5 Size-Selective Filtration and Enzymatic Digestion

1. Prepare two Petri dishes with 80 mL of digest mix in each. Label one dish as “PBEC 60’s” and the other as “PBEC 150’s” (*see* **Note 5**).
2. Assemble the filter holder unit. Place a sterile 150 μm nylon mesh on the filter holder unit and secure tightly. Filter 200 mL of homogenate through the unit and rinse with 20 mL of isolation medium (Fig. 1F) (*see* **Note 6**).
3. Using sterile curved forceps, carefully remove the mesh from the filter unit and transfer to the Petri dish labeled as PBEC 150’s. Carefully transfer the filtrate into a sterile beaker.

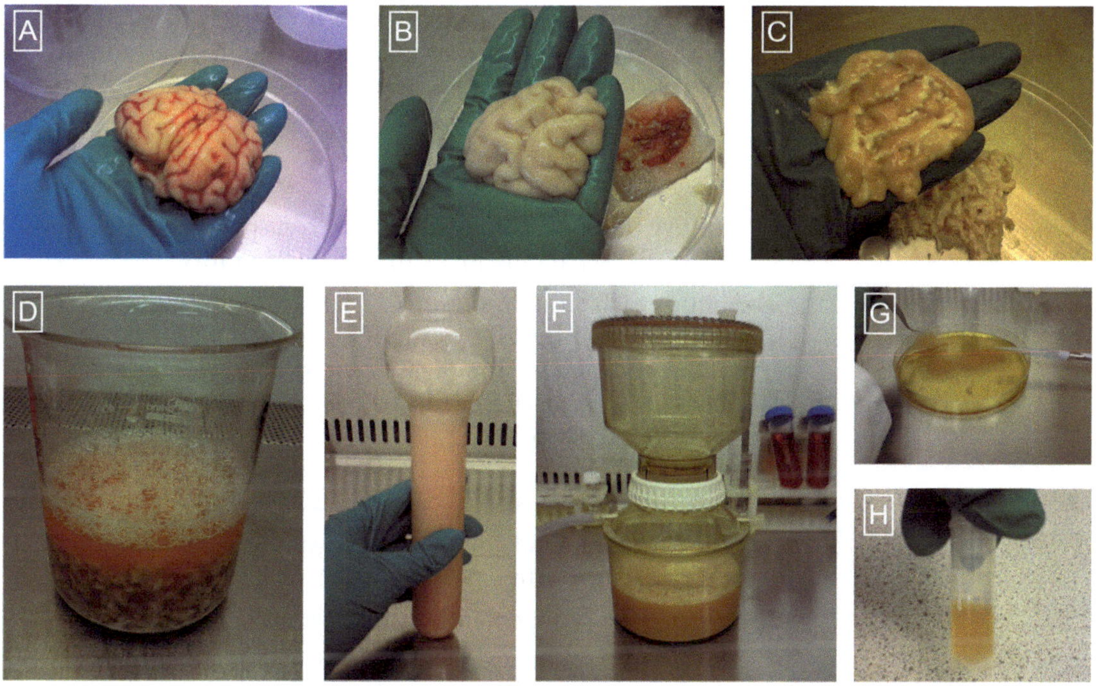

Fig. 1 Procedure for isolating porcine brain microvessels from fresh porcine brains. In brief, porcine brains (**A**) are cleared of meninges (**B**), white matter (**C**), and cut into small pieces (**D**) before homogenizing (**E**) and size-selective filtration through two nylon meshes (60 μm and 150 μm pore size) (**F**). The microvessel fragments that are retained on the meshes are enzymatically digested and collected (**G**) and then centrifuged to remove traces of enzymes and frozen in 1 mL aliquots (**H**)

4. Repeat **steps 2** and **3** for the remaining 150 μm nylon meshes. Then transfer the PBEC 150's Petri dish to the incubator for 1 hour at 37 °C on a rocker.
5. Using the filtrate, repeat **steps 2** and **4** for the 60 μm nylon meshes.
6. Transfer the PBEC 60's Petri dish containing the 60 μm nylon meshes to the incubator for 1 hour at 37 °C on a rocker.

3.6 Centrifugation

1. Remove the Petri dishes from the incubator and transfer to the class II biological safety cabinet. Some microvessel fragments will be attached to the filters, and these can be released using a 1000 μL pipette (Fig. 1G) (*see* **Note** 7).
2. Transfer the contents from each Petri dish to two labeled 50 mL centrifuge tubes. Ensure to keep the two fractions (60's and 150's) separately at all times.
3. Centrifuge at 4 °C for 5 minutes at 240 g.

4. Carefully remove the supernatant from each tube and resuspend with 10 mL of fresh isolation medium. Then add 20 mL of isolation medium and gently triturate. Centrifuge at 4 °C for 5 minutes at 240 g.
5. Repeat **step 4**.

3.7 Freezing

1. Remove the supernatant and gently resuspend each pellet (60's and 150's) separately in 12 mL of freezing mix.
2. Aliquot each suspension into 12 labeled cryovials (1 mL each) (Fig. 1H) and freeze overnight at −80 °C in a freezing container. Transfer to liquid nitrogen storage the next day. The cells can be resuscitated up to 5 years after freezing without losing culture quality.

3.8 Coating Culture Plastics

1. Coat two T75 flasks per vial of PBEC.
2. Add 2 mL of rat tail collagen, type I solution per flask. Incubate at room temperature for 2 hours.
3. Remove the collagen and wash twice with 5 mL of PBS.
4. Add 2 mL of human fibronectin per flask. Incubate at room temperature for 2 hours (*see* **Note 8**).
5. Remove the fibronectin and wash twice with 5 mL of PBS.
6. Leave a small volume of PBS in the flask to avoid drying of the surface until ready for seeding.
7. Transwell inserts can also coated using the same procedure. Use 200 μL of extracellular matrix solution per insert, and wash with 500 μL of PBS.

3.9 Thawing and Growth

1. Prepare 20 mL of PBEC growth medium and warm to 37 °C in a water bath.
2. Carefully remove a vial of PBEC from liquid nitrogen and thaw in a 37 °C water bath for 1–2 minutes, swirling gently.
3. Add the thawed suspension dropwise into the growth medium using a 1 mL plastic Pasteur pipette.
4. Mix gently using a 10 mL serological pipette and transfer equally to the coated T75 flasks.
5. Transfer the flasks to the incubator and leave undisturbed for 3 days.

3.10 Passaging and Seeding onto Transwell Inserts

1. The cells will be 60–80% confluent (Fig. 2) and ready for passaging and seeding onto Transwell inserts 3 days after thawing (*see* **Note 9**).
2. Prepare 50 mL of PBEC culture medium and warm to 37 °C in a water bath.

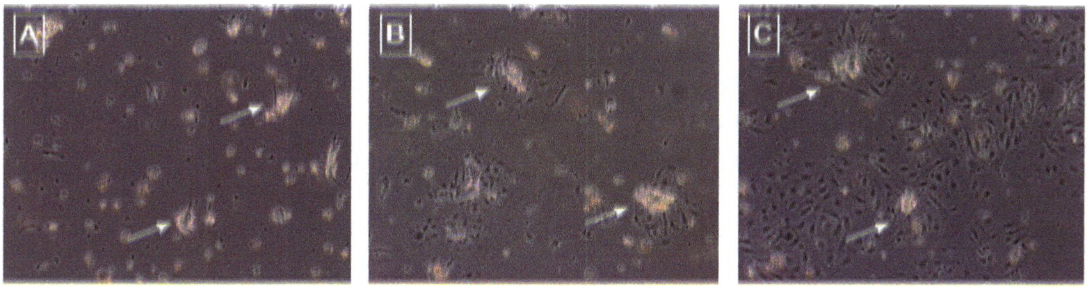

Fig. 2 Porcine brain endothelial cell growth pattern. Porcine brain endothelial cells (PBEC) were grown in medium containing 4 μg/mL puromycin for 3 days to remove contaminating cells such as pericytes that lack P-glycoprotein. Within 1 day of culture, PBEC start to migrate from the microvessel fragments (**A**). PBEC then start forming islands around the microvessel fragments by day 2 (**B**). By 3 days in culture, significant growth around the islands can be seen (**C**). The cultures will be 60–80% confluent by this time and must be passaged

3. Remove PBEC growth medium from the flasks and wash twice with PBS without Ca^{2+}, Mg^{2+}.
4. Add 2 mL of Trypsin-EDTA and transfer to the incubator for 2–5 minutes.
5. Remove the flasks from the incubator and observe under the microscope to assess cell detachment. Once the cells are rounded up, gently tap the flask to aid detachment.
6. Add 8 mL of PBEC culture medium per flask and resuspend gently. Transfer to a 15 mL centrifuge tube.
7. Centrifuge at room temperature for 5 minutes at 360 g.
8. Remove supernatant and resuspend pellet in 1 mL of PBEC culture medium.
9. Use the trypan blue exclusion method to count the cells with a hemocytometer.
10. Calculate the volume required to seed the cells at 1×10^5 cells per insert.
11. Resuspend the cells in a total volume of 12 mL of PBEC culture medium.
12. Add 1.5 mL of PBEC culture medium into each well of a 12-well plate.
13. Add 0.5 mL of cell suspension to each coated Transwell insert.
14. Transfer the well plate to the incubator for 48 hours.

3.11 PBEC Differentiation

1. Observe the cells under the microscope 48 hours after seeding. The cells must be 100% confluent to proceed to the next step (*see* **Note 10**).
2. Prepare 50 mL of serum-free PBEC switch medium and warm to 37 °C in a water bath. Warm an aliquot of BPDS to 37 °C in a water bath.

3. Carefully remove PBEC culture medium from the wells and then the inserts, making sure not to touch the cells.
4. Add 1.5 mL of serum-free PBEC switch medium into the wells and 0.5 mL of medium to the inserts.
5. Mix pCPT-cAMP and RO-20-1724 in the proportion 20:1. Immediately add 5.3 μL of the mixture above and 15.7 μL below the filter. Then add 50 μL of BPDS to each insert (*see* **Note 11**).
6. Transfer the plate to the incubator.
7. The PBEC model will be ready for permeability experiments in 24 hours (*see* **Note 12**).
8. TEER must be measured after 24 hours of this media change to assess the tightness of the monolayer (see [4, 6] for TEER measurement method).

4 Notes

1. Ideally pig brains must be extracted from the skull within 1–2 hours of death. This does not require sterile technique, and an abattoir worker can be instructed to extract the brain at the time of death. It is not necessary to extract the whole brain in one piece. Hemispheres are fine as long as they are intact and not bruised. Only the cerebrum is required, and the cerebellum can be cut off. Once the brain is extracted from the skull, wash thoroughly with ice-cold transport medium to remove any blood, hair, or skull fragments. The transport medium contains 1% penicillin (100 U/mL)/streptomycin (100 μg/mL) to limit possible bacterial contamination.
2. Removing the meninges is easier if the brains are kept ice-cold. This is much like removing wallpaper; sometimes it comes off in fragments and sometimes all at once. As this is a very important step, it is worth taking some time to remove as much of the meninges as possible. Shortening the time to remove meninges usually makes no difference to culture quality.
3. White matter is very sticky, and therefore, removing white matter from the brains can also be time-consuming and difficult. Please note that it is impossible to remove all white matter (Fig. 1C). However, taking time to remove as much as possible will lead to improved culture quality.
4. Do not take the pestle out of the solution while homogenizing to avoid air bubble formation. Air bubbles will make the homogenization process difficult. The homogenate should resemble strawberry milkshake.
5. It is essential to keep the two microvessel fractions (PBEC 60's and 150's) separate. The purity and the quality of the PBEC

cultures obtained from the two fraction differs significantly due to the size of the microvessels retained on the nylon meshes during the filtration process [4].

6. Make sure that the nylon mesh does not get clogged. If it gets clogged, this indicates incomplete removal of meninges and other cells during the initial dissection.
7. It is easier to remove microvessel fragments stuck to the mesh by using sterile forceps to hold the mesh on the side of the Petri dish and using a 1000 μL pipette to discharge the fragments with some digest mix. Pinch off any white matter stuck to the mesh at this stage.
8. If human fibronectin is not available, flasks and Transwell inserts can be coated with rat tail collagen, type I for 3 hours at 37 °C.
9. PBEC growth medium contains puromycin, which is toxic to contaminating cells such as pericytes that do not express the efflux transporter, P-glycoprotein [7]. Culturing the microvessels in the first 3 days with PBEC growth medium leads to a reduction in contaminating cells. Assess the confluency of the cells using an inverted microscope 3 days after seeding. They are ready for passaging if 60–80% confluent. However, do not allow the cells to become over confluent as this will lead to an increase in contaminating cell growth.
10. If cells are not 100% confluent, a change of medium is required. Feed the cells with PBEC culture medium and assess cell growth the next day. When the cells are 100% confluent, serum-free PBEC culture medium is required.
11. Treatment with pCPT-cAMP and RO-20-1724 (increases intracellular calcium levels) in addition to changing PBEC culture medium to the serum-free PBEC switch medium (contains hydrocortisone) leads to a significant increase in the barrier properties of PBEC cultures. Addition of glucocorticoids and agents for increasing intracellular cAMP to the culture medium has been demonstrated to increase in TEER in several BBB models [8–10]. In addition, using BPDS in the luminal (insert) side but not on the abluminal (well) side for 24 hours leads to a further increase in TEER.
12. PBEC 60's fraction gives an average TEER of ~800 $\Omega.cm^2$ (maximum TEER = 1300 $\Omega.cm^2$), and the 150's fraction gives TEER of ~200 $\Omega.cm^2$ (maximum TEER = 400 $\Omega.cm^2$) [4]. Cells cultured from the 60's fraction are uniform as they are derived from smaller vessels and are ideally suited for permeability screening experiments, while cells derived from the 150's are derived from larger vessels and are suitable for immunocytochemistry studies and extraction of RNA and proteins of interest.

Acknowledgements

This work was funded by the UK Department for Trade and Industry. Support and useful discussions with Joan Abbott and Rob Skinner during the initial stages of the development of the model is acknowledged. AP also acknowledges support by the NSW Ministry of Health, Australia, under the NSW Health Early-Mid Career Fellowships Scheme and Edge Hill University, UK. All content is solely the responsibility of the author and do not reflect the views of the NSW Health Entity. Some of this work is derived from Patabendige et al. [5] and is used under a Creative Commons Attribution License (https://creativecommons.org/licenses/by/3.0/). The original version can be found here (https://doi.org/10.1016/j.brainres.2013.04.006).

References

1. Galla H-J (2018) Monocultures of primary porcine brain capillary endothelial cells: still a functional in vitro model for the blood-brain-barrier. J Control Release 285:172–177
2. Patabendige A (2012) The value of in vitro models of the blood-brain barrier and their uses. Altern Lab Anim 40(6):335–338
3. Abbott N, Dolman D, Patabendige A (2008) Assays to predict drug permeation across the blood-brain barrier, and distribution to brain. Curr Drug Metab 9(9):901–910
4. Patabendige A, Skinner RA, Abbott NJ (2013) Establishment of a simplified in vitro porcine blood-brain barrier model with high transendothelial electrical resistance. Brain Res 1521: 1–15
5. Patabendige A, Skinner RA, Morgan L, Abbott NJ (2013) A detailed method for preparation of a functional and flexible blood-brain barrier model using porcine brain endothelial cells. Brain Res 1521:16–30
6. Patabendige A, Abbott NJ (2014) Primary porcine brain microvessel endothelial cell isolation and culture. Curr Protoc Neurosci 69:3.27.1–17
7. Perrière N, Yousif S, Cazaubon S, Chaverot N, Bourasset F, Cisternino S et al (2007) A functional in vitro model of rat blood-brain barrier for molecular analysis of efflux transporters. Brain Res 1150:1–13
8. Rubin LL, Hall DE, Porter S, Barbu K, Cannon C, Horner HC et al (1991) A cell culture model of the blood-brain barrier. J Cell Biol 115(6):1725–1735
9. Hoheisel D, Nitz T, Franke H, Wegener J, Hakvoort A, Tilling T et al (1998) Hydrocortisone reinforces the blood-brain properties in a serum free cell culture system. Biochem Biophys Res Commun 247(2):312–315
10. Förster C, Silwedel C, Golenhofen N, Burek M, Kietz S, Mankertz J et al (2005) Occludin as direct target for glucocorticoid-induced improvement of blood–brain barrier properties in a murine in vitro system. J Physiol 565(Pt 2):475–486

Chapter 8

Establishment of an In Vitro Model of Human Blood–Brain Barrier to Study the Impact of Ischemic Injury

Rais Reskiawan A. Kadir, Mansour Alwjwaj, and Ulvi Bayraktutan

Abstract

The blood–brain barrier (BBB), mainly composed of brain microvascular endothelial cells, astrocyte end-feet, and pericytes, serves as a physical and biochemical barrier that selectively limits the passage of circulating molecules into the brain parenchyma. The disruption of its integrity and function is a major cause of increased mortality and disability among ischemic stroke patients. Hence, scrutiny of the cellular and molecular mechanisms that alter BBB permeability following an ischemic injury remains of paramount importance. In this context, establishment of an in vitro model of BBB that closely simulates human cerebral barrier may offer an easy, inexpensive, and straightforward approach to identify signaling pathways involved in BBB breakdown and may help to discover new therapeutic targets to restore its damage. This chapter describes a sequential method pertaining to establishment of a triple culture model of human BBB consisting of the three main cellular components of the cerebral barrier. It also documents how the integrity and function of this barrier are evaluated through measurements of transendothelial electrical resistance (TEER) and paracellular flux of permeability marker and sodium fluorescein (NaF, 376 Da), respectively, both in normal and experimental conditions mimicking ischemic injury.

Key words Blood–brain barrier, Stroke, In vitro, Endothelial cells, Astrocytes, Pericytes, Experimental model, Cerebral barrier

1 Introductions

Stroke continues to be one of the leading causes of death and disability worldwide. In general, it is classified into two main categories, ischemic stroke and hemorrhagic stroke. Ischemic strokes occur due to an abrupt interruption of blood flow leading to or within the brain and constitute about 85% of all stroke events in Western countries. Hemorrhagic strokes, on the other hand, stem from the rupture of an artery within or on the surface of the brain and account for about 15% of all strokes in the Western world [1, 2]. Despite being the main cause of human cerebrovascular injury, thrombolysis with recombinant tissue plasminogen activator (rtPA) remains the only approved pharmacotherapy for ischemic

Nicole Stone (ed.), *The Blood-Brain Barrier: Methods and Protocols*,
Methods in Molecular Biology, vol. 2492, https://doi.org/10.1007/978-1-0716-2289-6_8,

stroke. However, due to a narrow therapeutic time window (within 4.5 h of stroke onset) and elevated risk of hemorrhage beyond this time point, globally <1% of patients at present receive this treatment [3, 4].

The blood–brain barrier (BBB) mainly consists of brain microvascular endothelial cells (BMECs), astrocyte end-feet, and pericytes. It is a physical and biochemical barrier that regulates the exchange of molecules between systemic circulation and the brain. Given that disruption of the BBB is associated with vasogenic edema, hemorrhagic transformation, and increased rates of mortality and disability in patients with ischemic stroke, a great deal of research interest has been channeled into the investigation of cellular and molecular mechanisms involved in BBB breakdown during and/or following an ischemic stroke-mediated injury. It is anticipated that better understanding of these mechanisms may open up new avenues for future therapeutic strategies for ischemic stroke [5, 6].

To this end, attempts have been made over the years to develop various in vitro and in vivo models of human cerebral ischemic injury. Although in vivo models, notably in larger species, more closely resemble to human physiology, in vitro models offer several unique advantages [7, 8]. Firstly, these models are easier, more straightforward, and less expensive to study the effect of ischemic injury. Secondly, via in vitro models, the contribution of each cell and the complex molecular pathways underlying ischemic stroke injury can be easily and extensively investigated. Thirdly, these models allow large-scale screening of compounds to study their cellular and molecular impacts in ischemic state [9]. In addition, considering the differences in immune responses across species, the use of human cells in in vitro models may somewhat replicate the inflammatory response in humans [10].

The most widely used in vitro BBB models are based on simultaneous culture of different cerebrovascular cells, namely, BMECs, astrocytes, and pericytes, on Transwell inserts. Since astrocytes cover 99% of brain capillaries, and there is an abundance of pericytes in the central nervous system (CNS), triple culture models consisting of human BMECs (HBMECs), human astrocytes (HAs), and human pericytes (HPs), i.e., EAP or EPA models, may anatomically and physiologically better represent human BBB than the co-culture models comprising HBMECs and HAs or HPs, i.e., EA or EP models, respectively (Fig. 1). Indeed, discovery of a crosstalk between cerebral endothelial cells and astrocytes in EAP model has proven this particular model as the best in vitro model with significantly higher barrier integrity and function compared to BMEC monoculture, EA, EP, and EPA models [11]. Although none of these models accommodate mechanically active factors, such as shear stress, as adopted in dynamic BBB models, the exposure of EAP model to 4 h of oxygen-glucose deprivation (OGD)

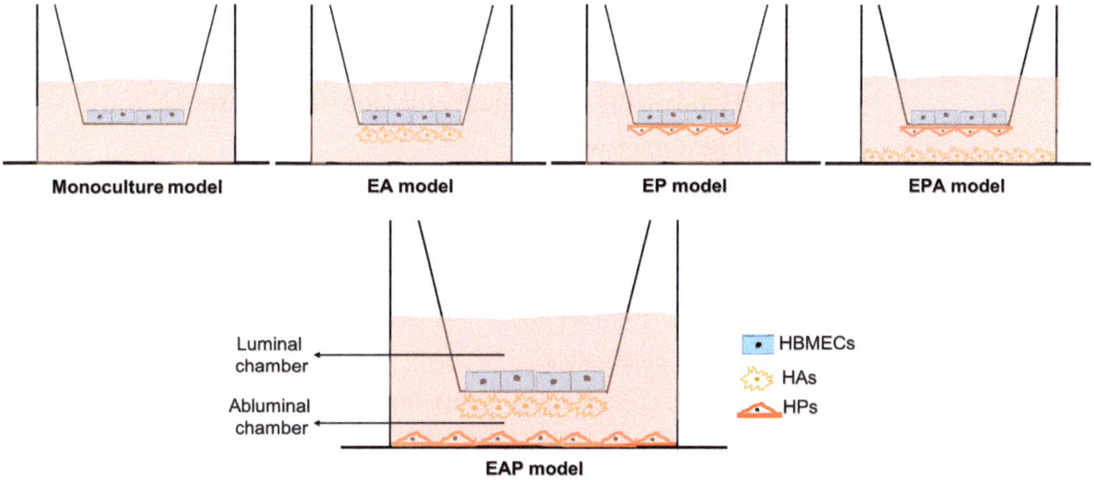

Fig. 1 Schematic representation of various in vitro models of human blood–brain barrier. With astrocyte end-feet covering all endothelial cells and the presence of pericyte at the bottom of the 12-well plate, the EAP model is anatomically and physiologically more representative of the human BBB and produces more electrical resistance than other models

alone or followed by 20 h of reperfusion (OGD±R) has been shown to closely mimic the cerebral ischemic injury observed in clinical settings [12, 13]. Hence, this chapter describes the experimental process leading to generation of the EAP model and reveals how OGD±R injury affects its integrity and function. Accumulating evidence demonstrated that this particular model has been very beneficial in evaluating the precise contribution of key cellular and molecular mechanisms to ischemic stroke-mediated cerebral barrier injury and in assessing the barrier-reparative or barrier-protective effects of a series of the so-called therapeutic agents [11, 14].

2 Materials

1. Astrocyte medium: 5% FBS, 100-units/mL penicillin, 100 μg/mL streptomycin, and growth factor.
2. Endothelial cell medium: 10% FBS, 100-units/mL penicillin, 100 μg/mL streptomycin, and growth factor.
3. Pericyte medium: 5% FBS, 100-units/mL penicillin, 100 μg/mL streptomycin, and growth factor.
4. EVOMX meter.
5. Fluorometer.
6. Hemocytometer.
7. Hank's balanced salt solution (HBSS).
8. Human astrocytes (HAs).

9. Human brain microvascular endothelial cells (HBMECs).
10. Human pericytes (HPs).
11. Hypoxic O_2/CO_2 incubator (for ischemic simulation).
12. Light microscope.
13. Phosphate-buffered saline (PBS).
14. RPMI 1640 glucose-free media.
15. Sodium fluorescein (NaF).
16. Sterile tweezer.
17. STX2 electrodes.
18. Tissue culture CO_2 incubator.
19. Tissue culture flasks—T25 or T75.
20. Tissue culture plates—12 and 96 wells.
21. Transwell inserts—12-well 0.4 μm pore polyester inserts.
22. Trypan blue.
23. Trypsin.
24. Water bath.

3 Methods

3.1 Cell Culture

3.1.1 Astrocyte Seeding

1. Thaw the frozen vial containing astrocytes and seed the cells in T25 or T75 flask in a humidified atmosphere (95% relative humidity) under normal conditions (70% N_2, 25% O_2, and 5% CO_2) at 37 °C.
2. Change the media the day after and subsequently every other day with astrocyte growth media.
3. Once HAs reach about 90% confluence, wash the cells twice with PBS and detach the cells by incubating with trypsin for 3–4 min at 37 °C. Thereafter, tap the flask gently until the cells lift off.
4. Resuspend the cells with astrocyte growth medium (*see* **Note 1**) and determine cell numbers by hemocytometer.
5. Open the 12-well plate and place them upside down (Fig. 2). The lid of 12-well plate acts as the "container," while the bottom of the cell culture plate acts as the "lid."
6. Open the Transwell inserts and carefully remove the insert using sterilized tweezer from its sterile packaging. Place the inserts upside down on the lid of the 12-well plate prepared in **step 5**.
7. Seed 1.5×10^5 HAs on the basolateral side of the inserts, now facing upward, and then add a total of 250 μL astrocyte growth media.

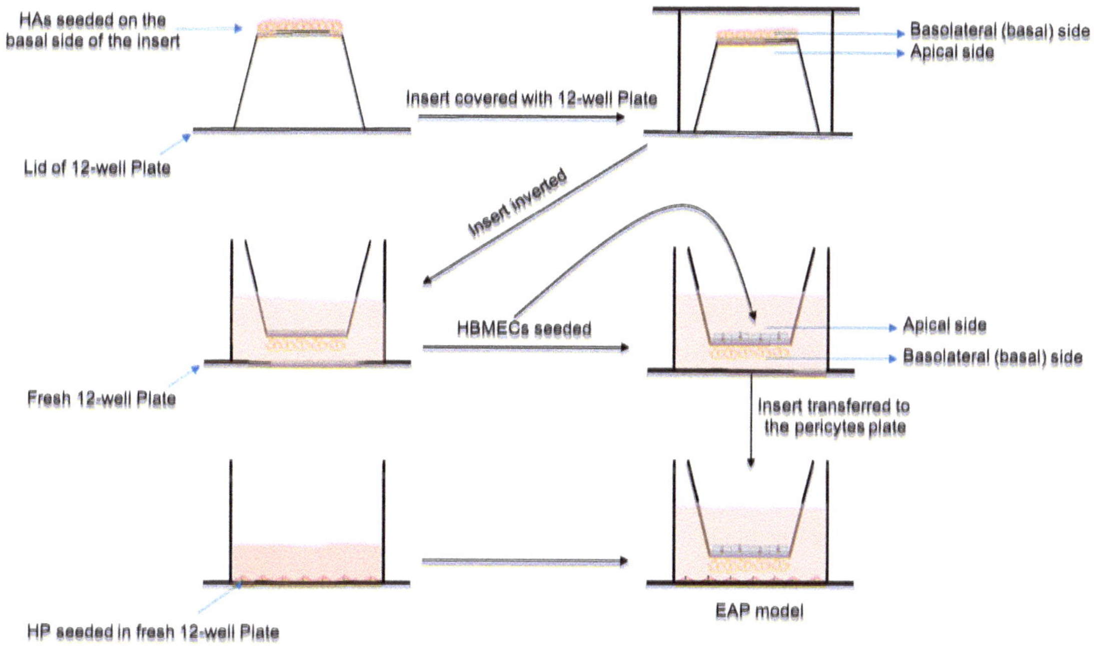

Fig. 2 Schematic of the establishment of an in vitro model of human BBB. The insert is placed upside down on the lid of a 12-well plate. The human astrocytes (HAs) are subsequently seeded onto the basolateral side of the insert, for which the 12-well plate acts as the "lid." After incubating for 3–4 h, the insert is transferred to a fresh 12-well plate. Once the HAs reach 80% confluence, human brain microvascular endothelial cells (HBMECs) are seeded onto the apical side of the insert, and human pericytes cells (HPs) are seeded to another 12-well plate. The next day, the insert is transferred to the 12-well plate containing HPs

8. Classify the inserts according to the group. To simulate ischemic stroke injury, at least three groups are needed: control, OGD, and OGD+R (*see* **Note 2**). Make each group in triplicate, to ensure reproducibility.
9. Incubate the cells for 3–4 h at 37 °C. Once the cells attach onto the surface of Transwell membrane, aspirate carefully the residual media and flip back the inserts to return to their original orientation, and then place them into a fresh 12-well plate.
10. Add 1 mL and 750 μL of HA pre-warmed astrocyte growth media in the abluminal and luminal chambers, respectively.
11. Check the cell attachment and uniform seeding through light microscope. At this point, the morphology of astrocyte is still a round shape, which will stretch and flatten later on (Fig. 3).
12. Incubate the inserts in similar conditions to **step 1** and change the media every day.
13. Astrocytes usually reach 80% confluence within 4 days (Fig. 4).

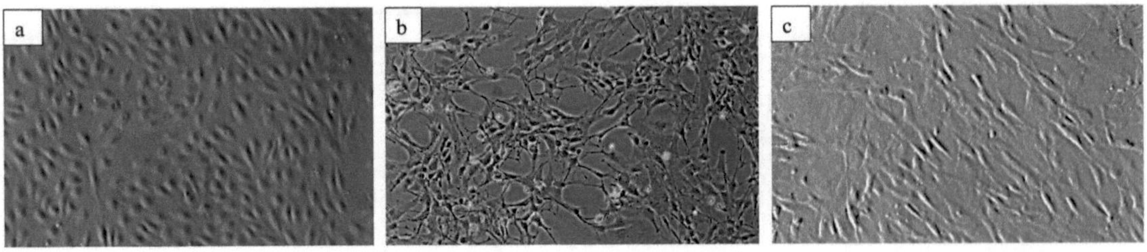

Fig. 3 The morphology of human brain microvascular endothelial cells (**a**), astrocytes (**b**), and pericytes (**c**) under normal conditions

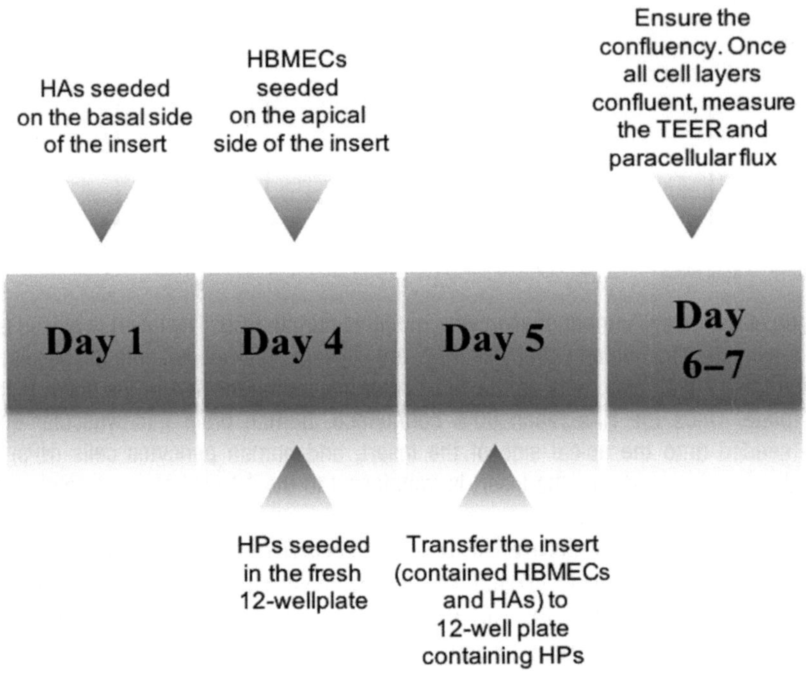

Fig. 4 Timeline for setting up an in vitro model of human BBB. On day 1, human astrocytes (HAs) are seeded on the basal side of the insert. Once the HAs reach about 80% confluence (usually within 4 days), seed the human brain microvascular endothelial cells (HBMECs) on the apical side of the insert. Human pericytes (HPs) are seeded in the fresh 12-well plate on the same day. On day 5, lift out the insert and transfer to the 12-well plate containing pericytes. Measure the electrical resistance and paracellular flux of sodium fluorescein once all cell layers attain full confluence (usually at days 6–7)

3.1.2 Seeding of Human Brain Microvascular Endothelial Cells and Pericytes

1. Culture HBMECs and HPs in two different flasks with similar conditions to **step 1** (*see* Subheading 3.1.1).
2. Change the media of HBMECs and HPs with their respective media on the following day and subsequently every other day.
3. Once the HBMECs and HPs reach 90% confluence, wash the cells with PBS twice, and detach the cells through incubating with trypsin for 3–4 min at 37 °C.

4. Resuspend the HBMECs and HPs in their specialized media and determine cell numbers (*see* **Note 3**).
5. Take out the 12-well plate that contain the inserts with astrocyte from the incubator and place in the hood.
6. Remove astrocyte growth medium from both luminal and abluminal chambers.
7. Seed 4 × 10^4 HBMECs onto the apical side of the insert and add a total of 750 μL of pre-warmed ECM growth media in to the luminal chamber (*see* **Note 4**).
8. Add 1 mL of astrocyte growth media in the abluminal chamber and incubate the plate in conditions similar to **step 1** in Subheading 3.1.1.
9. For HPs, seed 1.5 × 10^5 HPs into fresh 12-well plate and add the pre-warmed pericyte growth medium. Incubate the 12-well plate in those conditions similar to **step 1** in Subheading 3.1.1 (*see* **Note 5**).
10. Since co-culture of HBMECs and HAs in the luminal and abluminal chamber of Transwell insert leads to the difficulties to differentiate those cells through light microscope, it is a good idea to seed a similar number of HBMCEs (5 × 10^4 cells) in another insert to simultaneously monitor the cells' proliferation and confluency under light microscope.
11. The HBMECs and HPs usually reach full confluence within 2–3 days.

3.1.3 In Vitro Model of Human BBB

1. The next day, take out the 12-well plate that contain the insert (with HBMECs and HAs) and the other 12-well plate (containing HPs) from the incubator, and place them inside the hood.
2. Remove the endothelial, astrocyte, and pericyte growth media from both luminal and abluminal chambers.
3. Wash all the cell layers once with PBS.
4. Carefully transfer the inserts that contain HBMECs and HAs to 12-well plates containing HPs using a sterile tweezer.
5. Add 750 μL of pericyte growth media and another 750 μL of astrocyte growth media in the abluminal chamber, and 750 μL ECM growth media in the luminal chamber.
6. Change the media every day.

3.2 Exposure to OGD +R

The main principle of the pathophysiology of ischemic stroke is the cessation of blood flow within or to the brain, which leads to nutrient and oxygen deprivation. To mimic this in cell culture system, the in vitro model of human BBB is exposed to 4 h of oxygen-glucose deprivation (OGD); 4 h is selected considering the therapeutic window of approved rtPA pharmacological drugs for ischemic stroke. In addition, if the clot successfully dissolves and blood supply is restored after the administration of thrombolytic drug or the application of endovascular treatment, brain tissue may experience reperfusion injury. To simulate this, the inserts that have been exposed to OGD are returned to normal condition (similar condition to **step 1** in Subheading 3.1.1) and also to their specialized media for 20 h (OGD+R group).

1. The exposure of an in vitro model of human BBB to ischemia/reperfusion injury can only be performed if all layers attain full confluence (*see* **Note 6**).
2. To simulate the ischemic state, set the hypoxic O_2/CO_2 incubator into 95% N_2, 0.05% O_2, and 5% CO_2 at 37 °C. It usually takes about 1 h to achieve these parameters.
3. At about 15 min before reaching 1 h, take out the inserts for OGD and OGD+R group from normal incubator and place them inside the hood.
4. Remove the media from both the luminal and abluminal chambers, and wash both chambers once with PBS.
5. Add 1 mL and 750 μL of RPMI 1640 glucose-free media to the abluminal and luminal chambers, respectively.
6. Place the inserts for OGD and OGD+R group into the hypoxic O_2/CO_2 incubator once all parameters have been reached, as described earlier (Subheading 3.2, **step 2**).
7. Set the timer for 4 h.
8. 15 min before reaching 4 h, pre-warm the endothelial, astrocyte, and pericyte growth media in water bath, which will be used for the reperfusion group.
9. After exactly 4 h, take out all the inserts from the hypoxic O_2/CO_2 incubator and place them inside the hood.
10. For OGD+R group, replace the RPMI 1640 glucose-free media with fresh cell media and return the inserts into normoxic condition (Subheading 3.1.1, **step 1**).
11. Take out the control group from normal incubator and place it inside the hood, together with the OGD group.
12. The measurement of TEER and NaF paracellular flux for control and OGD group are now ready to be performed, while the measurement for OGD+R group will be performed the following day (after 20 h).

3.3 Integrity Measurement

Measurement of transendothelial electrical resistance (TEER), based on electrical resistance across a cellular barrier, has been extensively used to assess the integrity of BBB in live cells during various stages of their growth and development. This method offers advantages over functional assay (*see* Subheading 3.4), including being more rapid and noninvasive. In addition, it is also an excellent strategy to assess barrier integrity, as the electrical impedance across an endothelium barrier depends on the formation of robust tight junctions between neighboring cells. Hence, higher value of TEER readings indicate better tight junctional formation between endothelial cells [15].

1. Connect the STX2 electrode cable to the electrode port on the EVOMX resistance meter.
2. Wipe the electrode cable and EVOMX volt-ohm meter with a tissue dipped into 70% ethanol.
3. Sterilize the electrode by submerging in 70% ethanol for 15 min and allow it to dry for 15 s.
4. Place the probe vertically in a well containing the inserts.
5. There are two electrodes with uneven length, designed to easily access both compartments. The longer part of the electrode gently touches the bottom of the well, while the shorter one is placed slightly above the membrane of the insert, not directly making contact with the HBMEC layer (Fig. 5) (*see* **Note** 7).
6. Repeat the reading at least twice to ensure reproducibility.
7. Calculate the TEER value using the following formula:

$$\text{TEER value } \left(\Omega\ \text{cm}^2\right) = (A - B) \times C$$

A = TEER reading from each insert
B = TEER reading from insert just with cell culture medium
C = Area of the insert (1.12 cm^2).

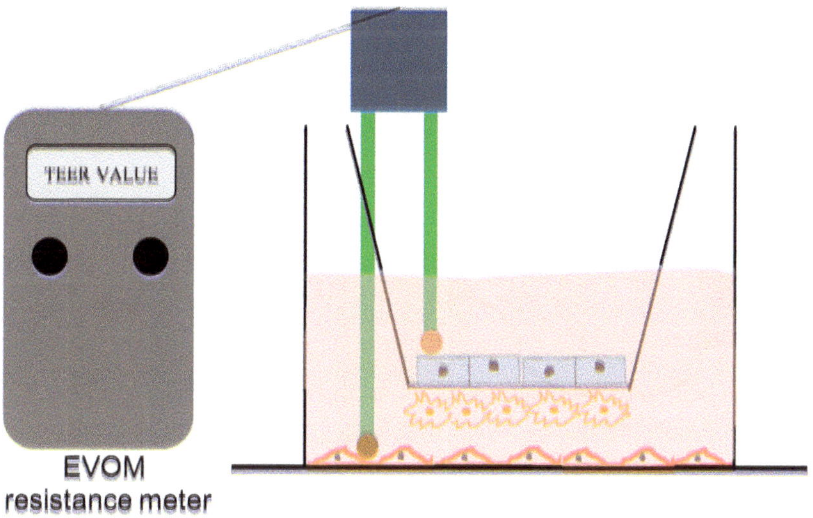

Fig. 5 Measurement of transendothelial electrical resistance (TEER) value in an in vitro model of BBB. The electrode probe should be maintained in upright position at all times to ensure consistent readings

3.4 Functional Assay

BBB integrity is based on impedance measurements reflecting the ionic conductance of the monolayer, which cannot be extrapolated to reflect the permeability of barrier. Validation assessment through a functional assay is thus needed in addition to electronic measurement of the ion flux. This assay basically measures the amount of solutes that can cross the cellular barrier (from luminal to abluminal compartment), thereby reflecting the functionality of the barrier to limit paracellular flux. Hence, both TEER measurement and paracellular flux assessment are equally needed for optimal characterization of an in vitro model of BBB [16].

1. After measuring TEER value, remove the media both from luminal and abluminal compartments.
2. Wash the cells once with HBSS.
3. Transfer the inserts into fresh 12-well plates containing 2 mL HBSS.
4. Carefully add NaF (10 μg/mL) solution in the luminal chamber (*see* **Note 8**).
5. Incubate the plate well containing the inserts into the conditions similar to **step 1** in Subheading 3.1.1 for 1 h.
6. After 1 h, take out the plate well from incubator and carefully transfer 400 μL of NaF solution and HBSS from the luminal and abluminal chambers, respectively, into two different Eppendorf tubes (*see* **Note 9**).
7. From those 400 μL solutions, transfer 100 μL into a 96-well plate in triplicate (Fig. 6).
8. Measure the amount of NaF paracellular flux by fluorometry (excitation/emission: 440/525 nm).
9. Calculate the paracellular flux of NaF using the following formula:

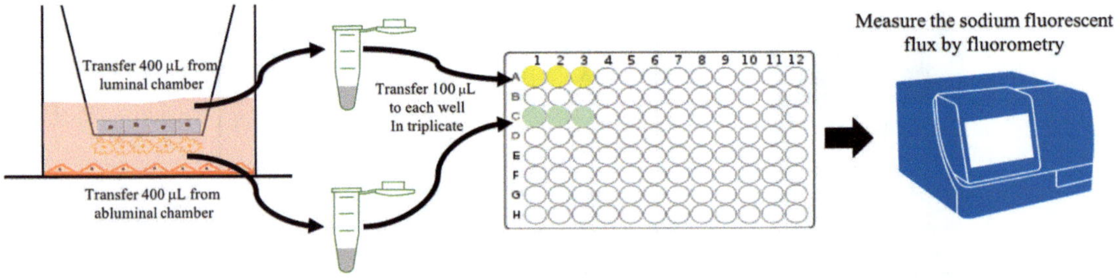

Fig. 6 Measurement of sodium fluorescein (NaF) flux across the in vitro model of human BBB. After incubation for 1 h at 37 °C, 400 μL from luminal and 400 μL from abluminal compartment are transferred into two different Eppendorf tubes. 100 μL from each compartment is subsequently transferred to a 96-well plate in triplicate. Finally, the 96-well plate is transferred to a fluorometer to measure the paracellular flux of NaF

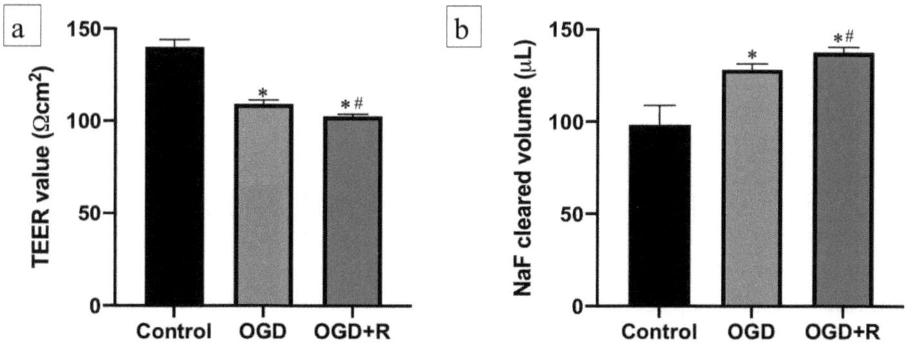

Fig. 7 The effect of oxygen-glucose deprivation (OGD) and OGD followed by reperfusion (OGD+R) mimicking ischemia and ischemia/reperfusion injury, respectively, on the integrity and function of the blood–brain barrier as assessed by measurements of (**a**) transendothelial electrical resistance (TEER) and (**b**) paracellular flux of sodium fluorescein (NaF). Data are expressed as mean ± SEM from three independent experiments. $^{*}p < 0.05$ vs controls, $^{\#}p < 0.05$ vs OGD

$$\text{NaF paracelullar flux} = \frac{\text{Reading from abluminal chamber}}{\text{Reading from luminal chamber}} \times 500$$

10. The exposure of the BBB to the OGD±R will impair its integrity and function as observed by the decreases of TEER value and concomitantly increases of NaF flux (Fig. 7).

4 Notes

1. As the basolateral side of the insert can only accommodate a maximum of 250 μL media in total, resuspend the HAs in a smaller volume (less than 100 μL), then carefully top up with 150 μL astrocyte growth media until reaching 250 μL in total.
2. If intending to investigate the effect of some particular compounds on the integrity and function of in vitro BBB model during OGD±R injury (for instance, the effect of protein kinase-C-β inhibitor to such condition, see Fig. 8), add such compounds in the glucose-free media (for OGD+drug group) and in the ECM growth media (for OGD+R+drug group).
3. The maximum capacity of luminal chamber is 750 μL, so make sure the volume of HBMECs suspension is less than 750 μL.
4. The seeding of HBMECs on the insert is performed when the astrocytes reach 80% confluence, which usually occurs within 4 days. The seeding of pericytes to the fresh 12-well plate is performed on the same day of the seeding of HBMECs on the insert. As the timing for setting up the model is crucial, the timeline for the establishment of the model is outlined in Fig. 4.

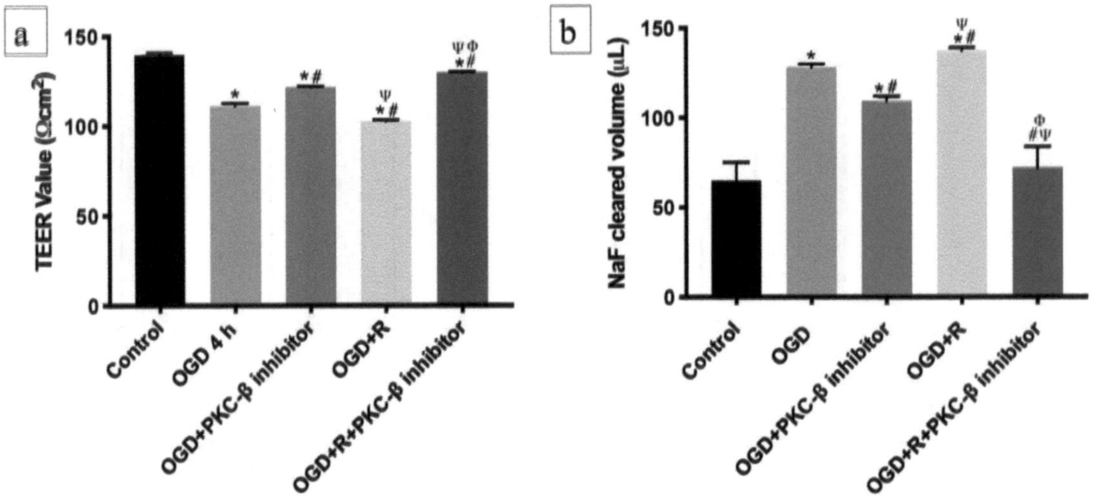

Fig. 8 The effect of an inhibitor for protein kinase C-β (PKC-β) on the (**a**) integrity and (**b**) function of an in vitro model of human blood–brain barrier during 4 h oxygen glucose deprivation alone or followed by 20 h reperfusion (OGD±R) injury. Data are expressed as mean ± SEM. $^{*}p < 0.05$ vs control, $^{\#}p < 0.05$ vs OGD, $^{\Psi}p < 0.05$ vs OGD+PKC-β inhibitor and $^{\Phi}p < 0.05$ vs OGD+R

5. As each group (control, OGD, OGD+R) goes through a different process, the plating of HPs should be into three different 12-well plates. If performing OGD+drug, HPs can be seeded for both OGD and OGD+drug group in the same plate (similar to seeding for the OGD+R and OGD+R+drug group).
6. The critical step for setting up the triple culture model of BBB is the confluency of each cell line. Initiating the treatment with non-confluent cells results in lower TEER values and higher NaF paracellular flux than anticipated. Hence, the seeding of HBMECs into another insert (Subheading 3.1.2, **step 10**) is really helpful to monitor cell confluency.
7. Since TEER values are very susceptible to change, it is really crucial to maintain the electrode in a 90° perpendicular position, avoiding tilting or touching the wall of the plate well or Transwell insert, which could lead to fluctuating TEER values.
8. To make 10 μg/mL of NaF solution, first make NaF solution at a concentration of 1 mg/mL by dissolving 1 mg of NaF into 1 mL HBSS (stock solution). Thereafter, dilute 100 μL of the stock solution into 9.900 μL of HBSS. Wrap the bottle with foil to avoid light and keep in the fridge (4 °C) for a maximum of 6 months.
9. Great care is needed as the Transwell membrane is vulnerable to breakage. Carelessness in sucking up the NaF solution can also prompt the NaF to cross the membrane, which can markedly increase the amount of paracellular flux above anticipated levels.

References

1. Sacco Ralph L, Kasner Scott E, Broderick Joseph P, Caplan Louis R, Connors JJ, Culebras A, Elkind Mitchell SV, George Mary G, Hamdan Allen D, Higashida Randall T, Hoh Brian L, Janis LS, Kase Carlos S, Kleindorfer Dawn O, Lee J-M, Moseley Michael E, Peterson Eric D, Turan Tanya N, Valderrama Amy L, Vinters Harry V (2013) An updated definition of stroke for the 21st century. Stroke 44(7):2064–2089. https://doi.org/10.1161/STR.0b013e318296aeca
2. Hisham NF, Bayraktutan U (2013) Epidemiology, pathophysiology, and treatment of hypertension in ischaemic stroke patients. J Stroke Cerebrovasc Dis 22(7):e4–e14. https://doi.org/10.1016/j.jstrokecerebrovasdis.2012.05.001
3. Kadir RRA, Bayraktutan U (2020) Urokinase plasminogen activator: a potential thrombolytic agent for ischaemic stroke. Cell Mol Neurobiol 40(3):347–355. https://doi.org/10.1007/s10571-019-00737-w
4. Powers William J, Rabinstein Alejandro A, Ackerson T, Adeoye Opeolu M, Bambakidis Nicholas C, Becker K, Biller J, Brown M, Demaerschalk Bart M, Hoh B, Jauch Edward C, Kidwell Chelsea S, Leslie-Mazwi Thabele M, Ovbiagele B, Scott Phillip A, Sheth Kevin N, Southerland Andrew M, Summers Deborah V, Tirschwell David L (2018) 2018 Guidelines for the early management of patients with acute ischemic stroke: a guideline for healthcare professionals from the American Heart Association/American Stroke Association. Stroke 49(3):e46–e99. https://doi.org/10.1161/STR.0000000000000158
5. Jiang X, Andjelkovic AV, Zhu L, Yang T, Bennett MVL, Chen J, Keep RF, Shi Y (2018) Blood-brain barrier dysfunction and recovery after ischemic stroke. Prog Neurobiol 163–164:144–171. https://doi.org/10.1016/j.pneurobio.2017.10.001
6. Li Y, Zhong W, Jiang Z, Tang X (2019) New progress in the approaches for blood–brain barrier protection in acute ischemic stroke. Brain Res Bull 144:46–57. https://doi.org/10.1016/j.brainresbull.2018.11.006
7. Canazza A, Minati L, Boffano C, Parati E, Binks S (2014) Experimental models of brain ischemia: a review of techniques, magnetic resonance imaging, and investigational cell-based therapies. Front Neurol 5:19–19. https://doi.org/10.3389/fneur.2014.00019
8. Stone NL, England TJ, O'Sullivan SE (2019) A novel Transwell blood brain barrier model using primary human cells. Front Cell Neurosci 13:230. https://doi.org/10.3389/fncel.2019.00230
9. Sivandzade F, Cucullo L (2018) In-vitro blood–brain barrier modeling: a review of modern and fast-advancing technologies. J Cereb Blood Flow Metab 38(10):1667–1681. https://doi.org/10.1177/0271678X18788769
10. Holloway PM, Gavins FNE (2016) Modeling ischemic stroke in vitro: status quo and future perspectives. Stroke 47(2):561–569. https://doi.org/10.1161/STROKEAHA.115.011932
11. Shao B, Bayraktutan U (2013) Hyperglycaemia promotes cerebral barrier dysfunction through activation of protein kinase C-β. Diabetes Obes Metab 15(11):993–999. https://doi.org/10.1111/dom.12120
12. Allen CL, Bayraktutan U (2009) Antioxidants attenuate hyperglycaemia-mediated brain endothelial cell dysfunction and blood–brain barrier hyperpermeability. Diabetes Obes Metab 11(5):480–490. https://doi.org/10.1111/j.1463-1326.2008.00987.x
13. Abdulkadir R, Alwjwaj M, Othman O, Rakkar K, Bayraktutan U (2020) Outgrowth endothelial cells form a functional cerebral barrier and restore its integrity after damage. Neural Regen Res 15(6):1071–1078. https://doi.org/10.4103/1673-5374.269029
14. Mathur M, Bayraktutan U (2017) Therapeutic hypothermia protects in vitro brain barrier from ischaemic damage through attenuation of inflammatory cytokine release and apoptosis. Stroke Res Ther 2(1:4):1–10
15. Srinivasan B, Kolli AR (2019) Transepithelial/transendothelial electrical resistance (TEER) to measure the integrity of blood-brain barrier. In: Barichello T (ed) Blood-brain barrier. Neuromethods, vol 142. Humana Press, New York. https://doi.org/10.1007/978-1-4939-8946-1_6
16. Helms HC, Abbott NJ, Burek M, Cecchelli R, Couraud P-O, Deli MA, Förster C, Galla HJ, Romero IA, Shusta EV, Stebbins MJ, Vandenhaute E, Weksler B, Brodin B (2016) In vitro models of the blood-brain barrier: an overview of commonly used brain endothelial cell culture models and guidelines for their use. J Cereb Blood Flow Metab 36(5):862–890. https://doi.org/10.1177/0271678X16630991

Chapter 9

A Novel Dynamic Human In Vitro Model for Studying the Blood–Brain Barrier

Patricia Miranda-Azpiazu and Sikha Saha

Abstract

Constructing a reliable in vitro blood–brain barrier (BBB) model using human primary cells has been considered a major challenge during the past decades. These systems could provide valuable information regarding the effect of therapeutic compounds on different BBB cell types (endothelial cells, astrocytes, pericytes) and their ability to cross the barrier in order to reach the brain. Several attempts have been made to develop in vitro BBB models, but these studies mainly used rat, bovine, and porcine cells rather than human primary cells. Genetically modified cell lines have also been used, but they do not appear to maintain physiological properties of the BBB. Here, we describe a detailed protocol for co-culturing and maintaining human brain primary endothelial cells, pericytes, and astrocytes under flow to create an in vitro human BBB model, which can be used for toxicity testing and for studying cross-interaction among different cell types involved in the BBB formation.

Key words Blood–brain barrier (BBB), Human primary cell, Shear stress, Quasi-Vivo system, Cell culture, Conditioned media, Endothelial cells, Astrocytes, Pericytes, Multicellular organ flow system

1 Introduction

The blood–brain barrier (BBB) is a brain-specific capillary barrier critical for preventing entry of toxic substances into the central nervous system [1]. The BBB has a dual function as it (a) maintains the homeostasis of the brain microenvironment and (b) regulates the entry of nutrients and neurotransmitters into the brain [2]. It is well established that brain endothelial cells conform complex intercellular tight junctions that restrict the paracellular permeation of molecules through the junctional cleft [3].

The BBB has always been a constant boulder against efficient drug delivery [4]. Due to the difficulties associated with studying the complex BBB in animals, in vitro BBB models play an important role in understanding the structure and function of the BBB both in health and disease [4–6]. Indeed, several cell culture models

Nicole Stone (ed.), *The Blood-Brain Barrier: Methods and Protocols*,
Methods in Molecular Biology, vol. 2492, https://doi.org/10.1007/978-1-0716-2289-6_9,

of the BBB are available, but there is currently no "gold standard" model [7]. Several in vitro BBB models were designed as endothelial monocultures [8–11], brain endothelial cells co-cultured with astrocytes [2, 12–21], and/or pericytes [3, 22–26] or with astrocytes's conditioned media (ACM) to help maintain the BBB phenotype in vitro [2]. Most rat, bovine, and human in vitro BBB models require co-culture of endothelial cells, astrocytes, and pericytes together to develop functional tight junctions with high transendothelial electrical resistance (TEER) and significantly lower permeability [25–27]. There are also many attempts to establish BBB under flow [28], as endothelial cell in vivo are in contact with blood flow. Therefore, it is of high importance for future research to develop a BBB model to mimic physiological conditions using human primary cells grown under flow, as it can be found in the brain.

Herein, we describe a protocol for culturing primary human brain endothelial cells, astrocytes, and pericytes and further establishing a dynamic BBB model using Kirkstall Quasi-Vivo (QV) system (QV500). This model will allow us to perform further studies on the effects of different compounds on each cell type separately, as well as understand possible mechanisms of interactions between different BBB cell types.

2 Materials

1. Primary human astrocytes (HAs).
2. Primary human pericytes (HBVP).
3. Primary human brain microvascular endothelial cells (HMBECs).
4. Class II vertical laminar-flow biological safety cabinet.
5. Fetal bovine serum or fetal calf serum (FBS/FCS).
6. L-glutamine.
7. Penicillin–streptomycin (P/S).
8. Incubator.
9. Optical microscope.
10. Gilson pipettes (P20, P200, and P1000) and filter tips.
11. Hemocytometer.
12. Trypan blue.
13. T75 flasks.
14. 12-well plates, 24-well plates.
15. Glass coverslips (12 mm diameter).

16. Fine curved forceps.
17. Poly-D-lysine.
18. Cell culture grade water (ready-to-use).
19. Rat tail collagen type I.
20. Ethanol.
21. Trypsin-EDTA solution (ready-to-use).
22. Penicillin–streptomycin solution (ready-to-use).
23. Phosphate-buffered saline (PBS) sterile for tissue culture (ready-to-use).
24. DMEM medium with and without phenol red.
25. 15 mL and 50 mL Falcon centrifuge tubes.
26. Refrigerated centrifuge.
27. Stripettes (5, 10, and 25 mL).
28. Cryovials.
29. Astrocyte and pericyte culture medium: Dulbecco's Modified Eagle's Medium/Ham's Nutrient mixture F12, (DMEM), 1 L, 100 mL fetal bovine serum or fetal calf serum (FBS/FCS), 10 mL L-glutamine, 10 mL penicillin–streptomycin (P/S).
30. HMBEC culture medium: Endothelial Cell Growth Medium MV (ready-to-use).
31. Kirkstall QV500 system.
32. 3 QV500 silicone chambers with tubing and connectors.
33. 1 reservoir bottle.
34. 100 glass coverslips, 12 mm diameter.
35. 6 × 22 cm extension tubing with luer connectors.
36. Parker pump.
37. MTT solution for cell viability measurement: prepare a 1 mg/mL solution by adding 5 mL of DMEM without phenol red to the 5 mg of MTT.
38. 0.22 μm sterile filters.
39. Horse serum.
40. Mounting medium with DAPI.
41. Antibodies of interest (GFAP, CD31, ZO1, α actin).
42. Confocal microscope.
43. Paraformaldehyde (PFA).

3 Methods

3.1 Cell Culture

3.1.1 Culture and Passage of Primary Human Astrocytes

1. Add 2 mL poly-D-lysine solution per T75 flask (*See* **Note 1**).
2. Incubate 1 h at 37 °C.
3. Remove poly-D-lysine solution and wash thoroughly twice with 5 mL Sigma sterile water.
4. The flask is ready to be seeded with cells.
5. Warm up astrocyte culture medium to 37 °C in water bath or incubator.
6. Thaw a vial of primary human astrocytes in 37 °C water bath or incubator for about 2 min, swirling gently to speed up the process.
7. Using a 1 mL Pasteur pipette, add thawed astrocytes dropwise into 10 mL astrocyte culture medium in a 15 mL Falcon tube. Mix gently using 10 mL stripette.
8. Pipette 10 mL into a coated T75 flask and transfer the flask to incubator (37 °C).
9. Change the medium next day, then every 2 days until the cells are 90% confluent.
10. When astrocytes are >90% confluent, remove culture medium from flask and wash twice with PBS.
11. Add 2 mL trypsin-EDTA solution per T75 flask and transfer the flask to incubator for 2 min.
12. Remove the flask from incubator and observe under an inverted microscope.
13. The astrocytes should be rounded up by this stage. Gently shake or tap the flask to aid detachment of cells from the culture surface.
14. Add 8 mL astrocyte culture medium per flask and mix gently using 10 mL stripette.
15. Transfer contents into 15 mL Falcon tube and centrifuge for 4 min at 1400 × *g* at room temperature.
16. Resuspend the astrocyte pellet in 10 mL astrocyte culture medium.
17. Add 8 mL astrocyte culture medium into new poly-D-lysine coated flasks.
18. Seed astrocytes and transfer to incubator.

3.1.2 Seeding Astrocytes on Coverslips

1. Seed 10^5 cells in 12- or 24-well plates with 12 mm coverslips (previously sterilized with 70% ethanol and UV light) and coated accordingly with poly-D-lysine as described above.

2. Normally, 10^5 cells correspond to a 500 μL volume. Add DMEM up to 1.5 mL in each well and place in the incubator at least 24 h.
3. Two days after seeding, remove half of the medium from all flasks (never remove all the medium) and place into a 50 mL Falcon tube. This constitutes the astrocyte-conditioned medium (ACM).
4. Feed astrocytes adding 8 mL fresh astrocyte culture medium to the "older" medium and return flasks to incubator.
5. Filter the collected ACM using a 0.2 μm syringe filter to remove cellular debris.
6. Aliquot into 15 mL Falcon tubes and snap freeze in dry ice or liquid nitrogen.
7. Transfer tubes to −80 °C freezer for storage.
8. Repeat the process every 2 days to collect more ACM.

3.1.3 Culture of Human Brain Endothelial Cells (HBMECs)

1. Add 2 mL rat tail collagen type I solution (from a 300 μg stock, *see* **Note 2**). Incubate at 37 °C for at least 3 h.
2. Remove collagen (*See* **Note 3**).
3. Warm up HMBECs cell culture medium to 37 °C in a water bath or incubator.
4. Thaw a vial of HMBECs in a 37 °C water bath or incubator for about 1–2 min, swirling gently to speed up the process.
5. Using a 1 mL pipette, add the thawed suspension cells dropwise into 15 mL falcon tube which contains 13 mL of culture medium previously warmed up. Mix gently using the same pipette and use an extra 2 mL of warmed media to take every possible rest of cells from the vial.
6. Add the resultant 15 mL of cells plus media into a coated T75 flask and transfer the flask to the incubator for 3 days. Up to 80% confluent cultures can be expected after 4 or 5 days post seeding.
7. Ensure that the media is changed next day of thawing to remove DMSO, adding fresh HMBECs media plus astrocytes conditioned media in a proportion of 1/5 and 4/5, respectively (see next section for more details).

3.1.4 Passage and Expansion of Human Brain Endothelial Cells

1. Warm up HMBECs culture medium to 37 °C in water bath or incubator.
2. When HMBECs are >90% confluent, remove culture medium from flask and wash once with 5–6 mL of PBS.
3. Add 2 mL trypsin-EDTA solution per T75 flask and transfer flasks to incubator for 1 min (*See* **Note 4**).

4. Remove flask from incubator and observe under an inverted microscope. The HMBECs should be rounded up by this stage. Gently shake or tap the flask to aid detachment of cells from the culture surface (*See* **Note 5**).
5. Add 8 mL HMBECs culture medium per flask to stop trypsin-EDTA reaction and mix gently using 10 mL stripette.
6. Transfer contents into 15 mL Falcon tube and centrifuge for 4 min at 1400 × *g* at room temperature.
7. Resuspend the pellet in 10 mL HMBECs media or 5 mL fetal bovine serum (FBS), depending on the next step. Resuspend HMBECs in media when expanding the culture is needed and in FBS when freezing for storage in liquid nitrogen is wanted.
8. Add 8 mL HMBECs medium into new collagen coated flasks.
9. Seed 2 mL of HMBECs on each flask and transfer to incubator.
10. Incubate cells to let them attach to the collagen-coated T75 flask at least overnight.

3.1.5 Seeding Human Brain Endothelial Cells in Coverslips

1. Following the same steps as for passage and expansion of last section, seed 10^5 cells in 12-well plates with 12 mm coverslips (previously sterilized with 70% ethanol and UV light, if possible) coated with collagen as described before.
2. Normally, 10^5 cells correspond to a 500 μL volume. Add endothelial cells medium up to 1.5 mL in each well and place in the incubator at least 24 h.
3. Two days after seeding, remove all the medium from all flasks and place into a 50 mL Falcon tube. This constitutes the endothelial cells conditioned medium.
4. Feed endothelial cells adding 10 mL of fresh HMBECs culture medium plus 5 mL of astrocytes conditioned media (*see* Sect. 3.2 for details) and return flasks to incubator.
5. Filter the collected medium using a 0.2 μm syringe filter to remove cellular debris.
6. Aliquot into 15 mL Falcon tubes and snap freeze in dry ice or liquid nitrogen.
7. Transfer tubes to −80 °C freezer for storage.
8. Repeat the process every 2 days to collect more conditioned medium.

3.1.6 Culture and Passage of Primary Human Pericytes (HBVPs)

1. Warm up pericytes culture medium to 37 °C in water bath or incubator.
2. Thaw a vial of primary human a pericytes in 37 °C water bath or incubator for about 2 min, swirling gently to speed up the process.

3. Using 1 mL Pasteur pipette, add thawed astrocytes dropwise into 10 mL pericytes culture medium in a 15 mL Falcon tube. Mix gently using 10 mL stripette.
4. Pipette 10 mL into a coated T75 flask and transfer to incubator.
5. Change medium the next day, then every 2 days until 90% confluent.
6. When pericytes are >90% confluent, remove culture medium from flask and wash twice with PBS. Add 2 mL trypsin-EDTA solution per T75 flask and transfer flask to incubator for 2 min.
7. Remove flask from incubator and observe under an inverted microscope. The astrocytes should be rounded up by this stage. Gently shake or tap the flask to aid detachment of cells from the culture surface.
8. Add 8 mL pericytes culture medium per flask and mix gently using 10 mL stripette.
9. Transfer contents into 15 mL Falcon tube and centrifuge for 4 min at 1400 × g at room temperature.
10. Resuspend the astrocyte pellet in 10 mL astrocyte culture medium.
11. Add 8 mL pericytes culture medium into new poly-D-lysine coated flasks.
12. Seed pericytes at 1:4 and transfer to incubator.

3.1.7 Seeding Pericytes in Coverslips

1. Seed 10^5 cells in 12- or 24-well plates with 12 mm coverslips (previously sterilized with 70% ethanol and UV light) and coated accordingly with poly-D-lysine.
2. Normally, 10^5 cells correspond to a 500 μL volume. Add medium up to 1.5 mL in each well and place in the incubator at least 24 h.
3. Two days after seeding, remove all the medium from all flasks and place into a 50 mL Falcon tube.
4. Filter the medium and freeze it. This constitutes the pericyte-conditioned medium.
5. Feed pericytes adding 15 mL fresh DMEM medium to the cells and return flasks to incubator.
6. Filter the collected pericytes conditioned medium using a 0.2 μm syringe filter to remove cellular debris.
7. Aliquot into 15 mL Falcon tubes and snap freeze in dry ice or liquid nitrogen.
8. Transfer tubes to −80 °C freezer for storage.
9. Repeat the process every 2 days to collect more pericyte-conditioned medium.

3.2 Preparation of the Improved Media for Astrocytes, Pericytes, and Endothelial Cells Co-culture

1. Collect fresh or frozen (−80 °C) vials of astrocytes, pericytes, and endothelial cells conditioned media, as well as fresh DMEM and endothelial cells fresh media.
2. For a tube of improved media of a final volume of 50 mL, place 20 mL of fresh DMEM, 10 mL of fresh endothelial cells media, 7 mL of astrocytes conditioned media, 7 mL of pericytes conditioned media, and 7 mL of endothelial cells conditioned media. Mix with the pipette up and down and warm it up in case it is going to be used immediately.
3. For long periods of storage, freeze this conditioned or improved media (IM) using dry ice or liquid nitrogen and place it at −80 °C till use. Avoid repeated thaw and freeze cycles.
4. This specific combination of mediums was obtained after different trials (data not shown), giving the best results in terms of cell growth (cells grew three times faster, especially in case of pericytes).

Representative images for the three cell types are shown in Fig. 1. Differences obtained growing cells in normal and improved media are shown in Fig. 2, and the differences for each cell type in normal and improved media during day 1, day 2, and day 3 are shown in Figs. 3, 4, and 5 (astrocytes, pericytes, and endothelial cells, respectively).

When these cells are seeded into 12-well coverslips (using different coating materials depending on the cell line, as explained earlier), cell growth was improved after 3 days when they were fed with the improved media compared to the normal media used under static conditions (Figs. 3, 4, and 5).

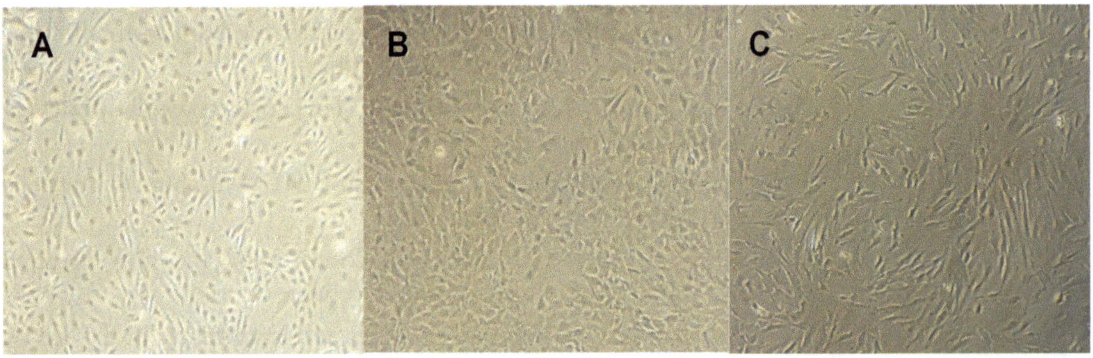

Fig. 1 Representative images of the three cell types cultured independently, endothelial cells (**a**), pericytes (**b**), and astrocytes (**c**) at early stages (passage 2). Images were taken from an optical microscope at 20× magnification

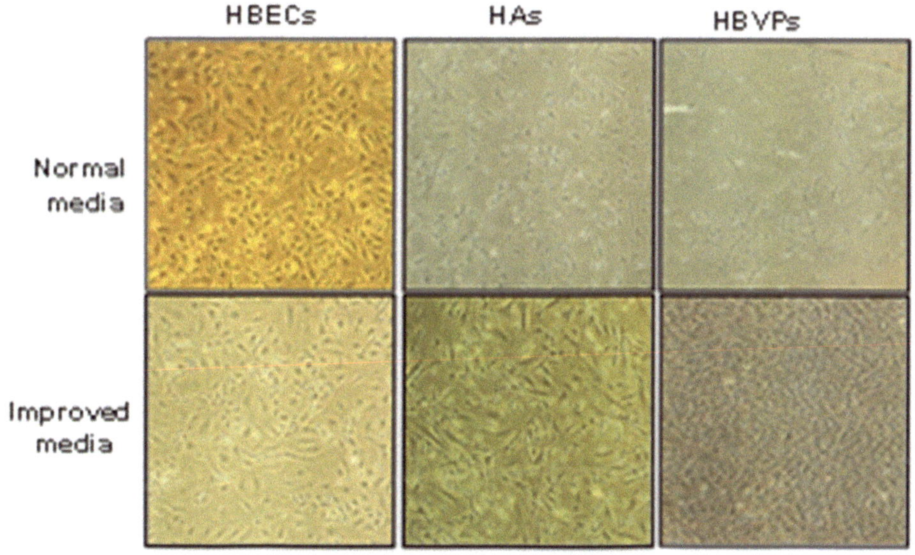

Fig. 2 Representative images obtained by an optical microscope of the three cell types after been cultured 3 days using the normal media (upper panel) and the improved conditioned media (lower panel). Images were taken from an optical microscope at 20× magnification

Fig. 3 Representative images obtained by an optical microscope of astrocytes after been cultured 3 consecutive days using the improved media (upper panel, IM) and the normal media (lower panel, normal). Images were taken from an optical microscope at 10× magnification

Fig. 4 Representative images obtained by an optical microscope of pericytes after been cultured 3 consecutive days using the improved media (upper panel, IM) and the normal media (lower panel, normal). Images were taken from an optical microscope at 10× magnification

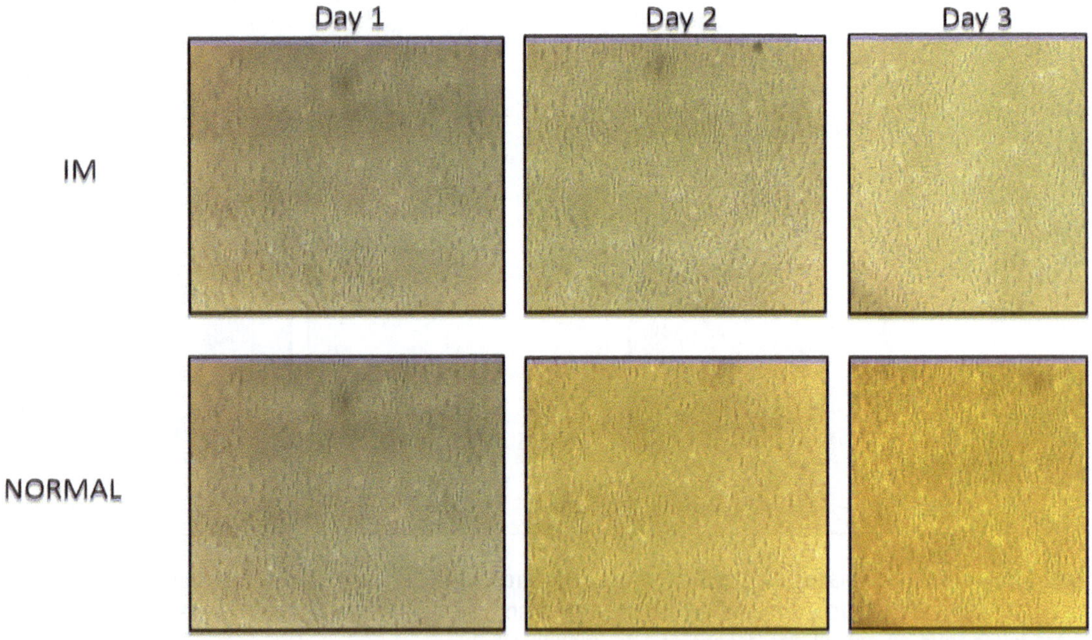

Fig. 5 Representative images obtained by an optical microscope of endothelial cells after been cultured 3 consecutive days using the improved media (upper panel, IM) and the normal media (lower panel, normal). Images were taken from an optical microscope at 10× magnification

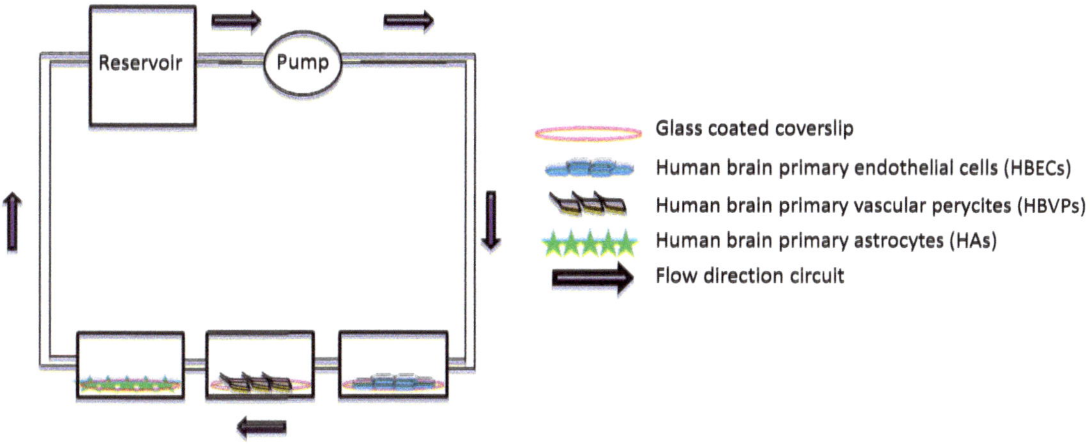

Fig. 6 Schematic representation of Quasi-Vivo QV500 system (Kirkstall Ltd.). (Adapted from Miranda-Azpiazu et al. [29]. http:/creativecommons.org/licenses/by/4.0/)

3.3 Co-culture of Astrocytes, Pericytes, and Endothelial Cells Using Kirkstall Quasi-Vivo (QV500) System

The Kirkstall QV500 system allows communication between different cells placed in the inter-connected chambers (Fig. 6). This design allows the cells to share the same media and experience a desired shear stress.

3.3.1 Preparation of the System for Sterilization and First Use

1. Sterilize the components of the model to be assembled by autoclaving before use.
2. Connect the whole system under aseptic conditions. Connect all the components of the QV500 system according to the manufacturer's instruction. This step takes approximately 15 min.
3. To sterilize the assembled system, place 25 mL of 70% ethanol inside the reservoir and apply a 100–200 μL/min flow during 24 h to sterilize the system.
4. Remove the ethanol and place sterile PBS to wash it and remove all the possible rests of ethanol.
5. Let the system run with PBS and 1% antibiotics (penicillin–streptomycin) for at least 24 h.
6. Remove the PBS and the antibiotics. The model is ready to run. The model can be autoclaved a maximum of three times if necessary.

3.3.2 Characterization and Setup of the Model to Determine the Optimal Flow

The interconnected Quasi-Vivo (QV500) modular chambers used for establishing BBB cells together under flow described herein were purchased from Kirkstall Ltd. The advantage of Kirkstall QV500 modular system is that it allows the possibility of communication between different types of cells as achieved by a continuous flow of culture media between the chambers. After the establishment of the optimal cell culture conditions for the three human brain cell types that form BBB, the conditions were optimized for a

12 or 24 plate format in order to seed the cells in coverslips able to be inserted in the QV500 system. The present in vitro BBB model gives the advantage of having all three cell types forming the barrier derived from human primary cells.

1. Grow cells until confluency in 12 mm coverslips (inside 12 or 24 well plates, as desired) coated with poly-D-lysine (for HA and HBVPs) or collagen (for HMBECs) (*See* **Note 6**).
2. Open the QV500 chambers inside a class II hood and place one by one the coverslips with the confluent cells (*See* **Note 7**).
3. Seed each confluent coverslip inside a QV500 chamber, independently (three replicates of a single cell type, Fig. 7a–c), or together (two or three cell types together, Fig. 7d–f). See Fig. 7a–f for all the combinations possible to be carried out (*See* **Note 8**).
4. Add 1 mL of improved media to each coverslip with the cells facing up inside the chamber (*See* **Note 9**).
5. Add also 30 mL of improved media to the tank and close properly.

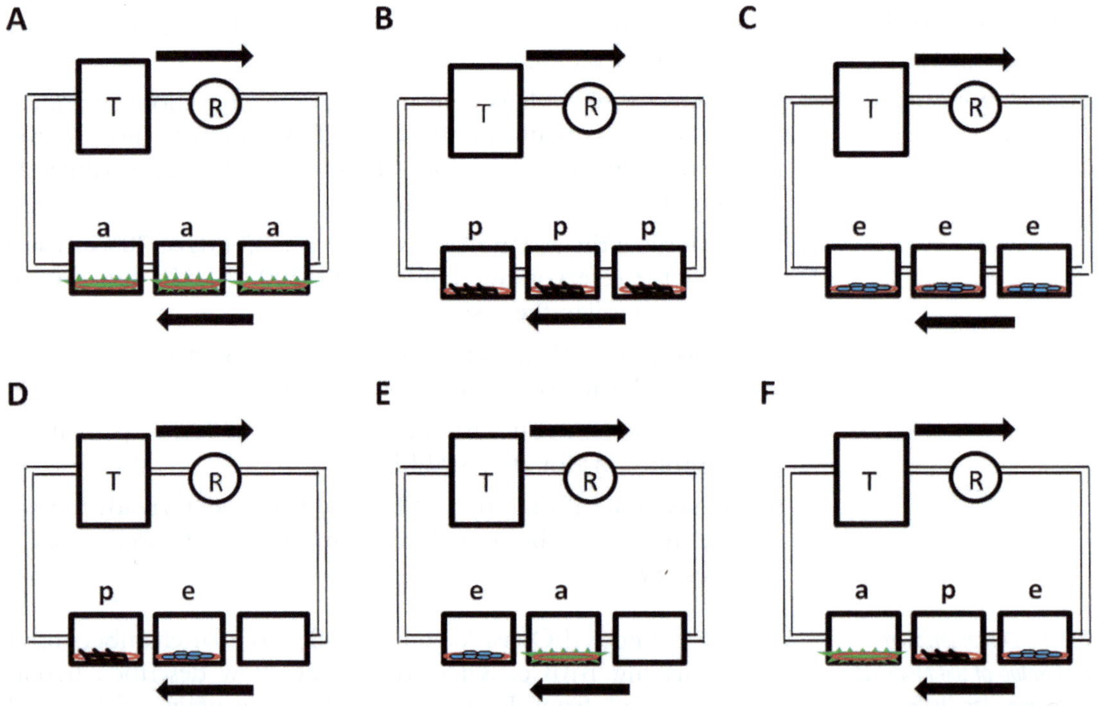

Fig. 7 Diagrammatic representation of the different combinations of cells inside the Kirkstall QV500 interconnected chambers. Each independent diagram shows the three cell lines cultured alone (**a–c**, astrocytes (a), pericytes (p), and endothelial cells (e), respectively) and together (**d**, pericytes and endothelial cells; **e**, endothelial cells and astrocytes; **f**, astrocytes, pericytes, and endothelial cells). T, tank (media reservoir); R, rotor (peristaltic pump)

6. The three QV500 chambers are interconnected between them and sharing the media (*See* **Note 10**).
7. Connect the tubing to the Parker pump and select the desired flow rate (from 0 to 100 μL/min).
8. Place the QV500 system in the incubator (pump included) and press "on" (*See* **Note 11**).
9. Let the cells stay under flow for 72 h (*See* **Note 12**).
10. Assess the effect of flow by studying cell viability (determined by MTT assay and immunohistochemistry assays).

3.3.3 Immunohistochem-istryImmunohistochemistry

To evaluate if the primary cells chosen are resembling the expected morphology of selective features, immunohistochemistry studies should be carried out. Figure 8 [29] shows astrocytes selectively express glial fibrillary acidic protein (GFAP, Fig. 8a), pericytes express α-actin (Fig. 8b), and endothelial cells express CD31 (Fig. 8c). Moreover, tight junction formation in case of endothelial cells under static conditions are also studied by using antibody to *zonula occludens* (ZO1, a tight junction protein), showing that they are able to express this marker even in the absence of flow (Fig. 8d).

1. Grow cells to confluency using 12 mm coverslips conveniently coated for each specific cell type.
2. In a fume hood, wash coverslips with PBS and fix the cells in 4% PFA for 10 min in the dark.
3. The PFA is discarded into a hazardous waste container and the coverslips are washed three times for 5 min with PBS.
4. Place the coverslips into a 12-well plate filled with 200 μL per well of blocking solution (Horse serum 1:10) for 1 h at room temperature.
5. Remove the blocking solution and replace it with the primary antibody solution (200 μL/well) for 2 h at room temperature with gentle agitation.

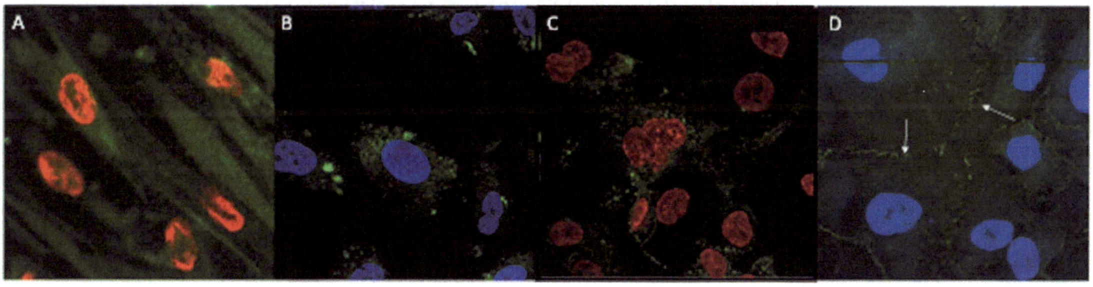

Fig. 8 Representative immunohistochemistry image obtained using a Zeiss LSM700 confocal microscope. (**a**) Astrocytes stained with GFAP, (**b**) pericytes stained with α-actin (**c**) endothelial cells stained with CD31, and (**d**) endothelial cells forming tight junctions (*zonula occludens*, ZO1) (white arrows). (Adapted from Miranda-Azpiazu et al. [29]. http:/creativecommons.org/licenses/by/4.0/)

6. Remove the primary antibody and wash the samples three times, for 5 min each, with PBS (300 μL/well).
7. Add fluorescent conjugated secondary antibody (200 μL/well) for 1 h at room temperature with gentle agitation (*See* **Note 7**).
8. Discard the secondary antibody and wash the samples twice for 5 min each with PBS (300 μL/well).
9. Invert the coverslips (cells facing down) into a drop of mounting medium containing DAPI on a microscope slide.
10. View the samples immediately or store at 4 °C for several days. The slides should be viewed under confocal or fluorescence microscope with the appropriated excitations for each fluorescent conjugated antibody used.

3.3.4 MTT Assay

1. Prepare 1 mg/mL solution of MTT in DMEM without phenol red.
2. Rinse cells (on coverslips) with 300 μL PBS.
3. Pipette 400 μL of the 1 mg/mL solution of MTT into each of the wells on the coverslips and incubate for 4 h at 37 °C inside the incubator avoiding light.
4. Rapidly remove the well contents with a pipette.
5. Add 200 μL of propan-2-ol to the cells to dissolve the dark blue crystals. Pipette up and down to dissolve the crystals formed.
6. Read on a microplate reader with a test wavelength of 560 nm.
7. To analyze the results, compare the absorbance of test cells (inside the model) with control cells (same conditions as under the model, but without flow), extracting the blank value (without cells). Transform the absorbance values into percentage relative to control viability (considered as 100%).

As it is shown in Fig. 9 [29], no statistically significant differences are observed when comparing the cells' viability using MTT assay for each cell line alone and together when using a 40 μL/min flow rate after 72 h inside the QV500 system. Even though higher flow rates are tried, astrocytes and pericytes do not survive to them (data not shown), and that is the main reason why this speed was chosen to co-culture the three cell types together. Notably, endothelial cells survive up to 100 μL/min flow rates. As commented, this flow rate was not chosen due to the cell death in case of the other two cell types.

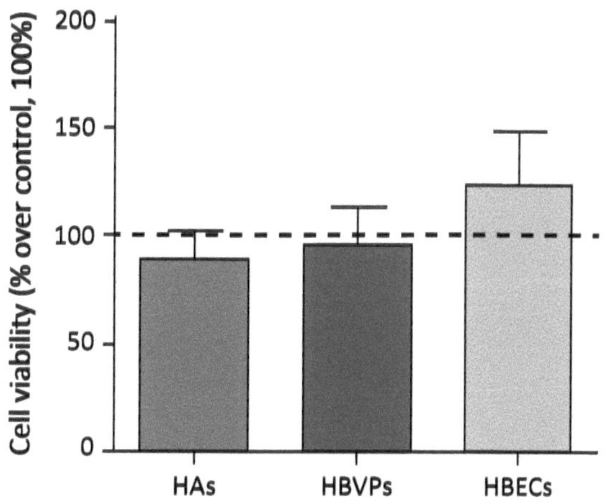

Fig. 9 Cell viability by MTT assay comparing the three cell types under flow (40 μL/min) after 72 h inside the QV500 chambers with the static condition control (black dotted line, 100%) from 3 to 8 independent experiments. No significant differences were obtained when one-way ANOVA test was performed. (Adapted from Miranda-Azpiazu et al. [29]. http:/creativecommons.org/licenses/by/4.0/)

4 Notes

1. To prepare poly-D-lysine solution (stock solution of 250 μg/mL), add 20 mL of graded tissue culture water to the 5 mg of commercial poly-D-lysine and mix.
2. To prepare the collagen stock (300 μg/mL), check the concentration of the crude collagen in the vial. For example, 100 mg/vial for every 22–25 mL would correspond to 4255.32 μg/mL. Thus, to prepare 50 mL of collagen stock (300 μg/mL), mix 3.524 mL of crude collagen with tissue culture grade water up to 50 mL in a Falcon tube. Keep this stock in the fridge.
3. Do not allow to dry out until ready for seeding. Try to use it fresh.
4. Avoid spending more than 1 min inside the incubator, HMBECs are very sensitive.
5. Try to do it as quick as possible.
6. Avoid using cells with a passage higher than seven to avoid loss of cell features. This is particularly important in case of HBMECs.
7. Avoid growing cells to a confluency higher than 95%. This is particularly important in case of HBMECs.
8. Avoid start culturing the three cell types at different times. To achieve a great confluence at the same time, please seed the three primary cell types at the same time.

9. Remember to keep as much conditioned media as possible and keep it at −80 °C.
10. The steps for QV500 preparation should be carried out at the same time as the three cell lines are seeded in the previously coated coverslips (collagen for endothelial cells and poly-D-lysine for astrocytes and pericytes) to have the system ready to place the confluent cells inside the QV500 chambers. The system must be sterilized in advance, and it is recommended that the system should be autoclaved only three times.
11. Plan your experiments to have everything prepared (confluency of cells, flasks coated, QV500 system sterilized, and so on) at the expected time.
12. As a control, replicates of those placed inside QV500 system will be maintained inside the incubator in the 12-well plate, corresponding to the static conditions.
13. All steps from here onward should be carried out with dimmed lights/under aluminum foil to avoid fading of fluorescent intensity.

Acknowledgements

This work was supported by the funding from the Biotechnology and Biological Sciences Research Council (BBSRC) and the Innovate UK. We acknowledge the support from Dr Kelly Davidge and Dr Malcolm Wilkinson at Kirkstall Ltd. for technical help with the QV500 system used in this study.

References

1. Takeshita Y, Kanda T (2015) The blood-brain barrier (BBB) and in vitro BBB models. Brain Nerve 67(8):1035–1042
2. Patabendige A, Abbott NJ (2014) Primary porcine brain microvessel endothelial cell isolation and culture. Curr Protoc Neurosci 69: 3.27.1–3.27.17
3. Wilhelm I, Krizbai IA (2014) In vitro models of the blood-brain barrier for the study of drug delivery to the brain. Mol Pharm 11(7): 1949–1963
4. Panagiotou S, Saha S (2015) Therapeutic benefits of nanoparticles in stroke. Front Neurosci 9:182
5. Patabendige A (2012) The value of in vitro models of the blood-brain barrier and their uses. Altern Lab Anim 40(6):335–338
6. Patabendige A (2012) Toward a humanised alternative to the use of laboratory animals for blood-brain barrier research. Altern Lab Anim 40(5):P12-3
7. Abbott NJ (2013) Blood-brain barrier structure and function and the challenges for CNS drug delivery. J Inherit Metab Dis 36(3): 437–449
8. Cucullo L, Hossain M, Tierney W et al (2013) A new dynamic in vitro modular capillaries-venules modular system: cerebrovascular physiology in a box. BMC Neurosci 14: 18
9. Edwards-Smallbone J, Pleass RJ, Khan NA et al (2012) Acanthamoeba interactions with the blood-brain barrier under dynamic fluid flow. Exp Parasitol 132(3):367–372
10. Yusof SR, Avdeef A, Abbott NJ (2014) In vitro porcine blood-brain barrier model for permeability studies: pCEL-X software pKa(FLUX) method for aqueous boundary layer correction

and detailed data analysis. Eur J Pharm Sci 65: 98–111

11. Zhang Y, Li CS, Ye Y et al (2006) Porcine brain microvessel endothelial cells as an in vitro model to predict in vivo blood-brain barrier permeability. Drug Metab Dispos 34(11): 1935–1943
12. Abbott NJ, Dolman DE, Drndarski S et al (2012) An improved in vitro blood-brain barrier model: rat brain endothelial cells co-cultured with astrocytes. Methods Mol Biol 814:415–430
13. Arisaka T, Mitsumata M, Kawasumi M et al (1995) Effects of shear stress on glycosaminoglycan synthesis in vascular endothelial cells. Ann N Y Acad Sci 748:543–554
14. Niego B, Medcalf RL (2013) Improved method for the preparation of a human cell-based, contact model of the blood-brain barrier. J Vis Exp (81):e50934
15. Hembury A, Mabondzo A (2008) Endothelin-1 reduces p-glycoprotein transport activity in an in vitro model of human adult blood-brain barrier. Cell Mol Neurobiol 28(7):915–921
16. Siddharthan V, Kim YV, Liu S et al (2007) Human astrocytes/astrocyte-conditioned medium and shear stress enhance the barrier properties of human brain microvascular endothelial cells. Brain Res 1147:39–50
17. Cucullo L, Hossain M, Rapp E, Manders T et al (2007) Development of a humanized in vitro blood-brain barrier model to screen for brain penetration of antiepileptic drugs. Epilepsia 48(3):505–516
18. Parkinson FE, Friesen J, Krizanac-Bengez L et al (2003) Use of a three-dimensional in vitro model of the rat blood-brain barrier to assay nucleoside efflux from brain. Brain Res 980(2):233–241
19. Krizanac-Bengez L, Hossain M, Fazio V et al (2006) Loss of flow induces leukocyte-mediated MMP/TIMP imbalance in dynamic in vitro blood-brain barrier model: role of pro-inflammatory cytokines. Am J Physiol Cell Physiol 291(4):C740–C749
20. Ment LR, Stewart WB, Scaramuzzino D et al (1997) An in vitro three-dimensional coculture model of cerebral microvascular angiogenesis and differentiation. In Vitro Cell Dev Biol Anim 33(9):684–691
21. Patabendige A, Skinner RA, Morgan L et al (2013) A detailed method for preparation of a functional and flexible blood-brain barrier model using porcine brain endothelial cells. Brain Res 1521:16–30
22. Katyshev V, Dore-Duffy P (2012) Pericyte coculture models to study astrocyte, pericyte, and endothelial cell interactions. Methods Mol Biol 814:467–481
23. Al Ahmad A, Taboada CB, Gassmann M et al (2011) Astrocytes and pericytes differentially modulate blood-brain barrier characteristics during development and hypoxic insult. J Cereb Blood Flow Metab 31(2):693–705
24. Bai Y, Zhu X, Chao J et al (2015) Pericytes contribute to the disruption of the cerebral endothelial barrier via increasing VEGF expression: implications for stroke. PLoS One 10(4): e0124362
25. Nakagawa S, Deli MA, Kawaguchi H et al (2009) A new blood-brain barrier model using primary rat brain endothelial cells, pericytes and astrocytes. Neurochem Int 54(3–4): 253–263
26. Thomsen LB, Burkhart A, Moos T (2015) A triple culture model of the blood-brain barrier using porcine brain endothelial cells, astrocytes and pericytes. PLoS One 10(8):e0134765
27. Deli MA, Abraham CS, Kataoka Y et al (2005) Permeability studies on in vitro blood-brain barrier models: physiology, pathology, and pharmacology. Cell Mol Neurobiol 25(1): 59–127
28. Neuhaus W, Lauer R, Oelzant S et al (2006) A novel flow based hollow-fiber blood-brain barrier in vitro model with immortalised cell line PBMEC/C1-2. J Biotechnol 125(1):127–141
29. Miranda-Azpiazu P, Panagiotou S, Jose G et al (2018) A novel dynamic multicellular co-culture system for studying individual blood-brain barrier cell types in brain diseases and cytotoxicity testing. Sci Rep 8:8784. https://doi.org/10.1038/s41598-018-26480-8

Chapter 10

A Dynamic, In Vitro BBB Model to Study the Effects of Varying Levels of Shear Stress

Gemma Molins Gutiérrez, Jordi Martorell, Antonio G. Salazar-Martin, and Mercedes Balcells

Abstract

The blood–brain barrier (BBB) consists of a tight network of blood capillaries in the brain that separate the circulatory system from the central nervous system. Its particular properties are based on the dynamic interaction between cerebral endothelial cells and other surrounding cells, especially astrocytes. We have designed and synthesized a three-dimensional scaffold that recapitulates the main hallmarks of the BBB extracellular matrix and serves as a platform to co-culture human brain microvascular endothelial cells and human cortical astrocytes. The scaffold can be exposed to flow, thereby allowing the study of flow-mediated pathways at the BBB.

Key words Blood–brain barrier, Cerebrovascular disease, Vascular biology, Fluid device, Flow loops, Shear stress

1 Introduction

The blood–brain barrier (BBB) consists of a tight network of blood capillaries in the brain that separate the circulatory system from the central nervous system. Its particular properties are based on the dynamic interactions between cerebral endothelial cells and other surrounding neurocytes such as astrocytes, pericytes, microglia, and neurons. These interactions give cerebral endothelial cells a unique phenotype capable of forming a morphological barrier of inter-endothelial junctions, the enzymatic and metabolic barriers, and the uptake and efflux transport systems [1]. Microvascular endothelial cells line brain capillaries, forming the core of the BBB, exhibit a unique phenotype that distinguishes them from endothelial cells in other vascular beds [2]. Endothelial cells are connected by tight junctions, which are composed of transmembrane proteins such as occludins and claudins. One of the hallmarks of the BBB phenotype is the highly restrictive paracellular permeability

Nicole Stone (ed.), *The Blood-Brain Barrier: Methods and Protocols*,
Methods in Molecular Biology, vol. 2492, https://doi.org/10.1007/978-1-0716-2289-6_10,

regulated by inter-endothelial tight junctions that maintains and regulates brain homeostasis, metabolism, and neuronal activity [3]. Astrocyte cell projections, called astrocytic feet, surround the endothelial cells of the BBB, providing them with biochemical support. The BBB protects the brain from an assortment of potential risks by restricting the passage of cytotoxic solutes, harmful proteins, and infectious agents into the CNS. Impairment of the BBB is involved in many neurological disorders and infections like meningitis, epilepsy, multiple sclerosis, and Alzheimer's disease [2, 4, 5].

The number of in vitro BBB models has increased notably since they emerged in the early 1990s. However, no in vitro model fully recapitulates all the functional and physiological properties of the BBB [6]. The cellular environment changes significantly, as many biochemical and biomechanical signals from both brain and blood are lost. Despite this, in vitro BBB models have many benefits when compared to in vivo ones: they are more reproducible, cost-effective, and versatile and can allow for high-throughput screening [6, 7]. The main characteristics that an ideal in vitro model should fulfill are stimulating endothelial cell differentiation into a mature phenotype, expressing tight junction markers and relevant transporters and responsive to shear stress [2, 8].

In this chapter, we present the design and synthesis of a three-dimensional scaffold that encompasses the main attributes of the extracellular matrix of the BBB in vitro. Our platform is flexible enough to modulate the level of interaction between endothelial cells and astrocytes. Our model allows for direct immunofluorescence of the samples and functional assays such as permeability tests, using fluorescent probes or lymphocytes to study immune cell migration. We have investigated the impact of different compositions in the synthesis of the scaffold by testing different polymers, concentrations, adhesion motifs, and buffer conditions. We include herein a combination optimized to achieve proper endothelial and astrocytic functionality and morphology using human cerebral microvascular endothelial cells (HCMEC/D3) and human cortical astrocytes (HCA).

2 Materials

2.1 Hydrogel Preparation

1. Collagen type I, rat tail high: 9.43 mg/mL (#354249, Corning).
2. Bovine fibronectin, 1 mg/mL (#03-090-1-05, Biological Industries).
3. Sodium hydroxide (NaOH): 1 M prepared in purified water.
4. USP sterile purified water (#46-000-CV, Corning).

5. Astrocyte medium, complete (#1801, ScienCell).
6. Trypsin EDTA, 0.05% (#25300, ThermoFisher Scientific).
7. Phosphate buffer saline 10X, Molecular Biology Grade (#46-013-CM, Corning).
    - A dilution of PBS 1X is prepared with deionized water and subsequently sterilized by autoclave.
8. 225 cm^2 cell culture flask, angled neck, tissue culture treated (#431082, Corning).
9. Cellometer disposable counting chambers (#SD100, Nexcelom Bioscience).
10. Cellometer Cell Counter Auto T4 (Nexcelom Bioscience).

2.2 Basement Membrane

1. Bovine fibronectin, 1 mg/mL (#03-090-1-05, Biological Industries).
2. Laminin, mouse, 1 mg (#354232, Corning).
3. Phosphate buffer saline (PBS) 10X, Molecular Biology Grade (#46-013-CM, Corning).
    - A dilution of PBS 1X is prepared with deionized water and subsequently sterilized by autoclave.
4. Sterile syringe filter, 0.22-um cellulose acetate (#28145-477, VWR).

2.3 Cells and Medium

1. Human cortical astrocytes, HA (#1800, ScienCell).
2. Human cerebral microvascular endothelial cells, HCMEC/D3 (#CLU512, CELLutions Biosystems Inc.).
3. Astrocyte medium, complete (#1801, ScienCell).
4. Endothelial cells medium is prepared as described below:
    - 500 mL of EBM-2 basal medium (#CC-3156, Lonza).
    - From EGM-2 MV SingleQuots (#CC-4147, Lonza), only the following supplements are added to the basal medium:
        - 0.2 mL of hydrocortisone.
        - 2 mL of hFGF-B.
        - 0.5 mL of ascorbic acid.
        - 25 mL of fetal bovine serum (FBS).
    - 5 mL of HEPES 1M (#15630080, ThermoFisher Scientific).
    - 5 mL of chemically defined lipid concentrate, CDLC (#11905031, ThermoFisher Scientific).
    - 5 mL of penicillin streptomycin, Pen Strep (#15140-122, ThermoFisher Scientific).
5. Trypsin EDTA, 0.05% (#25300, ThermoFisher Scientific).

6. Phosphate buffer saline 10X, Molecular Biology Grade (#46-013-CM, Corning).
   - A dilution of PBS 1X is prepared with deionized water and subsequently sterilized by autoclave.
7. Filter system, 0.22 μm, sterilizing, low protein binding (#430769, Corning).
8. 75 cm^2 cell culture flask, canted neck, treated (#430641U, Corning).
9. 225 cm^2 cell culture flask, angled neck, tissue culture treated (#431082, Corning).

2.4 Flow Loop

1. Proprietary customized microfluidic device.
2. Digital peristaltic pump (Model C.P.78002-00, Ismatec).
3. PharMed BPT Tubing, 2.79 mm ID (#95723-48, Ismatec).
4. Silastic Laboratory Tubing, 0.188 in ID × 0.312 in OD (#2415615, Dow Corning).
5. Silastic Laboratory Tubing, 0.130 in ID × 0.250 in OD (#2415607, Dow Corning).
6. Silastic Laboratory Tubing, 0.078 in ID × 0.125 in OD (#2415577, Dow Corning).
7. Masterflex Fitting Connector, Straight, Hose Barb Reducer, 3/16″ ID × 3/32″ ID (#UX-30703-47, Cole Parmer).
8. Masterflex Fitting Connector, Straight, Hose Barb Reducer, 1/8″ ID × 3/32″ ID (#UX-40703-42, Cole Parmer).
9. Masterflex Fitting Connector, Straight, Hose Barb Reducer, 1/8″ ID × 5/32″ ID (#UX-30703-45, Cole Parmer).
10. Elbow Luer Connector Male (#10802, Ibidi).
11. Cassette with occlusion lever (#IS 3820, Ismatec).
12. Customized glass reservoirs (James Glass Company).
13. Forma Series II Water Jacket CO_2 Incubator (Model 3851/3110M, ThermoFisher Scientific).

2.5 Immunostaining

1. Paraformaldehyde (PFA) powder (#158127, Sigma-Aldrich).
   - A solution of PFA 4% is prepared with deionized water and subsequently filtered through a 0.22 μm cellulose acetate syringe filter.
2. Phosphate buffer saline 10X, Molecular Biology Grade (#46-013-CM, Corning).
   - A dilution of PBS 1X is prepared with deionized water and subsequently sterilized by autoclave.
3. Triton X-100 (#T8787, Sigma-Aldrich).
   - A dilution of 0.1% Triton X-100 is prepared with PBS 1X.

4. Goat Serum (#16210064, ThermoFisher Scientific).
5. Blocking buffer (BB) solution.
   - BB solution is prepared by diluting Goat Serum in PBS 1X to a final concentration of 5%.
6. Claudin 5 Monoclonal Antibody, 4C3C2 (#35-2500, Invitrogen).
   - A 1:50 dilution of Claudin 5 Monoclonal Antibody is prepared with BB solution.
7. ZO-1 Polyclonal Antibody (#40-2200, Invitrogen).
   - A 1:50 dilution of ZO-1 polyclonal antibody is prepared with BB solution.
8. Rb pAb to GFAP (#ab7260, Abcam).
   - A 1:50 dilution of GFAP antibody is prepared with BB solution.
9. Anti-rabbit IgG Fab2 Alexa Fluor® 594 (#8889S, Cell Signaling).
   - A 1:200 dilution of Anti-rabbit Alexa Fluor® 594 is prepared with BB solution.
10. Anti-mouse IgG Fab2 Alexa Fluor® 488 (#4408S, Cell Signaling).
    - A 1:200 dilution of Anti-mouse Alexa Fluor® 488 is prepared with BB solution.
11. Anti-mouse IgG Fab2 Alexa Fluor® 594 (#8890S, Cell Signaling).
    - A 1:200 dilution of Anti-mouse Alexa Fluor® 594 is prepared with BB solution.
12. Anti-rabbit IgG Fab2 Alexa Fluor® 488 (#4412S, Cell Signaling).
    - A 1:200 dilution of Anti-rabbit Alexa Fluor® 488 is prepared with BB solution.
13. DAPI solution, 1 mg/mL (#62248, Thermo Scientific).
    - A 1:1000 dilution of DAPI is prepared with either BB solution or PBS 1X.

3 Methods

Carry out all procedures in the biological sterile hood and under sterile conditions unless otherwise specified.

3.1 Hydrogel Preparation with Embedded Human Astrocytes

Different hydrogel compositions have been investigated, and the final formulation has been optimized based on cellular morphology, proliferation, phenotype expression, and mechanical properties, namely, resistance to shear stress. In this case, the preparation of 245 μL of a 5 mg/mL collagen hydrogel preparation containing 10^5 astrocytes is described.

1. Clean and sterilize the work area of a biological hood with 70% ethanol. Let it evaporate.
2. Warm up astrocyte medium, trypsin, and PBS 1X in a water bath at 37 °C.
3. Place collagen type I, bovine fibronectin, NaOH 1M solution, a 15 mL Falcon tube with 2 mL of astrocyte medium, and a 1.5 mL Eppendorf tube on ice (*see* **Note 1**).

3.1.1 Astrocyte Dilution Preparation

1. Take a T225 flask with confluent astrocytes and carefully aspirate the medium with vacuum. Rinse the cells with pre-warmed PBS 1X and carefully aspirate with vacuum. Add 15 mL of pre-warmed trypsin and incubate at 37 °C for 1–2 min. Check cell detachment under microscope.
2. Stop trypsinization by addition of pre-warmed astrocyte medium in equal or greater volume than trypsin and transfer the cell suspension to a Falcon tube. Centrifuge the suspension for 5 min at 1400 rpm.
3. Aspirate the supernatant of the falcon tube gently, without disturbing the cell pellet. Resuspend the cell pellet in 1–5 mL of astrocyte medium (*see* **Note 2**).
4. Homogenize the cell suspension thoroughly and transfer 20 μL to a cell counter chamber. Perform this step twice (*see* **Note 3**).
5. Count the cells using a cell counter. Make three measurements per side and calculate the average of the six counts. The cell concentration should be equal or greater than 10^6 cells/mL.
6. Prepare a dilution of astrocytes from the cell suspension in **step 7** at a final concentration of 10^6 cells/mL. Keep the cell dilution inside the hood (*see* **Note 4**).

3.1.2 Hydrogel Preparation and Polymerization

1. Take the Eppendorf tube from the ice container and add 100 μL of the astrocyte dilution (10^6 cells/mL) prepared in **step 9**. Place on ice for 5 min.
2. Take the same Eppendorf tube from the ice container and add 12.3 μL of fibronectin. Vortex for 5 s and place it back on ice for 2 min.
3. Meanwhile, take a P200 pipette and set the volume to 130 μL. Cut the tip of a 200 μL pipette tip (*see* **Note 5**).

4. Take the Eppendorf tube from the ice container. Take collagen type I from the container and immediately pipette 130 μL using the previously cut pipette tip. Make sure to pipette slowly as the collagen stock is very viscous. Add collagen in the Eppendorf tube and vortex for 5 s. You should see the medium changing from red/pink to yellowish. Place it back on ice.
5. Meanwhile, take a P10 pipette and set the volume to 3.3 μL. Take a P1000 and set the volume to 220 μL. Cut a P1000 pipette tip to facilitate the aspiration of the final viscous hydrogel solution.
6. Prepare the chamber where the hydrogel will be placed before proceeding with next steps (*see* **Note 6**).
7. Take the NaOH 1M solution from the ice container. Take the Eppendorf tube from the ice container and immediately add 3.3 μL of the NaOH solution. Vortex for 5 s and without pause, gently pipette with the P1000 previously cut tip and place it in the hydrogel chamber/reservoir to be used as indicated in Fig. 1. Avoid formation of bubbles (*see* **Note** 7).
8. Take the hydrogel chamber/reservoir and place it in an incubator at 37 °C, 5% CO_2 and 80% humidity. If possible, use an orbital shaker to improve hydrogel homogeneity. Wait 1 h to ensure complete polymerization.
9. Finally, take the chamber with polymerized hydrogel from the incubator and add enough astrocyte medium to cover the hydrogel surface (about 1 mL). Put it back in the incubator for 2 days. Add medium every day to make sure the hydrogel does not dry out.

3.2 Hydrogel Coating with Basement Membrane Formulation

Prior to culturing brain endothelial cells on top of the hydrogel matrix, the surface is coated with a solution of fibronectin and laminin simulating the basement membrane found in vivo in brain capillaries. This step enhances the adhesion of endothelial cells and prevents them from detaching during application of flow.

1. Clean and sterilize the work area of a biological hood with 70% Ethanol. Let it dry.
2. Warm up astrocyte medium and PBS 1X in a water bath at 37 °C.
3. Take bovine fibronectin from the fridge and laminin from the freezer and bring them to the hood.
4. Take an Eppendorf tube and add 471 μL of pre-warmed astrocyte medium. Add 25 μL of fibronectin and vortex for 5 s.
5. Add 3.75 μL of laminin to the Eppendorf tube and vortex for 5 s.

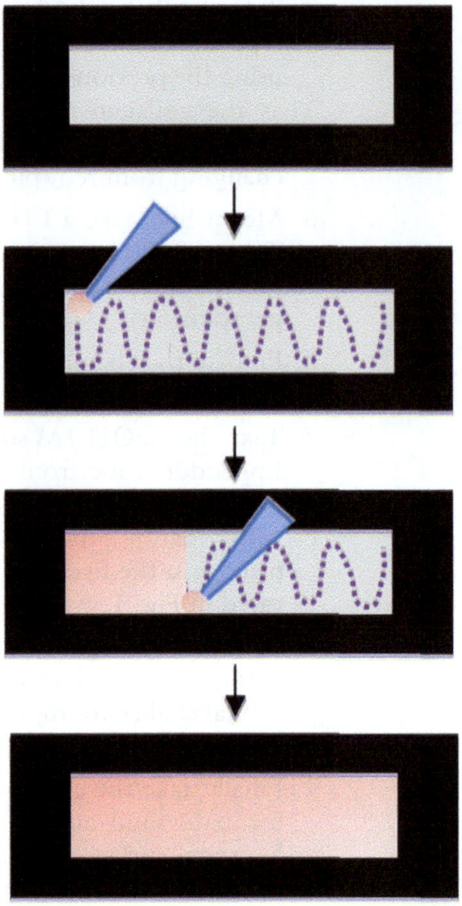

Fig. 1 Hydrogel placing representation. To avoid the formation of bubbles and ensure the complete filling of the device's reservoir, place the hydrogel solution from side to side throughout the length of the reservoir until obtention of a homogeneous flat hydrogel matrix

6. Take the chamber from the incubator and gently aspirate the medium using a pipette tip (*see* **Note 8**).
7. Rinse the hydrogel and chamber with pre-warmed PBS 1X and aspirate with pipette tip.
8. Add 500 μL of the basement membrane solution and incubate for at least 2 h at 37 °C, 5% CO_2 and 80% humidity.
9. Finally, gently aspirate the basement membrane solution and add enough astrocyte medium to cover the hydrogel. Put it back in the incubator.

3.3 Culture of Human Brain Microvascular Endothelial Cells

In this section, we describe the procedure to culture brain microvascular endothelial cells on top of the hydrogel (Fig. 2). Two types of endothelial cells have been investigated and used. After optimization, HCMEC/D3 have been selected for these type of flow

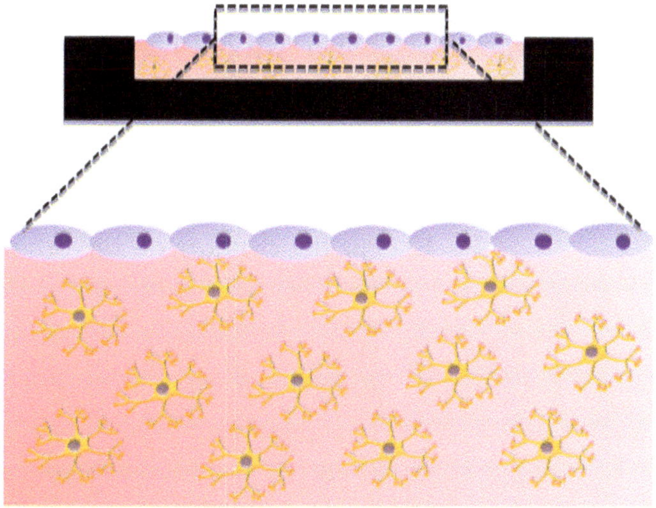

Fig. 2 Three-dimensional hydrogel scaffold with embedded star-shaped astrocytes covered with a confluent monolayer of brain endothelial cells

experiments. Given the hydrogel volume, surface area, and the chamber configuration, a dilution of 10^6 cells/mL is used. However, the number of cells should be optimized in each case according to the hydrogel volume and surface.

1. Clean and sterilize the work area of a biological hood with 70% ethanol. Let it dry.
2. Warm up endothelial cells medium, trypsin, and PBS 1X in a water bath at 37 °C.
3. Take a T225 flask with confluent HCMEC/D3 cells and carefully aspirate the medium with vacuum. Rinse the cells with pre-warmed PBS 1X and carefully aspirate with vacuum. Add 15 mL of pre-warmed trypsin and incubate at 37 °C for 1–2 min. Check cell detachment under microscope.
4. Stop trypsinization by addition of pre-warmed endothelial cell medium in equal or greater volume than the volume of trypsin and transfer the cell suspension to a falcon tube. Centrifuge the suspension for 5 min at 1400 rpm.
5. Aspirate the supernatant of the falcon tube gently, without disturbing the cell pellet. Resuspend the cell pellet in 1–5 mL of endothelial cell medium.
6. Homogenize the cell suspension thoroughly and transfer 20 μL to a cell counter chamber. Perform this step twice (*see* **Note 3**).
7. Count the cells using a cell counter machine. Make three measurements per side and calculate the average of the six counts. The cell concentration should be equal or greater than 10^6 cells/mL.

8. Prepare a dilution of HCMEC/D3 cells from the cell suspension in **step 7** at a final concentration of 10^6 cells/mL.
9. Place an orbital shaker in the incubator.
10. Take the chamber from the incubator and gently aspirate the medium.
11. Add 1000 μL of the HCMEC/D3 cell dilution to the chamber.
12. Put the chamber on the orbital shaker in the incubator and incubate overnight on slow shaking to ensure that endothelial cells fully cover the surface of the hydrogel (*see* **Note 9**).
13. The next day, warm astrocyte and endothelial cell medium in a water bath at 37 °C.
14. Take the chamber from the orbital shaker and gently aspirate using a pipette the endothelial cells medium.
15. Finally, add a mixture of 50% astrocyte and 50% endothelial cell medium to the chamber and put it back in the incubator.
16. Check cell growth under the microscope every day and change medium every other day. Let endothelial cells grow for 7 days (*see* **Note 10**).

3.4 BBB Flow Experiments

After culturing astrocytes and endothelial cells on the collagen hydrogel for 7 days, the in vitro model is connected to a flow loop system (Fig. 3) that allows simulation of physiological and

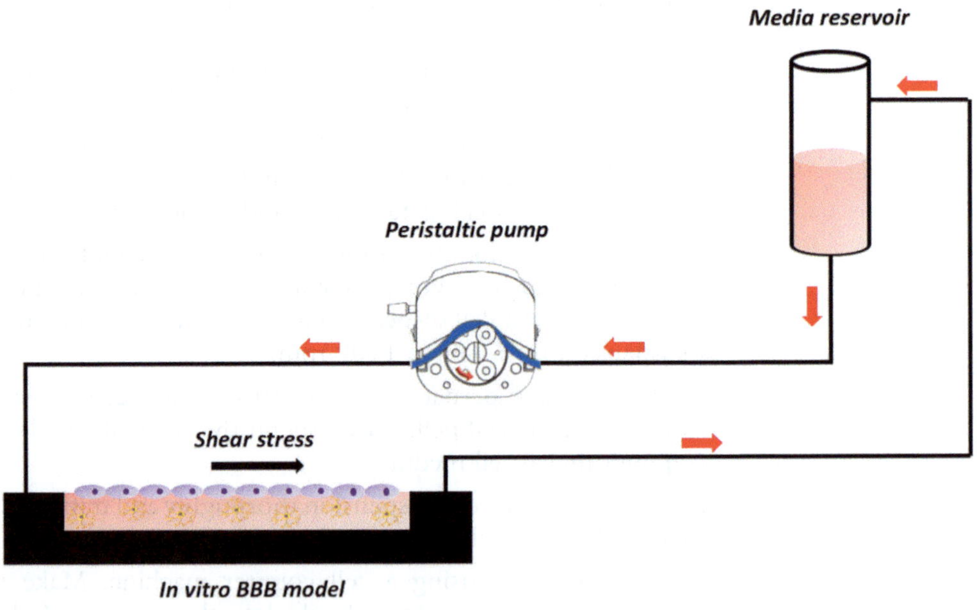

Fig. 3 In vitro BBB model flow system. Flow device with BBB model is connected to a flow loop consisting of a peristaltic pump connected to a medium reservoir. The peristaltic pump is set to specific flow rates and regimens simulating both cerebrovascular physiological and pathological conditions

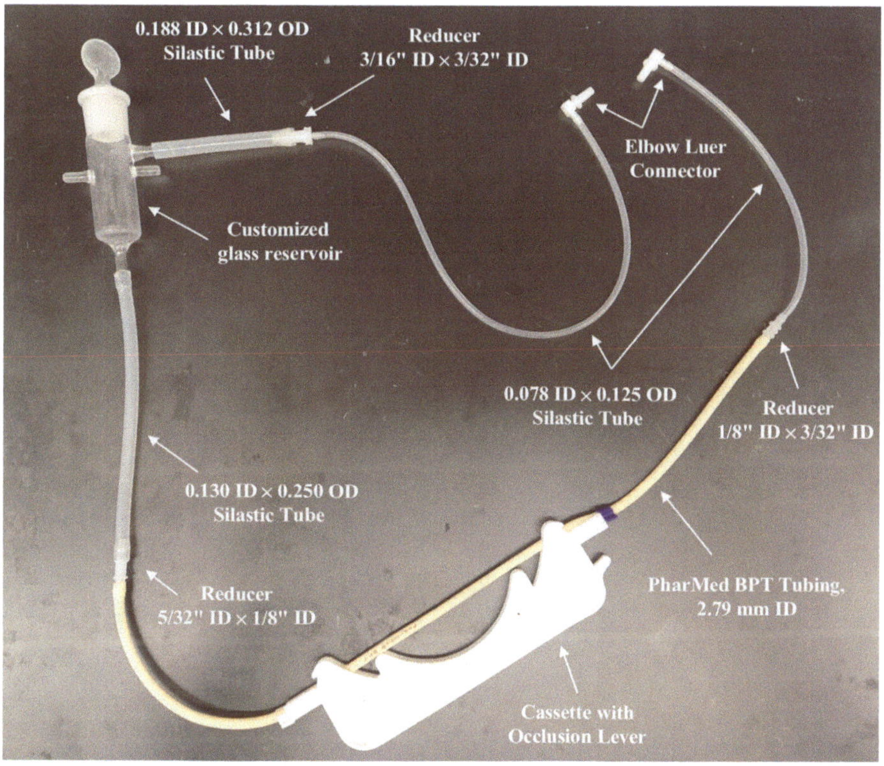

Fig. 4 Flow loop system construct

pathological flow by applications of different shear stresses and flow types [2].

1. Prepare flow loop constructs with sterile silicone tubing (0.188 in ID × 0.312 in OD, 0.125 in ID × 0.250 in OD, 0.078″ ID × 0.0125″ OD, 0.104 in ID × 0.192 in OD), connectors (barbed fittings, reducing connector, 1/8″ × 3/32″ ID, Cole Parmer, Vernon Hills, IL, USA) and customized glass reservoirs (Fig. 4).
2. Autoclave flow loop constructs.
3. Warm up endothelial and astrocyte medium in a water bath at 37 °C.
4. Take the chamber from the incubator and gently aspirate the medium with a pipette.
5. Add 1 mL of a mixture of 1:1 (v/v) of astrocyte and endothelial cell medium to the chamber.
6. Connect flow system to perfusion bioreactor (Fig. 5).
7. Fill in the glass reservoirs with 10–15 mL of a mixture of 1:1 (v/v) of astrocyte and endothelial cell medium to the chamber.

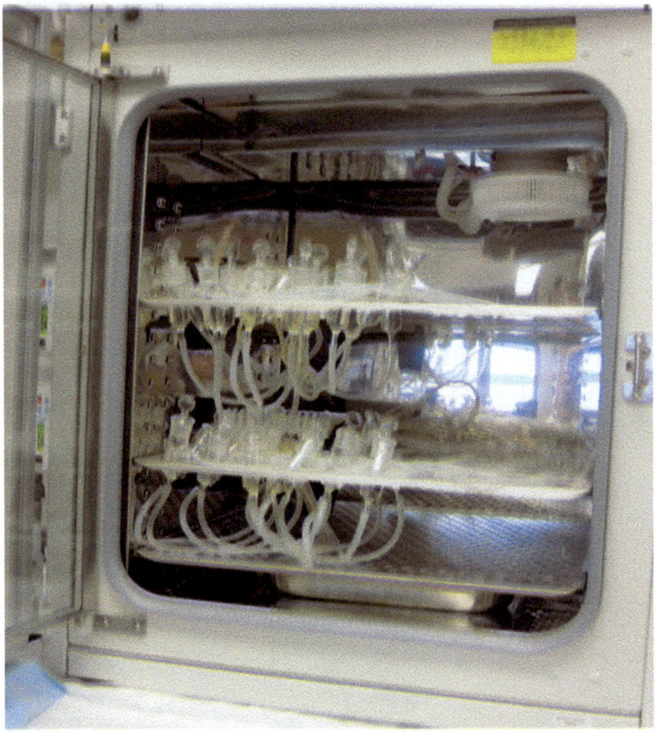

Fig. 5 Flow loop systems connected to the perfusion bioreactor

8. Once the flow system is connected to the peristaltic pump, slowly increase the pump velocity to the desired flow value. Make sure there is no leakage (*see* **Note 11**).
9. Keep cell cultures under dynamic conditions for 3 days prior to further analysis. Cells are exposed to laminar flow with shear stress ranging from 0 to 40 dyn/cm^2, under steady or pulsatile regime.

3.5 Immunostaining

After application of flow, analysis of cell morphology (Fig. 6) and tight junctions (Fig. 7) is performed by immunostaining.

1. Slowly decrease the flow rate of the peristaltic pump and disconnect the chambers from the flow loop.
2. Gently aspirate the remaining medium in the chamber with a pipette and rinse with PBS 1X.
3. Add 4% PFA in the chamber and incubate at RT for 20 min.
4. Gently aspirate the remaining solution in the chamber with a pipette and rinse three times with PBS 1X, 10 min each time.
5. If necessary, permeabilize the cells with a dilution of 0.1% Triton-X in PBS 1X and incubate at RT for 10 min (*see* **Note 12**).

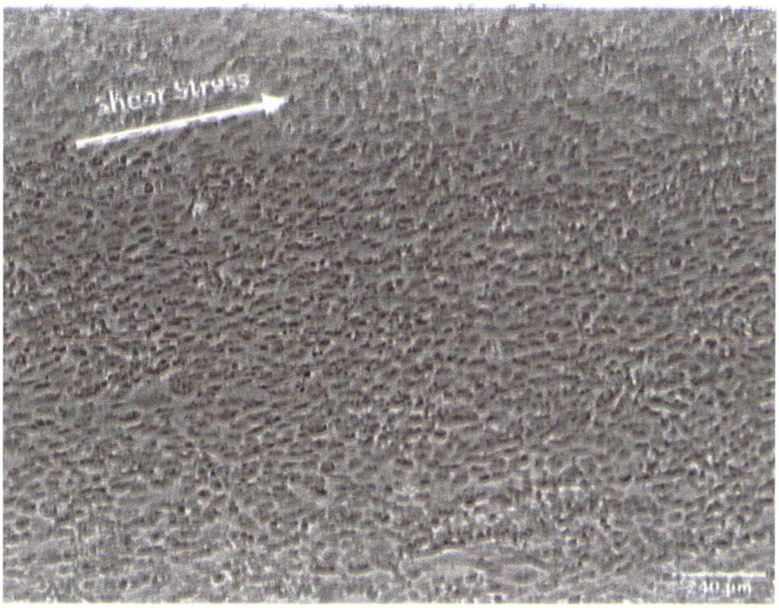

Fig. 6 Bright-field image of a brain endothelial cell confluent monolayer cultured on top of the hydrogel after exposure to 10 dyn/cm^2 of shear stress for 3 days. Notice stretch and alignment of cells in the direction of flow

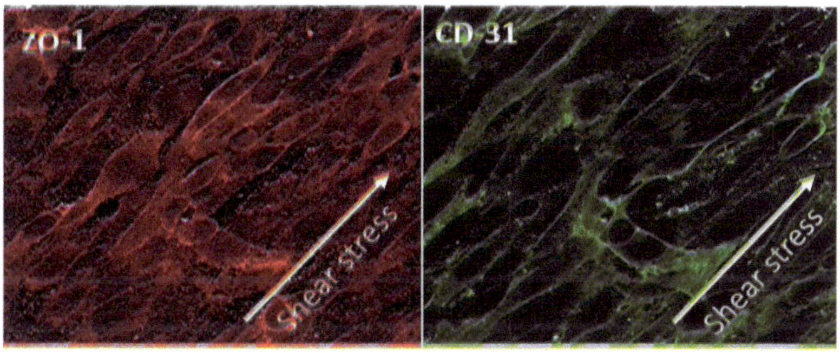

Fig. 7 Confocal microscopy images of immunostaining of tight junction protein ZO-1 (red) and endothelial cell marker CD-31 (green) on a confluent monolayer of HCMEC/D3 cultured on an astrocyte-embedded collagen hydrogel and exposed to steady flow of 10 dyn/cm^2 for 3 days

6. Gently aspirate the remaining solution in the chamber with a pipette and rinse five times with PBS 1X, 10 min each time.
7. Add blocking buffer in the chamber and incubate 1 h at 37 °C or overnight at 4 °C.
8. Prepare the desired primary antibody dilution in blocking buffer and add it to the chamber. Incubate overnight at 4 °C.
9. Gently aspirate the remaining solution in the chamber with a pipette and rinse five times with PBS 1X, 10 min each time.

10. Prepare the desired secondary antibody dilution in blocking buffer and add it to the chamber. Incubate for 2 h at 37 °C.
11. Gently aspirate the remaining solution in the chamber with a pipette and rinse three times with PBS 1X, 10 min each time.
12. Finally, prepare a 1:1000 dilution of DAPI in PBS 1X and incubate for 10 min at RT.
13. Gently aspirate the remaining solution in the chamber with a pipette and rinse three times with PBS 1X, 10 min each time.
14. Proceed with microscope imaging. Keep the chamber with PBS 1X at 4 °C.

4 Notes

1. Collagen stock solution used in these experiments has a high concentration. Therefore, it can polymerize very easily if it is not kept cold. When using collagen stock solution for hydrogel preparation, make sure it is always kept on ice. Do the same with the rest of materials that will be used to prepare the hydrogel solution, as detailed in the methodology description.
2. The number of astrocytes to be embedded in the hydrogel (considering the volume used in the protocol described here) is fairly high. Moreover, because a high collagen concentration is needed to form the hydrogel scaffold in which astrocytes will be embedded, a minimal cell dilution volume is required to keep the collagen concentration high. Therefore, when preparing the cell suspension, ensure the cell concentration is at least 1×10^6 astrocytes/mL.
3. Bubbles in the cell counter chamber will interfere with the cell count. Therefore, it is important to avoid formation of bubbles when introducing 20 μL of the cell suspension in the cell counter chamber. Place the tip of the pipette so that it faces the space to be filled. Additionally, make sure you fill the chamber with the appropriate volume, 20 μL in the case described here. To get the best results, perform three counts of each side of the chamber slide, and average the six counts to obtain the final cell concentration.
4. The astrocyte cell suspension will be used in the preparation of the hydrogel. As mentioned in **Note 1**, the collagen stock concentration as well as the final collagen concentration on the hydrogel matrix are high. That causes the polymerization to happen quickly at room temperature. Therefore, it is important to keep all the materials used in the preparation of the hydrogel on ice except for the cell suspension that is kept at room temperature in order to avoid cell damage.

5. The collagen stock is highly concentrated and therefore very viscous. Thus, it is very tricky to pipette it with a regular pipette tip. Cutting the pipette tip with regular sterilized scissors helps pipetting the viscous collagen solution more easily.
6. Polymerization of collagen takes place right after the addition of NaOH solution. Therefore, it is critical to have the device, pipette, and vortex prepared before adding NaOH solution and work as quickly as possible after its addition. Vortex right away and pipette and pour the resulting hydrogel solution in the device reservoir without waiting.
7. The hydrogel scaffold has astrocytes embedded and must be as flat as possible to ensure homogenous shear stress of the cells cultured on top of it. For that reason, formation of bubbles when adding the hydrogel solution in the reservoir must be avoided. Pipette gently and leave a drop of the hydrogel solution in the pipette tip. Pipetting all the way down will create a bubble at the end of the process.
8. When the hydrogel has polymerized in the device reservoir, proceed very gently with aspiration and addition of new medium to prevent disruption of the hydrogel scaffold.
9. Without proper shaking, endothelial cells can accumulate in specific areas of the hydrogel surface. Because these experiments require a homogeneous confluent monolayer of endothelial cells, orbital shaking for 24 h after addition of the endothelial cell suspension will ensure proper distribution of cells along the hydrogel surface.
10. After culture, endothelial cells should have formed a confluent monolayer of cells without empty spaces on top of the hydrogel scaffold. Likewise, astrocytes embedded in the hydrogel should be star-shaped all along the hydrogel.
11. When connecting the fluidic device in the flow loop, it is essential to increase the flow rate slowly and gradually on the peristaltic pump. If flow is increased too fast, cells can detach from the hydrogel surface. Likewise, if the final flow rate is too high, the hydrogel scaffold can disrupt. Thus, flow rates are key and require optimization, depending on the device used. Moreover, leakage of medium can occur when using mid to high flow rates. It is more common for connections between the tube and the device to leak, although it could also happen between tube-to-tube connections. Leakage can appear from minutes to hours after start of flow. Always check after 10 min, 1 h, and 3 h to ensure the system is not leaking.
12. Permeabilization is generally used in immunostaining protocols. In our protocols, we permeabilize the cell culture when staining intracellular proteins. However, eliminating permeabilization is interesting to minimize the intracellular

fluorescence noise when staining for membrane proteins like tight junctions, as long as the antigen receptor of the protein under study is located on the membrane and not intracellularly. This has helped us evaluate the differences in tight junction expression and localization in our endothelial cell cultures.

Acknowledgements

This work was supported in part by a grant from Spain Ministerio de Economía y Competitividad (SAF-2017-84773-C2, MB) and by the Global CoCreation Lab (GM, MB).

References

1. Abbott NJ (2013) Blood-brain barrier structure and function and the challenges for CNS drug delivery. J Inherit Metab Dis 36:437–449. https://doi.org/10.1007/s10545-013-9608-0
2. Garcia-Polite F, Martorell J, Del Rey-Puech P et al (2017) Pulsatility and high shear stress deteriorate barrier phenotype in brain microvascular endothelium. J Cereb Blood Flow Metab 37(7):2614–2625. https://doi.org/10.1177/0271678X16672482
3. Abbott NJ (2002) Astrocyte-endothelial interactions and blood-brain barrier permeability. J Anat 200:629–638. https://doi.org/10.1046/j.1469-7580.2002.00064.x
4. Singer J, Trollor JN, Crawford J et al (2013) The association between pulse wave velocity and cognitive function: the Sydney memory and ageing study. PLoS One 8:4–9. https://doi.org/10.1371/journal.pone.0061855
5. Weiss N, Miller F, Cazaubon S et al (2009) The blood-brain barrier in brain homeostasis and neurological diseases. Biochim Biophys Acta Biomembr 1788:842–857. https://doi.org/10.1016/j.bbamem.2008.10.022
6. Nicolazzo JA, Charman SA, Charman WN (2006) Methods to assess drug permeability across the blood-brain barrier. J Pharm Pharmacol 58:281–293. https://doi.org/10.1211/jpp.58.3.0001
7. Bicker J, Alves G, Fortuna A et al (2014) Blood–brain barrier models and their relevance for a successful development of CNS drug delivery systems: a review. Eur J Pharm Biopharm 87:409–432. https://doi.org/10.1016/j.ejpb.2014.03.012
8. Cucullo L, Hossain M, Puvenna V et al (2011) The role of shear stress in blood-brain barrier endothelial physiology. BMC Neurosci 12:40. https://doi.org/10.1186/1471-2202-12-40

Part IV

Microfluidic Chip Models of the BBB

Chapter 11

Novel, Emerging Chip Models of the Blood-Brain Barrier and Future Directions

Paul M. Holloway

Abstract

The use of microfluidic chips is now allowing for more advanced modelling of the blood–brain barrier (BBB) in vitro, recapitulating heterotypic interactions, 3D architecture, and physiological flow. This chapter will give an introduction to these new technologies and how they are being applied to model the BBB and neurovascular unit (NVU). A foundational understanding of the fluid dynamics germane to the effective use of these chips will be set and an overview of how physical phenomena at the microscale can be exploited to enable new possibilities to control the cell culture environment. The four main approaches to construct microfluidic blood vessel mimetics will be discussed with examples of how these techniques are being applied to model the BBB and more recently to study specific neurovascular disease processes. Finally, practical guidance will be given for researchers wishing to adopt these new techniques along with a summary of the challenges, limitations faced, and new opportunities opened up by these advanced cell culture systems.

Key words Microfluidics, Organ-on-chip, MPS, BBB, Neurovascular unit

1 Introduction

Microfluidic and "organ-on-chip" systems have emerged in recent years with the promise of offering more physiologically relevant in vitro microenvironments in which to study cell behaviors and the heterotypic interactions that make up the functional units of an organ. The blood–brain barrier (BBB) and neurovascular unit (NVU) can be considered a functional unit of the brain essential to maintaining brain homeostasis and the unique brain microenvironment. Disruption of the BBB is increasingly being recognized as an important feature of many neurological diseases, including Alzheimer's disease (AD), Parkinson's disease (PD), multiple sclerosis (MS), epilepsy, and stroke. On the other hand, the restrictive phenotype of the BBB is a significant obstacle in drug delivery, and overcoming this barrier has been the focus of a great deal of research. As such, having effective tools and model systems with

Nicole Stone (ed.), *The Blood-Brain Barrier: Methods and Protocols*,
Methods in Molecular Biology, vol. 2492, https://doi.org/10.1007/978-1-0716-2289-6_11,

which to dissect cellular mechanisms underlying BBB function and explore therapeutic opportunities and drug delivery strategies is greatly beneficial to tackling the global burden of neurological disease.

Microfluidic and "organ-on-chip" systems, from here on in referred to as microphysiological systems (MPS), offer a unique set of advantages over traditional in vitro cell cultures and are already being leveraged to gain novel mechanistic insights into the functioning of the BBB under physiological and pathophysiological conditions. Many of these advantages arise from the unique physics of fluid flow at the microscale and the ability to shape microenvironments at the cellular and even subcellular scale. However, with added complexity to the cell culture environment, there often also come new technical challenges for the end user. An appreciation of the underlying physics governing the chemical and physical microenvironment, along with an understanding of the benefits and limitation of MPS, provides an important foundation to effectively utilizing these tools which enable new insights into BBB function in health and disease.

This chapter gives an introduction to microfluidics, the technologies that have given rise to this rapidly advancing field, and a brief overview of fundamental physical phenomena germane to microfluidic cell culture. An overview of how these advanced microfluidic cell culture techniques are being applied to recapitulate the structure and function of the BBB and NVU will also be given, highlighting the strengths and limitations of each approach. Thus, this chapter will serve as an introduction to the new experimental possibilities provided by microfluidic technologies and how MPS may be used to study BBB function with enhanced physiological relevance.

2 Limitations of Established In Vitro Models: A Driving Force for Novel MPS

> All models are wrong, but some are useful. (George Edward Pelham Box 1979)

While there is a clear discrepancy, on many levels, between the high level of complexity of a human organ and the simplicity of an in vitro system that claims to model it, in vitro approaches do allow for simplification of a biological system and isolation of specific elements. This opens up a myriad of experimental possibilities for in-depth, and at times high-throughput, dissection of cellular and molecular mechanisms under defined conditions.

What might be considered as the primary advantage of in vitro systems is the ability to interrogate human cells. This benefit is particularly evident in the case of neurological disease where the possibilities to actively experiment on the human brain are extremely limited. While animal models have provided a great deal of insights into neurological diseases, particularly at the systems level, it should be noted that there are considerable interspecies differences, particularly with respect to the brain. In the BBB species-specific differences in efflux transporters, tight junctions and cell–cell signalling have been reported [1, 2]. Human astrocytes are three times larger and have ten times more processes than those of rodents [3, 4]. Astrocytes show a number of distinct genomic and functional traits across species [5], including differences in immune activation [6] and responses to extracellular glutamate [5], potentially having implications on studies of excitotoxicity and CNS disease. Homologous human and rodent neurons also show differences in electrical activity [7], gene expression, and morphology [8]. Immunological responses are increasingly being acknowledged to contribute to an array of neurological diseases. Along with substantial difference in the relative abundance of leukocyte subtypes [9], mouse and human leukocyte genomic responses to immune stressors have shown to be vastly different [10]. Even subtle differences in immune function may cause substantially altered responses, as became evident in the TGN1412 trial, where experiments in macaques failed to predict adverse immunological events. As with any model, it is therefore important to know the limitations of the system when interpreting results.

Despite the key advantage of enabling study in human cells, in vitro models often fall short in their physiological relevance. In a keynote given at the Organ-on-Chip World Congress 2018 in San Diego, John Wikswo articulated an observation that the majority of insights into cellular and molecular biology have up until now been made using "2D biology on plastic" and that:

> Many biological experiments are conducted on cells that: have cancer, are inbred, diabetic [couch] potatoes on a stiff plastic couch without exercise, . . . gorge themselves on sugar once a day . . . live in their own excrement, . . . take a complete or only partial bath every day or two, and talk only to cells of like mind. (Wikswo 2018)

While this is a somewhat facetious summary of traditional in vitro approaches, which have facilitated a myriad of fundamental insights into cell biology, it does highlight how often the cellular microenvironment has been overlooked in cell culture experiments. There is now an increasing awareness that not only the chemical and cellular makeup but also physical microenvironment has significant impacts on cellular phenotype.

The significance of the in vivo microenvironment in maintaining BEC properties was realized early on in BBB research, with observations that rat cerebral capillary fragments placed in culture gave rise to confluent monolayers but progressively lost their endothelial phenotype over time [11]. Since then numerous co-culture experiments have demonstrated that BBB function arises from complex heterotypic cellular interactions. For example, pericytes and astrocytes, or their conditioned media, can upregulate tight junction protein expression, apico-basal polarization of transporters and increased transepithelial/transendothelial electrical resistance (TEER) [12–14].

The mainstay of in vitro BBB models has been the use of Transwell models which allow for the partitioning of heterotypic cell cultures while providing moderate scalability and high-throughput screening capabilities. However, Transwell systems are limited by their planar architecture, with the presence of relatively thick (10–50 μm) membranes that prevent a large amount of direct heterotypic cellular interaction (with pores ranging from 0.4 to 3 μm diameter) and often display "edge effects" with high permeability at the walls [15]. In addition to being limited in their recapitulation of BBB 3D architecture, Transwell systems often fail to incorporate flow and the influence of wall shear stress (WSS), which has been shown to have a multifaceted influence on BEC genotype and phenotype.

With an increasing understanding of how the physical and chemical microenvironment shapes BBB function, microfluidic cell culture and MPS are now being developed to exploit advances in microfabrication techniques and shape BBB microenvironments at the cellular scale to provide in vitro models that promise greater physiological and pathophysiological relevance.

3 An Introduction to Microfluidics and Physical Phenomena at the Microscale

3.1 A Brief History

There's Plenty of Room at the Bottom. (Richard Reynman 1959)

This now famous lecture given by Richard Feynman in 1959 anticipated an exploration of engineering at the micro- and nanoscale, highlighting the possibilities afforded by miniaturization. While fluidics and biological implications of microscale technologies were not mentioned, microfluidics has a shared origin with the microelectronics that were beginning to be realized in 1959. Born out of the drive to miniaturize electrical circuits, the technique of photoengraving technique patented by Jay Andrus in 1957 and further developed by Jack Kilby at Texas Instruments in 1964 [16] (for which he was awarded the Nobel Prize in 2000) has allowed for the production of micro-/nanoscale architectures to route electrons, revolutionizing commuting and giving birth to the "silicone age." These techniques pioneered by the microelectronics industry have since been adopted to create miniature channels for the transport of liquids, molecules, and cells rather than electrons. As such microchips share their name with "lab-on-chip" and "organ-on-chip" technologies.

Initial developments in microfluidics were largely in the field of chemical analysis, such as microscale gas chromatography in 1979

[17] widely regarded as the first "lab-on-chip" system. In a seminal paper in 1990, Manz et al. laid out how the physics of fluid mechanics at the microscale could be leveraged to great effect in the conceptualization of a "miniaturized total analysis system" (μTAS) for automated sampling, transport, and analysis [18]. The development of microfluidic systems was subsequently accelerated by large-scale funding initiatives in the 1990's both from the Defense Advanced Research Projects Agency (DARPA) in the USA to develop portable analysis devices capable of detecting biological and chemical weapons [19] and the global Human Genome Project [20]. A critical development in the adoption of microfluidics for cell culture systems and in vitro modelling was the use of the elastomeric material poly(dimethylsiloxane) (PDMS) pioneered by George Whitesides' group at Harvard University [21]. Termed "soft lithography" these techniques allowed for replica molding of microstructures created using silicone chip photolithography techniques into a flexible, and crucially transparent, polymer that could be easily bonded to a glass slide to complete microscale channels. This process allowed for the rapid fabrication and prototyping of fluidic channels outside of a clean room. Furthermore, the gas permeability and relative biocompatibility of PDMS allowed for integration of cell culture. Elastomeric properties of PDMS were also exploited in the development of "Quake valves" by Stephen Quake's group at Stanford University [22] and more recently in "organ-on-chip" MPS to mimic the cyclic stretch experienced by cells in the lung alveoli, gut, and blood vessels. The results of this work pioneered by the Ingber Lab [23, 24] gave rise to the now commercially available Emulate chip.

A number of other fabrication techniques have since been developed including 3D printing and lasers for micromachining (reviewed elsewhere [25]). However, soft lithography remains the mainstay for rapid prototyping of microfluidic devices and can be easily used for the production of custom designs for novel application.

3.2 Advantages of Microfluidics over Macroscale Culture

Microfluidic cell culture techniques have garnered much attention for the development of in vitro models, due to a unique set of advantages that go beyond the reduced consumption of scarce or expensive reagents. Microfluidic devices provide the opportunity to precisely define micro-/nanoscale architecture to structure and pattern cells while providing control of fluid flows in the nano- to femtoliter range for temporospatial shaping of chemical and physical ques. For in-depth reviews of the advantages and limitations of microfluidic systems for general cell culture, *see* refs. [26, 27].

3.2.1 Defining Geometries at the Scale of the Cell

Microfluidic devices provide a physical architecture at cellular and subcellular scales, allowing for physical features to be used to pattern, confine, and direct cells. This enables precision investigation of cell–cell interactions, para-juxta and autocrine signalling, improved automation of image analysis, and the highly controlled study of chemotaxis, locomotion, axonal guidance, and trafficking.

Utilizing cell traps and physical confinement of cell researchers have been able to study contact and proximity critical cell–cell signalling such as immune synapse formation, myelination, and synaptic activity [28, 29]. This ability to grow cells in devices with precisely defined physical barriers has been particularly useful in the field of neuroscience where devices that guide axon growth, separating soma and axon to allow the formation of simplified circuits, have been widely exploited to garner a variety of new biological insights [30–38].

MPS geometries can also be used to define 3D microenvironments by retaining gels within set boundaries [39–41] or even aligning fibrous ECM nanostructure to guide cell growth [42]. A 3D environment and the mechanical properties of substrates are increasingly being recognized as influencing cellular phenotype [43–45]. Yet plastic and glass substrates typically used in traditional cell cultures have a high mechanical stiffness, in the gigapascal range, many orders of magnitude higher than the Young's modulus of the brain at ~1–2 kilopascals [46]. Even simply adding an extra dimension to cell culture has been shown to influence cell behaviors. Astrocytes cultured in 3D matrigel exhibit more in vivo-like ramified morphology responsive phenotype allowing more effective study of reactive astrogliosis such as in neuroinflammation when compared to cells grown on a 2D matrigel surface [47]. The native ECM however comprises a complex set of biophysical (topography, stiffness, and porosity) and biochemical (nutrients, matrix components, and sequestered signalling molecules) cues that far extends beyond a simple cell scaffold. Many microfluidic models are now incorporating 3D microenvironments that will allow further exploration of the specific roles the matrisome plays on cellular functions and may even reveal novel therapeutic targets.

3.2.2 Control of Physiologically Relevant Flow

While low volumes and high-density cultures mean metabolite buildup can be rapid in MPS, introduction of flow facilitates removal of waste and continuous exchange of nutrients, growth factors, and O_2, as is the case in vivo. As such, it is important that users of MPS acquaint themselves with the concepts of effective culture volume and a critical perfusion rate as described by Walker and Beebe [48, 49]. Introducing fluid flow to microfluidic cultures has furthermore enabled new insights into mechanotransduction, demonstrating how wall shear stress (WSS) can act as a pleiotropic modulator of BBB phenotype.

In human BEC, shear stress has been shown to significantly increase the expression of tight junctions [50–52], multidrug resistance transporters, ion channels, several p450 enzymes, and carrier-mediated transport systems [53]. Furthermore shear stress has been found to suppress glycolytic bioenergetic pathways in favor of aerobic respiration while inhibiting the cell cycle and apoptosis and reducing cell turnover in monolayer cultures [53, 54]. Interestingly unlike endothelial cells (ECs) of other vascular beds, brain endothelial cells (BECs) have been found not to elongate and align under flow, potentially as a mechanism to reduce junctional area [54, 55]. Crucially many studies have found that overall, BECs acquire more stringent BBB properties under flow than those observed in static platforms [56]. Indeed, side-by-side comparisons with bovine ECs cultured with or without shear stress (4 dynes/cm^2) in a hollow fiber microfluidic system showed an increase in TEER by over tenfold [57]. Similar results have been obtained in rat primary BMECs in a 3D microfluidic device, while hCMEC/D3 monolayers subjected to pulsatile flow in a capillary cartridge system showed highly restrictive TEER values of 1000–1200 $\Omega \cdot cm^2$, which was found to rapidly drop following flow cessation [58, 59]. The mere loss mechanical shear input as a result of flow can result in leukocyte-mediated cytokine release and BBB disruption [60]. Such findings have considerable implications to in vitro BBB mimetics. For instance, in modelling stroke the accompanying perturbations in fluid flows have been largely overlooked. Microfluidic systems offer the advantage of enabling precise control of flow and shear forces, owing to the laminar flow regime discussed in detail below; furthermore, flow can be maintained over relatively long timescales without the use of prohibitively large volumes of media.

The laminar flow that prevails in microfluidic devices also means that convective mixing is essentially absent and thus spatial and temporal chemical milieus can be shaped within the same cell culture platform. As examples, microfluidic devices have enabled the production of stable chemical gradients to study angiogenesis and vasculogenesis [61, 62] and microglial chemotaxis [63] and reveal mechanisms of axonal guidance [64].

3.2.3 Integration of Sensors for Real-Time Analytics

Many MPS are now also harnessing the potential to integrate electrical sensors for the real-time, longitudinal measurement of TEER, impedance, and electrophysiological parameters. There are now a number of examples of MPS which integrate electrodes above and below cells cultured on porous membranes (analogous to a Transwell) [65–69]. Owing to the planarity of the technology, MPS are increasingly leveraging the power of microelectrode arrays (MEAs) [70, 71]. When coupled with the ability to accurately pattern cells and shape chemical environments, MEAs can provide high-content, spatially resolved, noninvasive cellular and environmental monitoring such as barrier function or neuronal activity [72, 73]. Maoz et al. have even demonstrated the capability to simultaneously measure cellular electrical activity and barrier function integrating both multi-electrode array and transepithelial electrical resistance onto a single device [74]. Multi-parametric sensing, including electrochemical enzymatic sensing of environmental factors such as glucose and lactate, along with CO_2 production and O_2 consumption, is also rapidly advancing. Integration of such technologies could yield a detailed picture of cell metabolism within MPS [75]. The majority of these systems, however, rely on flat, rigid electrodes, and the movement of many MPS into 3D microenvironments presents a significant challenge—one that may in the future be met by the next generation of flexible polymeric and transparent sensors [76–78].

3.3 Understanding Physical Phenomena at the Microscale

Understanding a few simple mathematical descriptions of fluidic behaviors at the microscale enables an insight into microfluidic design and operation, allowing appropriate interpretation and analysis of results obtained from MPS. Physical phenomena that dominate at the microscale are laminar flow, diffusion, fluidic resistance, interfacial and surface tension, and surface area-to-volume ratio [79].

While microfluidics is a relatively young field, the physical principles of fluid flows that underpin our understanding were developed in the 1800s. While the first microfluidic device to take advantage of photoliographic microfabrication, the inkjet printer, was developed in the 1960s [80], this technology was based upon fundamental principles described by Walter Raleigh in 1879 [81]. Many of the advantages of microfluidics are due to the fact that microfluidic flows are not turbulent and lack the eddies and vortices that mix and fold fluids at larger scales [82]. At the microscale, flows are ordered in layers of fluid streams or "lamina," not only making these flows mathematically predictable but also allowing for the control of gradients of solutes, reduced diffusion times, and rapid reaction rates. This enables environmental perturbations to be matched to the time and spatial scales of biological system.

It is important to understand that at the microscale, viscous forces dominate over inertial forces, so much so that flows can even be considered as time reversible. As an example, an organism must utilize flagella and cilia that execute nonreciprocal motions to be able to swim at the microscale, since utilizing methods that are familiar at the macroscale, such as the back and forth of a fish's tail or the opening and closing of a scallop, would result in no net forward movement. Such macroscale methods of propulsion rely on inertia. This is known as Purcell's scallop theorem. The ratio of inertial forces to viscous forces in a fluidic system is described by the dimensionless Reynolds number (Re) which is given in Eq. 1.

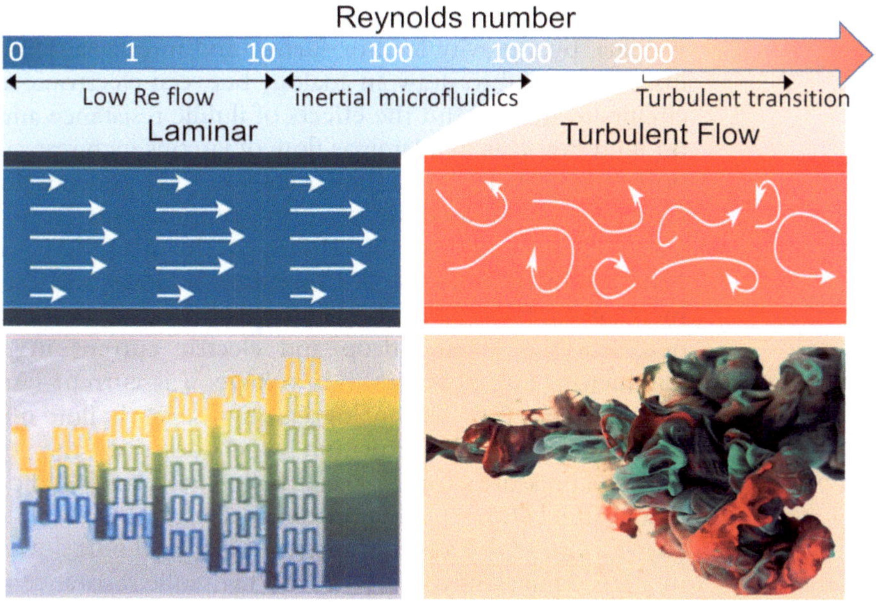

Fig. 1 Reynolds number of laminar flow. At low Re viscous forces dominate, and flows remain laminar so that fluidic streams do not mix by convective forces but only by molecular diffusion. At high Reynolds number at macroscales or high velocities, flows become turbulent, and inertial forces dominate creating chaotic mixing of fluid streams

$$\mathrm{Re} = \frac{(\text{inertial force})}{(\text{viscous force})} = \frac{\rho v\, D_{\mathrm{h}}}{\mu} \tag{1}$$

where ρ is the density of the fluid, ν is the velocity, μ is the dynamic viscosity, and D_{h} is the hydraulic diameter of the channel, as given by Eq. 2.

$$D_{\mathrm{h}} = \frac{2wh}{2w + 2h} \tag{2}$$

where w is the channel width and h the height. As the dimensions of the system are reduced, it is evident that the Reynolds number is also reduced. This number can be used as a descriptor of the flow regime. At high Re, (typically >2000) inertial forces dominate, and flow is turbulent, while at low Re (<2000), viscous forces are more important, and flow is laminar (Fig. 1). In microfluidic systems flow is typically at low Re (<10), well within the laminar flow regime. As such mathematical descriptions of flow are greatly simplified.

The Navier–Stokes equations are a set of partial differential equations which describe the velocity field of fluid flow. In microfluidic flows since inertial forces have little influence, these terms can be omitted, thus simplifying the nonlinear Navier–Stokes equations to the linear Stokes equation, as outlined by Oh et al. [83].

Within the microfluidic systems and MPS, it is often desirable to be able to predict fluid flow velocities based upon an applied pressure, either by a pump or, in the case of passive pumping systems, by a gravity feed or surface and interfacial forces. In this case, it is useful to draw an analogy between electronic and fluidic circuits to understand the effects of fluidic resistance and pressure on fluid flows, as in a laminar flow of viscous incompressible fluid, both systems behave in a similar manner [83]. Indeed, fluidic flows have even been used to perform computations from hydraulic algebraic machines in the 1900s to the MONIAC (Monetary National Income Analogue Computer) in 1949 and the development of fluidic logic in the 1960s [84]. In electronics Ohm's law describes the voltage drop and electric current in a resistive conductor (Eq. 3) where V is voltage, I is current (Amps), and R is resistance (Ω). The analogous physical law in fluid mechanics is the Hagen–Poiseuille equation which describes the relationship between pressure drop, volumetric flow rate, and fluidic resistance of a channel (Eq. 4), where pressure drop across the fluidic circuit (Δp) *measured in Pa* is equivalent to voltage, fluidic flow rate (Q) in m^3/S is equivalent to current, and hydraulic resistance (R_h) in Pa s^3/m is comparable to electrical resistance.

$$\text{Ohms law}: V = IR \tag{3}$$

$$\text{Hagen} - \text{Poiseuille's law}: \Delta P = Q\ R_h \tag{4}$$

The resistance of a circular channel can be calculated using the formula

$$R = \frac{8\mu L}{\pi\, r^4} \tag{5}$$

where L is the channel length and r is the channel radius. However, many microfluidic channels, due to limitations in fabrication, have a rectangular cross section. Determining the hydraulic resistance of a rectangular channel is more complex but can be approximated using the following equations:

$$\text{Where } w > 10h: \quad R \cong \frac{12\mu L}{wh^3} \tag{6}$$

$$\text{Where } w < 2h: \quad R \cong \frac{32\mu L}{wh^3} \tag{7}$$

In a simple microfluidic device where the fluid flow is driven passively by a Colum of liquid at the inlet, ΔP is given by subtracting the outlet pressure from the inlet pressure where P is

$$P = h \times g \times \rho \tag{8}$$

where h is the height of the fluidic head, ρ is the density of the fluid (water = 998 kg/m^3), and g is gravity (9.8 m/s^2). The above equations allow a researcher to estimate the laminar flow rates within an MPS. Under these conditions (low Re flow), there is no convective mixing, and mixing of dissolved molecules between fluid streams is dictated primarily by molecular diffusion, resulting in predictable kinetics [16]. As such flow rates can be tailored to effectively isolate fluidic streams that are in direct contact with one another, and concentration gradients can be precisely defined. The Péclet number (Eq. 9) is a dimensionless number that describes the relative importance of advective versus diffusive transport on a molecule.

$$\mathrm{Pe} = \frac{vL}{D} \tag{9}$$

where D is the experimentally defined diffusion coefficient of a molecule in a solution (m^2/s). Reducing the scale of a system results in a reduced Pe, making the kinetics more predictable. Diffusion can be approximated in a single dimension by the following equation:

$$x^2 = 2Dt \tag{10}$$

where x is the distance a molecule moves in a given dimension over a period of time t. Given that diffusion scales to the square power of distance, diffusion rates are dramatically different between macro- and microscale cell culture systems. The following example provided by Bebee et al. [79] illustrates this: hemoglobin ($D = 7 \times 10^{-7}$ cm^2/s) in water takes over 11 days to diffuse 1 cm in the absence of mixing; however, the same molecule under identical conditions only takes 1 s to diffuse 10 μm. This allows for rapid reaction times and exchange of nutrients and oxygen as well as cellular signalling molecules in MPS. While diffusion is rapid in MPS, under dynamic conditions flow rates can be used to tailor molecular transit times and thus shape the chemical environment with high precision.

Calculating flow rates and rates of diffusion allows for an appreciation of the delivery of nutrients and removal of waste products to and from cells in a micro-device, along with the influence and timescales of paracrine and autocrine signalling events. To understand how fluid flows may also exert physical force upon a cell, initiating mechanotransduction pathways and signalling events, it is useful to convert fluid flow rates into wall shear stress (WSS). Wall shear stress is the force per unit area exerted by a fluid on a solid boundary and in biological systems is typically expressed as dyne/cm^2 (1 Pa is 10 dyne/cm^2). WSS is calculated as follows:

$$\tau = 10\,\mu \frac{du}{dy} \tag{11}$$

where τ is WSS (in dyne/cm^2), μ is the fluid viscosity, and du/dy is wall shear rate (the rate at which fluid layers or laminae move past each other). In a channel of cylindrical cross section, the velocity profile is symmetrical, and wall shear rate can be calculated as follows given by

$$\frac{du}{dy} = 8\frac{v}{D} \tag{12}$$

where v is the mean fluid velocity (m/s, which can be calculated from the volumetric flow rate), D is the diameter of channel (m), and du/dy is wall shear rate (s^{-1}). However, in many MPS channels are rectangular. In this case du/dy can be approximated for h/w ratios up to 1 using Eq. 13 where V_{max} is the max velocity at the centerline (1.5 × mean velocity), y is the distance from the centerline flow to the wall, and h is the channel height. WSS can then be calculated by Eq. 11.

$$\frac{du}{dy} = 8\frac{v_{max}\,y}{h^2} \tag{13}$$

WSS plays an important role in shaping endothelial cell behavior and BBB phenotype [53, 54], along with intravascular adhesion events such as leukocyte capture, adhesion and transmigration [85], and platelet activation and thrombosis [86, 87]. Obtained WSS can thus be used to benchmark the physiological relevance of a system [88] and in understanding the thresholds for physiological and pathophysiological responses to shear. In the arterial tree, the wall shear stress ranges between 10 and 70 dyne/cm^2 [89, 90] with oscillations in blood pressure during the cardiac cycle, resulting in pulsatile arterial blood flow. However, these fluctuations are progressively ***damped*** owing to the elasticity of vessel walls. In the human bulbar conjunctiva, erythrocyte velocity has been used to estimate a mean wall shear stress of 1–6 dyne/cm^2 in capillaries and post-capillary venules ranging in size from 4 to 20 μm [88, 91]. Interestingly, aging has been shown to alter cerebral arterial elasticity and the dampening of pulsatile flow which has been postulated as a contributor to microvascular damage [92]. MPS BBB models could readily be applied to such questions, utilizing the precise control of fluid flow that can be obtained in microfluidic systems using electronically controlled pumps.

At the microscale surface effects of a liquid can have significant effects on fluid behavior. Surface tension (the tendency for a liquid to minimize its surface to air interface to reduce its free energy) and interfacial tension (the same phenomena but for two immiscible fluids, such as oil in water) have been widely exploited in microfluidics. Interfacial forces have notably been used in the rapidly growing field of droplet microfluidics [93] but also for the creation of virtual walls where fluidic circuits are confined by a stable interface with another liquid rather than a solid wall [94]. Surface tension can be utilized to form valves for automated liquid control, to pin liquids in a specified pattern (as in protocols where surface tension confines a cell-laden gel to a central channel [95–97]) as well as in passive pumping mechanisms based on Laplace pressure [98]. The pressure generated by a liquid surface with perpendicular radii of curvature *R*1 and *R*2 can be calculated with the Young–Laplace equation described in greater detail here [99]. Although it might seem counterintuitive at first, in the case of a passive pump, a small droplet at the inlet of a microchannel will force liquid through the channel to a larger droplet at the opposite end. This is occurs since a small drop has a higher internal pressure than a large drop. This phenomenon has been used in conjunction with Saffman–Taylor instability to pattern lumens through ECM hydrogels in the creation of blood vessel and BBB models [100, 101].

Capillary forces can also play a significant role in fluid behavior at the microscale and have been harnessed in many analytical devices such as in pregnancy tests and blood glucose meters and also widely within the field of paper-based microfluidics [102]. However, while liquid-surface interactions are important forces at the microscale, capillary forces are not heavily utilized in MPS and BBB models and are outside the scope of this chapter.

While computational fluid dynamics (CFD) modelling can be used to precisely predict more complex behaviors of fluids and dissolved species in MPS, the above examples provide a basic foundation for MPS users to build an understanding of fluid flows at the microscale.

4 Modelling the BBB Using Microphysiological Systems

> If the human brain were so simple that we could understand it, we would be so simple that we couldn't. (Emerson M. Pugh 1977)

The brain is arguably one of the most complex known structures, and as such modelling this system to recapitulate relevant features and functions is extremely challenging. Our incomplete understanding of the relative importance of structural features in the brain along with the exact molecular basis of neurological disease states further hinders the establishment of reliable models that are able to predict drug delivery, efficacy, and toxicity. Nevertheless, being able to distil complex system into distinct functional units that can be then dissected into its constitutive components has aided in developing our understanding.

The BBB provides a multimodal barrier, acting as (1) a gating barrier to prevent paracellular diffusion (tight junctions, etc.), (2) a transport barrier utilizing active efflux systems (such as multidrug resistance proteins), and (3) a metabolic/enzymatic barrier (such as monoamine oxidase and cytochrome P450 enzymes) which catalyze the breakdown of xenobiotic substances such as drugs and other potentially toxic chemicals [103, 104]. Expanded to include the surrounding cellular milieu (astrocytes, pericytes, neurons, and microglia), as well as extracellular components such as the ECM and basement membrane, it allows a more complete understanding of the BBB as a functional unit termed the "neurovascular unit" (NVU) [103]. This essential functional unit of the brain maintains a restrictive, selective, and dynamic barrier which maintains the unique brain microenvironment. Effective models of the BBB not only allow for investigation of the mechanisms and consequences of BBB breakdown and dysregulation in diseases but also provide a platform to explore strategies to overcome this significant obstacle to drug delivery to the brain. Various components of the NVU have been shown to influence BEC BBB properties, and as such many BBB models incorporate elements of the NVU to promote barrier properties and enhance physiological relevance. As such many of the models detailed below are best considered as partial NVU models, which can enable investigation of BBB function.

The application of microfluidic cell culture techniques to model specific functional units or organs has led to the development of various "organ-on-chip" models. Here these are referred to as microphysiological systems (MPS) to distinguish that these systems do not replicate the form or function of an entire organ, but enable the study of cell function in a quasi-physiological microenvironment.

A number of distinct approaches have been made to model the BBB in MPS, which can be broadly grouped into (1) devices that use membranes to separate compartments, (2) those that use surface tension to create membraneless interfaces, (3) those that use sacrificial molding to create truly 3D vessel mimetics, and (4) those that exploit cellular self-assembly (angiogenesis and vasculogenesis). Each approach presents its own benefits and limitations and should be benchmarked according to the parameters being studied [88].

4.1 Membrane-Based Compartmentalized MPS

Dual-layered microchambers separated by porous membranes (Fig. 2a) have enabled MPS to model a wide variety of compartmentalized biological systems, as pioneered by the Ingber group at Wyss Institute for Biologically Inspired Engineering at Harvard University with their lung-on-chip device [105]. This microfluidic device consists of stacked flow/culture channels separated by a thin porous membrane. These channels are also flanked by pneumatic channels that allow for controlled cyclic stretch to mimic breathing, peristalsis, or the cardiac cycle. This system is now available from Emulate and has further been applied to model the intestine [106], liver [107], kidney [108], and spinal cord [109]. Emulate has now recently released a commercially available "Brain-Chip" model of the BBB claiming 35% closer overlap compared to traditional Transwell culture using the same constituent cells. Other dual-layered MPS are also commercially available, such as the

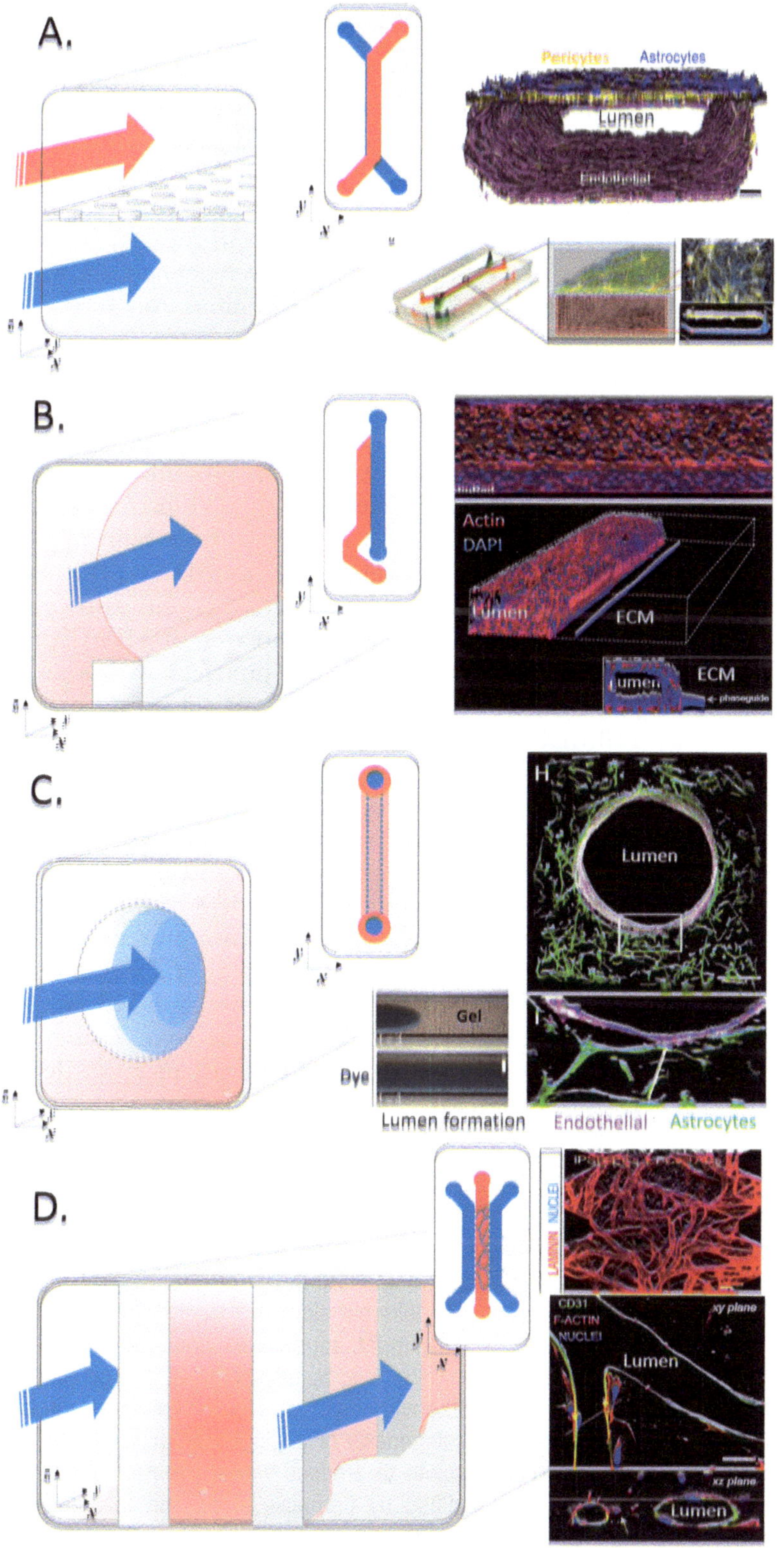

Fig. 2 Microfluidic approaches to modelling the BBB. (**a**) Membrane-based dual-layer systems such as the Emulate chip allow for flexibility and integration of TEER sensors. Examples of models taking this approach include Park et al. (2019) (top right, reproduced from [163] under Creative Commons License 4.0) and Maoz

BI/OND CHIP, while many other groups rely on in-house production for custom chip setups.

Such membrane-based multichannel flow systems mimic Transwell-like connectivity but show greater compatibility with microscopy (owing to the low-aspect-ratio channels and optical transparency) while also crucially allowing for individual control of perfusion both in the vascular and parenchymal channels [65, 66]. Compartmentalization in these systems allow neurovascular microenvironments to be well defined, allowing easy sampling of vascular and parenchymal regions for permeability and drug transport studies while also offering the possibility integration of electrodes to monitor TEER. Griep et al. have used such a system to demonstrate responsiveness of hCMEC/D3 BECs to physiological and pathological stimuli. This work demonstrated WSS to increase the expression of tight junction proteins and elevate TEER by a factor of 3, while challenge with TNF-α treatment reduced TEER by a factor of 10 [66]. Ahn et al. have built upon this multilayer approach by incorporating a 3D co-culture of Matrigel encapsulated human brain pericytes and astrocytes in the lower chamber, with an apical monolayer of HBMEC cultured on an extremely fine membrane of 7 μm thick with 8 μm-diameter pores at a density of 1×10^5 pores/cm^2. Flow was maintained over the HBMEC layer at 16 μL/min to give a physiologically relevant shear stress of 4 dyne/cm^2. In this system Ahn et al. demonstrate co-culture upregulates occludin, zonula occludens-1, and vascular endothelial cadherin, along with the expression and membrane transporters glucose transporter 1 and cholesterol efflux regulatory protein [47]. They additionally demonstrate polarized expression of aquaporin-4 in the BBB chip, not present in 2D cultures. Here AQP4 was localized to the astrocytic end feet in the vascular side of the channel and was significantly induced in the presence of HBVPs, as previously observed in vivo. Using this model Ahn et al. were furthermore able to study receptor-dependent transcytosis of nanoparticle coupled drug delivery [47].

Fig. 2 (continued) et al. [111] (bottom right, figure modified from [111] with permission from Springer Publishing). (**b**) Pinning of a cell-laden ECM gel using micro-pillars or PhaseGuides™ (illustrated here), as with the MIMETAS OrganoPlate, allows for easy incorporation of 3D parenchymal cell culture and direct vascular contact, as used by Wevers et al. [50] (right, reproduced from [50] under Creative Commons Attribution 4.0 International License). (**c**) Lumen patterning by displacement of a gel (here viscous fingering) allows for circular large-scale lumen in a fully 3D environment, as demonstrated by Herland et al. [101] (right, figure adapted from [101]). (**d**) Exploiting vasculogenesis or angiogenesis utilizing gradients of angiogenic factors or interstitial flow can direct vessel growth to produce vessel geometries that are defined by the physiochemical and cellular milieu as exemplified by Campisis et al. and the AIM Biotech chip (right, reproduced from [41] under Creative Commons license)

Recent work has also utilized iPSC technologies in the Emulate chip to create an isogenic NVU model, which demonstrated high, physiologically relevant TEER (>1000 ohms/cm^2) maintained over several days under physiological shear [110], along with the appropriate expression of ZO-1, occludin, claudin-5, and the BBB glucose transporter GLUT-1. Barrier function within this model was demonstrated to protect cultured neurons from plasma-induced toxicity upon whole blood perfusion, with IgG and albumin also being confined to the blood side, while transferrin accumulated on the brain side indicating a recapitulation of transferrin receptor-mediated transcytosis. Furthermore modelling vascular inflammation using TNF-α, IL-1β, or IL-8 was found to alter ZO-1 expression leading to dose-dependent barrier breakdown, as measured by fluorescent 3 kDa dextran [110].

Such dual-chamber chips can also be connected in series to provide flexibility of experimentation. Maoz et al. fluidically connected three separate chips: a BBB chip connected to a hES-derived neuronal culture "brain chip" upstream of a second BBB chip. This allowed perfused metabolites to be tracked across the BBB, into brain compartment along with their efflux out of the subsequent BBB chip. This work demonstrated a previously unknown metabolic coupling between BBB and neurons, whereby endothelial and perivascular metabolites were able to directly influence neuronal GABA and glutamatergic neurotransmitter synthesis [111]. Such systems may allow for an exploration of the links between neurometabolism and neuropathologies such as stroke, AD, and TBI.

Disadvantages of dual-layer compartmentalized MPS relate to the inclusion of an artificial barrier between the vascular and parenchymal side, limiting the extent of direct neurovascular interaction and mimicry of capillary-level vessel ensheathment. Non-physiological channel geometries also mean that BECs are grown in flat monolayers and show flow dead zones at the channel edges.

4.2 Exploiting Surface Tension for Membraneless Interfaces

One method to allow for direct cellular contact at interfaces without the artificial presence of a membrane is to make use of surface tension at the microscale to pin a hydrogel at a gel–air interface during polymerization. In such cases, abrupt changes in channel geometry using PhaseGuides™ [112] or micro-pillars [67] pin a cell-laden gel to a specified region as is the case in the B^{3}C BBB model [67] or the commercially available MIMETAS OrganoPlate® [50] and SynVivo SynBBB models [113]. More recently a method using local surface modification through directing laminar flow has been used to the same effect, without the need for PhaseGuides™ or pillars, but has not yet been employed for BBB modelling [114].

Membraneless interfaces allow for direct contact of BEC with perivascular cells and astrocyte end feet within well-defined channel geometries. Deosarkar et al. show that such an approach can be used to co-culture astrocytes (in 3D) and BEC, resulting in significantly enhanced barrier function as measured both by TEER and 40 kDa dextran leakage, when compared to equivalent Transwell models [67]. Adriani et al. have also used micro-pillars, in this case, to create two contacting central cell-laden hydrogels of separate neuronal and astrocyte cultures in direct contact (in collagen type I, 2.5 mg/mL and 7 mg/mL for neurons and astrocytes, respectively). The central gel channels were flanked by media channels, one of which was lined with human BECs, resulting in junctional VE-cadherin and ZO-1 expression [115]. Interestingly in this model HUVECs in co-culture show a more restrictive barrier to 10- and 70-kDa dextrans compared to monocultures while with a different EC cell source (hCMEC) resulted in barriers in co-culture showing higher permeability than monoculture, highlighting the importance of cell source. Additionally, it should be noted that this model mixed cells from different species, and while they were able to integrate an active neuronal network which allowed for investigation of the influence of perfused drugs on neuronal function, this model omitted pericytes.

While micro-pillars use constrictions in the *x/y* plane to pin fluids, PhaseGuides™ utilize a change in the *z* dimension of the channel (Fig. 2b) as incorporated into the MIMETAS OrganoPlate®. Wevers et al. use this system to develop a BBB model for the study of receptor-mediated transcytosis [50]. This system uses PhaseGuides™ to pin an ECM hydrogel in two- or three-lane microfluidic cell cultures in a 384-well plate format, containing 96 or 40 chips, respectively. Flow (producing ~ 1.2 dyne/cm^2) is provided by a gravity feed which is maintained by placing the plate on a timed rocking platform inside the incubator. In their study Wevers et al. use TY10 BEC culture separated by a collagen gel interface from a co-culture of primary human brain pericytes and astrocyte cell line hAst to demonstrate the potential of antibody targeting of the transferrin receptor as a means of drug delivery [50].

A drawback of approaches using PhaseGuides™ or micro-pillars to define a vascular channel is that only one side of the BEC culture is in contact with the parenchyma, and vessel cross sections are not circular.

4.3 Sacrificial Molding for 3D Hydrogel Models

The cylindrical geometry of microvessels means that cells experience curvature inversely proportional to the diameter, and, in brain capillaries, BECs even wrap entirely around and form tight junctions with themselves. Given BECs do not align to flow, this has been proposed to minimize the area of cell–cell junctions to maximize barrier function [88]. Three-dimensional cylindrical vessels also allow for ensheathment of astrocytes and pericytes, an important feature of the BBB that allows for neurovascular coupling.

One approach to create BBB MPS with circular lumens has been through the use of sacrificial molds or removable templates within a cast hydrogel (e.g., the removal of a needle to reveal a channel [116] or use of a dissolvable sugar template [117–119]). This technique has been used to study the role of pericytes in angiogenesis [120] and also to demonstrate that SARS-CoV-2 spike protein alters barrier function in primary human BEC grown under low shear (0.7 dyne/cm^2) in a gel comprised of type I collagen, hyaluronan, and Matrigel [121]. Another method to form cylindrical lumen within a hydrogel is to make use of Saffman–Taylor instability (also termed viscous-fingering) in which a flow of a lower viscosity fluid can replace a denser hydrogel, to drive out a fully circular cross-section lumen during gel polymerization (Fig. 2c). Herland et al. were able to use this technique in straight microchannels to reproducibly create cylindrical lumen with a diameter between 600 and 800 μm, in type I collagen gels, depending on the flow conditions used [101]. This approach allowed for endothelial lined vessel mimetics to be co-cultured with pericyte/astrocyte 3D cultures. Endothelial cells were shown to secrete their own underlying type IV collagen-containing basement membrane at the gel interface and expressed ZO-1-containing tight junctions. When compared with static monocultures and co-cultures of the same cells cultured in Transwell plates, this 3D model was shown to have a significantly more restrictive barrier phenotype (when measured using 3-kDa dextran), indicating that the 3D microenvironment promoted barrier function. Inflammatory cytokine release in response to TNF-α stimulation also differed between models, with the 3D system being much more responsive, with distinct secretion profiles detected depending on the presence of astrocytes or pericytes [101]. De Graaf et al. have since further refined techniques using Saffman–Taylor instability for vessel patterning technique, for scalable production of blood vessel models with lumen patterning success rates of over 90% giving a consistent lumen diameter averaging 336 ± 15 μm [122].

A limitation to this approach however is that the lumen produced is comparatively large relative to the average human brain capillary which has been reported to be ~6 μm [123, 124]. As a result, in the absence of high volumetric flow rates, physiological shear rates are difficult to sustain over long-term culture. Importantly, this scale prohibits pericyte or astrocyte ensheathment, which may limit this technique in recapitulating capillary-level BBB physiology. Multistep molding may also be used to produce more complex vasculature [125, 126]; however, this requires technically challenging fabrication techniques, and there are currently no commercially systems available that use these techniques.

4.4 Leveraging Cellular Self-Assembly

A method that is increasingly being explored to model a number of different vascular beds [96, 127, 128] is to promote EC cellular self-organization for the formation of 3D vascular networks with in vivo-like morphology and scale (Fig. 2d). By using meniscus pinning techniques to confine a hydrogel flanked by perfusion channels, gradients of growth factors or interstitial fluid flow can be employed to stimulate vasculogenesis or angiogenesis (depending on whether ECs are encapsulated within the gel, or outgrowth is stimulated from lined perfusion channels). Lumen formation and vessel stabilization in such models are a critical step and dependent on the quality of cells used and support from mural cells or fibroblast-derived factors [129]. Bang et al. use this approach to culture HUVECs in a central fibrin gel channel flanked by a vascular perfusion channel and a neuronal/astrocyte co-culture that allowed for a neuronal/astrocyte migration at the gel interface. A fourth channel on the vascular side confined a fibroblast culture whose vasoactive secretions stimulated in-gel vasculogenesis. This vascular network showed BBB permeability on par with in vivo values (20-kDa FITC-dextran, $0.45 \pm 0.11 \times 10^{-6}$ cm/s; 70-kDa FITC-dextran, $0.36 \pm 0.05 \times 10^{-6}$ cm/s) and a high degree of neurovascular interfacing, which promoted BBB characteristics in HUVECs [130]. However, fibroblasts are not a brain-resident cell type.

Campisi et al. have since demonstrated stable, perfusable vessel formation in the absence of fibroblast-mediated support, by culturing hiPSC-BEC with human primary brain pericytes and astrocytes (Fig. 2d) [41]. This approach encapsulated all three cell types within the central gel, allowing direct EC support, with networks forming within 4 days of culture. Due to the restriction of microfluidic channel height, the vessels produced were slightly elliptical with transverse diameters ranging between 10 and 40 μm, while the lateral diameters were influenced by cell culture conditions, with pericytes and astrocytes co-culture found to produce the minimal lateral diameters of vessels (25 and 50 μm) [41]. Co-cultures also elevated the deposition of basement membrane proteins laminin and collagen IV by approximately twofold while also increasing levels of ZO-1, occludin, and claudin-5 compared to monoculture. This resulted in a highly restrictive barrier phenotype (8.9×10^{-8} cm/s and 2.2×10^{-7} cm/s for 40-kDa and 10-kDa FTIC-dextran). The same microfluidic chip design has also been used in a model that combined human ES-derived motor neuron spheroids with iPSC-BECs in a collagen gel; in this system BEC formed vessels and promoted neuronal connectivity and synchronization through both paracrine and contact-dependent signalling [131]. This chip is now commercially available from AIM Biotech, with each chip containing three devices. Protocols have been developed using this chip for applications including cell invasion and migration, cancer research, and vasculogenesis and angiogenesis and as a BBB model (based upon the work of Campisi et al. [41]). While the protocols for using this microfluidic chip to create BBB models use a self-assembly approach, the same chip could conceivably be used to define parenchymal culture with flanking vascular channels as described in Subheading 4.2.

Such platforms are well suited to studies of the innate capacity of ECs to form vessels and thus could be used in conjunction with iPSC technologies to study cerebral small vessel disease such as COL4A1 mutations, cerebral autosomal dominant arteriopathy with subcortical infarcts and leukoencephalopathy (CADASIL), and cerebral autosomal recessive arteriopathy with subcortical infarcts and leukoencephalopathy (CARASIL). On the other hand, this might be seen as a limitation of this technique as it relies heavily on highly vasculogenic, low-passage ECs. Furthermore, while this approach arguably creates the highest level of physiological mimicry, the vessel geometry is not specifically defined, and thus for analysis of WSS, the vessel geometry and dimensions must be analyzed in each case for subsequent computation of flow and WSS. Additionally, integration of TEER in such systems is challenging.

5 Moving Toward Disease Models

> When you want to know how things really work, study them when they're coming apart. (William Gibson 2010)

The extent to which these novel BBB models are able to mimic pathological events in neurovascular diseases may serve to test their degree of physiological relevance. While a BBB or NVU model may be useful in ascertaining the effectiveness of new drug delivery strategies to overcome the BBB, drugs are given primarily to elevate a disease, and diseases where the brain is the target often have an influence on BBB function. As such disease modelling presents as an important step in the development of predictive BBB models.

MPS are increasingly being utilized for modelling of neurological disease and have been particularly beneficial in shedding light on neuronal transmission and mechanisms of disease propagation in Alzheimer's disease [132–134] and Parkinson's disease [135, 136]. Additionally MPS that allow precise control of chemical microenvironments have provided insights on chemotactic and inflammatory responses in neurological disease, including microglial, neuronal, and astrocytic responses to Aβ gradients [63, 137]. BBB MPS are now also beginning to be applied to disease modelling.

5.1 Neuroinflammation

A number of the BBB models detailed above have been used to explore neuroinflammatory states by application of inflammatory cytokines such as TNF-α [66, 101]. Achyuta et al. use a membrane-based BBB model to show that vascular application of TNF-α resulted in ~75% of resident microglia and astrocytes on the neural side to become activated [138]. Using a 3D vessel, rat brain endothelial cell monoculture, Cho et al. demonstrated that TNF-α treatment elevated the release of several cytokines, including VEGF, TIMP1, CINC1, CINC2–a/b, CX3CL1, CXCL-10, and CCL20, while EC ZO-1 protein levels were reduced and delocalized from the cellular boundary [139]. Using the same model, the authors were able to investigate neutrophil chemotaxis toward a flanking chemoattractant source channel connected by a capillary array. Neutrophils added to the vascular channel become activated; however, none were observed to cross the BBB [139]. Oxygen glucose deprivation, relevant to modelling stroke, was also assessed in this system significantly increasing reactive oxygen species and ROCK activation while also reducing ZO-1 expression, allowing for an investigation of the protective effects of prospective antioxidant therapies, edaravone and Y-27632 [139].

5.2 Neurotoxicity

MPS BBB models have also been applied to study the neurotoxic effect of organophosphate-based compounds found in pesticides and biological weapons. Using the MIMETAS OrganoPlate system, Koo et al. culture bEnd.3 immortalized murine BEC in a vascular channel in contact with a 3D culture of N2a murine brain neuroblastoma cells, C8-D1A immortalized murine astrocytes, and BV-2 immortalized murine microglia in a type I collagen gel [140]. Using this model Koo et al. confirmed that organophosphates infiltrated BBB and rapidly inhibited acetylcholinesterase activity and that in vitro toxicity correlated with in vivo toxicity [140]. Maoz et al. in the previous mentioned system of linked membrane-based chips were also able to study methamphetamine toxicity, showing reversible disruption of the BBB as has been shown in vivo [111], suggesting such models might prove useful in predicting neurotoxicity.

5.3 Neurodegenerative Disease

Microfluidic BBB models have also been employed to investigate pathological events underlying Alzheimer's disease. Using multi-compartment pinned gels, Shin et al. were able to produce a BBB model incorporating 3D cultures of human neural progenitors with mutations in the amyloid precursor protein gene, thus enabling an investigation of the influence of Aβ on BBB phenotype. This system recapitulated a number of hallmarks of AD, including extracellular deposition of amyloid plaques, developed neurofibrillary tangles, BBB disruption, and increased reactive oxygen species [141]. MPS based upon the AIM Biotech chip have also been used in BBB models to study how Alzheimer's patients' serum can result in vascular dysfunction [142]. Preliminary studies have also been made in the development two NVU disease models in the Emulate chip by using iPSC-derived cells from Huntington's disease patients and by genetic manipulation of the monocarboxylate transporter 8 (MCT8) gene as a model for MCT8 deficiency. Both these iPSC-derived models demonstrated BBB disruption as observed in the human disease context and in the future may be used to explore therapeutic strategies [110].

5.4 Brain Cancer and Metastasis

Vascularized MPS have allowed the study of anticancer therapies targeted toward the vasculature [96] as well as in BBB-specific models to assess strategies to enhance drug delivery [143]. Given that most malignant brain lesions are metastasized from other organs [144], BBB MPS may provide useful systems to model brain metastasis. MPS have already provided insights into the role of astrocytes in restricting extravasation of malignant cancer cell lines MDA-MB-231 and A549 [145]. Such MPS might be used in the future to provide a more detailed understanding of the mechanisms of brain tropism and BBB penetration displayed by certain tumor cells [146].

While there are still only a handful of examples of MPS BBB models being applied to the modelling of specific neurological pathologies, given that BBB dysfunction is a common factor in a wide variety of neurological diseases combined with the enhanced physiological relevance of MPS, it is likely that the development and application of such models in the coming years will enable new pathophysiological insights. Additionally a number of studies are now linking multiple MPS which model different organs; this may allow in the future for a dissection of key organ–organ interactions such as the gut–brain axis [147]. A review of the developing work on linked MPS for systems modelling can be found here [148].

6 Choosing the Right MPS

> Everything should be made as simple as possible, but not simpler. (Albert Einstein 1950)

The specific advantages of these systems are in some cases obvious, such as self-assembly approaches in studying cerebral small vessel disease or multi-chamber devices in studying drug transport and metabolism. However, in others, it is yet to be seen as to what exact extent that more faithfully mimicking the BBB structure provides greater physiological relevance and predictive power of actual BBB function over more simple and higher-throughput in vitro systems. So far, most studies have demonstrated that in vitro systems follow the rule of "function follows form" with respect to flow [53], 3D environments [149], and heterotypic cultures. However, for certain readouts simplified models may be sufficient to answer a specific question in a clear and reproducible manner and thus should be considered as preferential model systems. DeStefano et al. have produced guidelines for gauging the relevance of in vitro BBB models, describing 12 design criteria for tissue engineering the human BBB [88]. Benchmarks should, where possible, be not to animal models but to the human physiological system or disease that is being modelled. However, achieving greater physiological mimicry and enhancing complexity can come at the expense of throughput and can in some cases limit experimental possibilities. Thus, a model should be considered in terms of not only its relevance to the system being modelled but also its relevance to the question being asked of it, on a case-by-case basis.

7 Challenges and Limitations Faced by MPS

With a rapidly growing understanding of the importance of physiochemical, mechanical, and cellular environmental cues in recapitulating the in vivo-like functional characteristics of the BBB, there is a growing interest in the use of MPS to provide increase predictive power of in vitro studies. Integration of other technologies such as iPSC culture and label-free real-time monitoring furthermore shows great promise to maximize the utility of such systems. As is common in many rapidly emerging fields, MPS still face challenges with validation, standardization, and acceptance. However, while the field was pioneered by cross-disciplinary collaborations between engineers, chemists, and biologists, MPS are now increasingly opening up to biologist end users with no previous experience in microfluidics. Some groups at the forefront of MPS development now offer their cross-disciplinary expertise as disease modelling services (such as Aracari Biosciences Inc. and Hesperos Inc.). Other spinout companies have made MPS devices (and where appropriate their accompanying control systems) commercially available (e.g., Emulate, BI/OND, SynVivo, MIMETAS, and AIM Biotech), and a growing number of companies even offer bespoke device fabrication services for custom applications (e.g., *TissUse GmbH and uFluidic*). As such there is now a wealth of opportunities for researchers to begin taking advantage of the opportunities afforded by MPS for BBB modelling.

Barriers to entry include a requirement for new skill sets and knowledge base and, in some cases, the expense of new equipment such as pumps and pressure controls. Users should also have an understanding of the current limitations in MPS including those that are inherent with scaling and material properties. Important considerations for new MPS users are briefly detailed below.

7.1 Cell Origin

Self-assembly protocols serve as an example of where the importance of cell source and quality is immediately obvious; here the endothelial cells found to be most successful in establishing stable vessels were of primary or iPSC origin and need to be low passage to retain the necessary vasculogenic phenotype to allow establishment of the model. As with all in vitro models, the cells used are of critical importance. Low-quality cells with little relevance used in a MPS shown to be capable of producing a highly relevant microenvironment will still not yield results of physiological relevance. While providing an abundance of cells at low cost for high-throughput screening, immortalized cell lines can often lack physiological relevance, showing substantial differences in their transcriptional and epigenetic profiles when compared to native cells in the tissue of origin [150, 151]. On the other hand, primary cells while highly relevant can be scarce, and lot-to-lot variability can be an issue. iPSC technologies show a great promise with protocols having been developed for multiple NVU cell types [152–157] albeit often at considerable cost and time. However, questions have been raised as to the exact identity of some iPSC-derived cells, particularly with regard to BEC protocols which may in fact produce cells more representative of epithelial cells than ECs [158].

Undoubtedly cell environment can help to shape cell phenotype; this is particularly the case with regard to BEC where loss of environmental cues (such as bFGF, angiopoietin-1, and transforming growth factor-β) can result in rapid loss of BBB characteristics [159]. Thus, there is a complex interplay between the MPS microenvironment and its cellular continuance in shaping physiological relevance, and careful consideration and choice of the cellular input are thus of huge importance.

7.2 Cell Loading

In a number of devices, achieving even cell loading can be challenging and is a skill mastered with practice and patience. For devices that use hydrogels, time is of utmost importance. For instance, thrombin converts fibrinogen to fibrin gel in a matter of seconds, while with temperature-sensitive gels such as Matrigel, the high surface area-to-volume ratio means change in temperature exchange is rapid following injection into the device. Poor loading technique can result in uneven distribution, unwanted alignment of fibers, and disruption of growth patterns. Accurate seeding densities and precise pipetting are also critical to reproducible MPS cultures. Such systems typically use tens to thousands of cells as opposed to millions of cells used in macroscale culture; thus, small inaccuracies can have a large impact. Furthermore, aggregates of cells, trapped bubble, and debris can easily block microfluidic devices and result in poor reproducibility and abnormal cell growth.

7.3 Scaling Laws

As dimensions of a system are reduced, there is a dramatic increase in surface area with respect to the volume. This ratio is exploited to the benefit of certain MPS but may also enhance surface adsorption losses. PDMS is a commonly used polymer for MPS fabrication due to its high gas permeability, transparency, and flexibility. However, it should be noted that PDMS can absorb a variety of small molecules such as drugs, media components, and metabolites, which can have significant impacts on detection and bioavailability [160]. Surface treatments can be used to negate some of these effects, while some commercial MPS manufactures are now instead favoring the use of more traditional thermoplastics; others are investigating novel materials [161, 162]. It should be noted that as the surface area-to-volume ratio increases with miniaturization, surface effects and material properties play an increasing role. Ren et al. (2013) provide an in-depth discussion of MPS material properties [162]. Another consequence of scaling, previously alluded to, is that MPS are often critically reliant on perfusion due to rapid depletion of nutrients and waste buildup in static micron-scale systems. As such users should also be aware of effective culture volume and critical perfusion rates [48, 49].

7.4 Outlook

While there are a number of challenges faced by the field, MPS are increasingly being utilized to model the BBB. As a result of the unique opportunities provided to precisely control and manipulate the physical and chemical microenvironment, these systems are allowing for greater physiological relevance to be achieved in vitro, enabling human cells to be interrogated in a setting that more closely resembles their native environment. As such MPS show the potential to enhance the predictive power of preclinical studies and open up new possibilities to investigate BBB function in health and disease.

References

1. Warren MS et al (2009) Comparative gene expression profiles of ABC transporters in brain microvessel endothelial cells and brain in five species including human. Pharmacol Res 59(6):404–413
2. Uchida Y et al (2011) Quantitative targeted absolute proteomics of human blood–brain barrier transporters and receptors. J Neurochem 117(2):333–345
3. Oberheim NA et al (2006) Astrocytic complexity distinguishes the human brain. Trends Neurosci 29(10):547–553
4. Oberheim NA et al (2009) Uniquely hominid features of adult human astrocytes. J Neurosci 29(10):3276–3287
5. Zhang Y et al (2016) Purification and characterization of progenitor and mature human astrocytes reveals transcriptional and functional differences with mouse. Neuron 89(1):37–53
6. Tarassishin L, Suh HS, Lee SC (2014) LPS and IL-1 differentially activate mouse and human astrocytes: role of CD14. Glia 62(6): 999–1013
7. Beaulieu-Laroche L et al (2018) Enhanced dendritic compartmentalization in human cortical neurons. Cell 175(3):643–651.e14
8. Hodge RD et al (2019) Conserved cell types with divergent features in human versus mouse cortex. Nature 573(7772):61–68

9. Haley PJ (2003) Species differences in the structure and function of the immune system. Toxicology 188(1):49–71
10. Seok J et al (2013) Genomic responses in mouse models poorly mimic human inflammatory diseases. Proc Natl Acad Sci U S A 110(9):3507–3512
11. Panula P, Joó F, Rechardt L (1978) Evidence for the presence of viable endothelial cells in cultures derived from dissociated rat brain. Experientia 34(1):95–97
12. Nielsen SSE et al (2017) Improved method for the establishment of an in vitro blood-brain barrier model based on porcine brain endothelial cells. J Vis Exp (127):56277
13. Abbott NJ, Rönnbäck L, Hansson E (2006) Astrocyte–endothelial interactions at the blood–brain barrier. Nat Rev Neurosci 7(1): 41–53
14. DeBault LE, Cancilla PA (1980) Gamma-Glutamyl transpeptidase in isolated brain endothelial cells: induction by glial cells in vitro. Science 207(4431):653–655
15. Hudecz D et al (2014) Reproducibility in biological models of the blood–brain barrier. Eur J Nanomed 6:185–193
16. Convery N, Gadegaard N (2019) 30 years of microfluidics. Micro Nano Eng 2:76–91
17. Terry SC, Jerman JH, Angell JB (1979) A gas chromatographic air analyzer fabricated on a silicon wafer. IEEE Trans Electron Devices 26(12):1880–1886
18. Manz A, Graber N, Widmer HÁ (1990) Miniaturized total chemical analysis systems: a novel concept for chemical sensing. Sensors Actuators B Chem 1(1–6):244–248
19. Whitesides GM (2006) The origins and the future of microfluidics. Nature 442(7101): 368–373
20. Woolley AT, Mathies RA (1994) Ultra-high-speed DNA fragment separations using microfabricated capillary array electrophoresis chips. Proc Natl Acad Sci U S A 91(24): 11348–11352
21. McDonald JC et al (2000) Fabrication of microfluidic systems in poly (dimethylsiloxane). Electrophoresis 21(1):27–40
22. Unger MA et al (2000) Monolithic microfabricated valves and pumps by multilayer soft lithography. Science 288(5463):113–116
23. Huh D et al (2010) Reconstituting organ-level lung functions on a chip. Science 328(5986):1662–1668
24. Villenave R et al (2017) Human gut-on-a-chip supports polarized infection of coxsackie B1 virus in vitro. PLoS One 12(2):e0169412
25. Wu WI et al (2013) 1 – Materials and methods for the microfabrication of microfluidic biomedical devices. In: Li X, Zhou Y (eds) Microfluidic devices for biomedical applications. Woodhead Publishing, pp 3–62
26. Halldorsson S et al (2015) Advantages and challenges of microfluidic cell culture in polydimethylsiloxane devices. Biosens Bioelectron 63:218–231
27. Tehranirokh M et al (2013) Microfluidic devices for cell cultivation and proliferation. Biomicrofluidics 7(5):51502–51502
28. Frimat J-P et al (2011) A microfluidic array with cellular valving for single cell co-culture. Lab Chip 11(2):231–237
29. Vu TQ, de Castro RMB, Qin L (2017) Bridging the gap: microfluidic devices for short and long distance cell-cell communication. Lab Chip 17(6):1009–1023
30. Taylor AM et al (2003) Microfluidic multi-compartment device for neuroscience research. Langmuir 19(5):1551–1556
31. Taylor AM et al (2005) A microfluidic culture platform for CNS axonal injury, regeneration and transport. Nat Methods 2(8):599–605
32. Park J et al (2009) Microfluidic compartmentalized co-culture platform for CNS axon myelination research. Biomed Microdevices 11(6):1145–1153
33. Shi M et al (2013) Glia co-culture with neurons in microfluidic platforms promotes the formation and stabilization of synaptic contacts. Lab Chip 13(15):3008–3021
34. Robertson G, Bushell TJ, Zagnoni M (2014) Chemically induced synaptic activity between mixed primary hippocampal co-cultures in a microfluidic system. Integr Biol (Camb) 6(6): 636–644
35. Frimat JP et al (2010) The network formation assay: a spatially standardized neurite outgrowth analytical display for neurotoxicity screening. Lab Chip 10(6):701–709
36. Nagendran T et al (2018) Use of pre-assembled plastic microfluidic chips for compartmentalizing primary murine neurons. J Vis Exp (141):10.3791/58421
37. MacKerron C et al (2017) A microfluidic platform for the characterisation of CNS active compounds. Sci Rep 7(1):15692
38. Fantuzzo JA et al (2020) Development of a high-throughput arrayed neural circuitry platform using human induced neurons for drug screening applications. Lab Chip 20(6): 1140–1152
39. Huang CP et al (2009) Engineering microscale cellular niches for three-dimensional multicellular co-cultures. Lab Chip 9(12): 1740–1748

40. Sharma AD et al (2019) Engineering a 3D functional human peripheral nerve in vitro using the Nerve-on-a-Chip platform. Sci Rep 9(1):8921
41. Campisi M et al (2018) 3D self-organized microvascular model of the human blood--brain barrier with endothelial cells, pericytes and astrocytes. Biomaterials 180:117–129
42. Kim SH et al (2017) Anisotropically organized three-dimensional culture platform for reconstruction of a hippocampal neural network. Nat Commun 8:14346–14346
43. Buxboim A, Ivanovska IL, Discher DE (2010) Matrix elasticity, cytoskeletal forces and physics of the nucleus: how deeply do cells "feel" outside and in? J Cell Sci 123(3): 297
44. Engler AJ et al (2006) Matrix elasticity directs stem cell lineage specification. Cell 126(4): 677–689
45. Rauti R, Renous N, Maoz BM (2019) Mimicking the brain extracellular matrix in vitro: a review of current methodologies and challenges. Isr J Chem 59(12):10–20
46. Budday S et al (2015) Mechanical properties of gray and white matter brain tissue by indentation. J Mech Behav Biomed Mater 46: 318–330
47. Ahn SI et al (2020) Microengineered human blood–brain barrier platform for understanding nanoparticle transport mechanisms. Nat Commun 11(1):175
48. Walker GM, Zeringue HC, Beebe DJ (2004) Microenvironment design considerations for cellular scale studies. Lab Chip 4(2):91–97
49. Young EWK, Beebe DJ (2010) Fundamentals of microfluidic cell culture in controlled microenvironments. Chem Soc Rev 39(3): 1036–1048
50. Wevers NR et al (2018) A perfused human blood–brain barrier on-a-chip for high-throughput assessment of barrier function and antibody transport. Fluids Barriers CNS 15(1):23
51. Seebach J et al (2000) Endothelial barrier function under laminar fluid shear stress. Lab Investig 80(12):1819–1831
52. Siddharthan V et al (2007) Human astrocytes/astrocyte-conditioned medium and shear stress enhance the barrier properties of human brain microvascular endothelial cells. Brain Res 1147:39–50
53. Cucullo L et al (2011) The role of shear stress in blood-brain barrier endothelial physiology. BMC Neurosci 12:40–40
54. DeStefano JG et al (2017) Effect of shear stress on iPSC-derived human brain microvascular endothelial cells (dhBMECs). Fluids Barriers CNS 14(1):20
55. Reinitz A et al (2015) Human brain microvascular endothelial cells resist elongation due to shear stress. Microvasc Res 99:8–18
56. Cucullo L et al (2008) Immortalized human brain endothelial cells and flow-based vascular modeling: a marriage of convenience for rational neurovascular studies. J Cereb Blood Flow Metab 28(2):312–328
57. Santaguida S et al (2006) Side by side comparison between dynamic versus static models of blood–brain barrier in vitro: a permeability study. Brain Res 1109(1):1–13
58. Weksler B, Romero IA, Couraud P-O (2013) The hCMEC/D3 cell line as a model of the human blood brain barrier. Fluids Barriers CNS 10(1):16–16
59. Poller B et al (2008) The human brain endothelial cell line hCMEC/D3 as a human blood–brain barrier model for drug transport studies. J Neurochem 107(5):1358–1368
60. Krizanac-Bengez L et al (2006) Loss of shear stress induces leukocyte-mediated cytokine release and blood–brain barrier failure in dynamic in vitro blood–brain barrier model. J Cell Physiol 206(1):68–77
61. Kuzmic N et al (2019) Modelling of endothelial cell migration and angiogenesis in microfluidic cell culture systems. Biomech Model Mechanobiol 18(3):717–731
62. Barkefors I et al (2008) Endothelial cell migration in stable gradients of vascular endothelial growth factor A and fibroblast growth factor 2: effects on chemotaxis and chemokinesis. J Biol Chem 283(20):13905–13912
63. Cho H et al (2013) Microfluidic chemotaxis platform for differentiating the roles of soluble and bound amyloid-beta on microglial accumulation. Sci Rep 3:1823
64. Romano NH et al (2015) Microfluidic gradients reveal enhanced neurite outgrowth but impaired guidance within 3D matrices with high integrin ligand densities. Small 11(6): 722–730
65. van der Meer AD et al (2010) Analyzing shear stress-induced alignment of actin filaments in endothelial cells with a microfluidic assay. Biomicrofluidics 4(1):11103–11103
66. Griep LM et al (2013) BBB on chip: microfluidic platform to mechanically and biochemically modulate blood–brain barrier function. Biomed Microdevices 15(1):145–150
67. Deosarkar SP et al (2015) A novel dynamic neonatal blood-brain barrier on a chip. PLoS One 10(11):e0142725

68. Brown JA et al (2015) Recreating blood--brain barrier physiology and structure on chip: a novel neurovascular microfluidic bioreactor. Biomicrofluidics 9(5):054124
69. Appelt-Menzel A et al (2017) Establishment of a human blood-brain barrier co-culture model mimicking the neurovascular unit using induced pluri- and multipotent stem cells. Stem Cell Reports 8(4):894–906
70. Müller J et al (2015) High-resolution CMOS MEA platform to study neurons at subcellular, cellular, and network levels. Lab Chip 15(13):2767–2780
71. Soscia D et al (2017) Controlled placement of multiple CNS cell populations to create complex neuronal cultures. PLoS One 12(11): e0188146
72. Kanagasabapathi TT et al (2011) Dual-compartment neurofluidic system for electrophysiological measurements in physically segregated and functionally connected neuronal cell culture. Front Neuroeng 4:13–13
73. van de Wijdeven R et al (2019) A novel lab-on-chip platform enabling axotomy and neuromodulation in a multi-nodal network. Biosens Bioelectron 140:111329
74. Maoz BM et al (2017) Organs-on-Chips with combined multi-electrode array and transepithelial electrical resistance measurement capabilities. Lab Chip 17(13):2294–2302
75. Prill S, Jaeger MS, Duschl C (2014) Long-term microfluidic glucose and lactate monitoring in hepatic cell culture. Biomicrofluidics 8(3):034102
76. Kalmykov A et al (2019) Organ-on-e-chip: three-dimensional self-rolled biosensor array for electrical interrogations of human electrogenic spheroids. Sci Adv 5(8):eaax0729
77. Pitsalidis C et al (2018) Transistor in a tube: a route to three-dimensional bioelectronics. Sci Adv 4(10):eaat4253
78. Curto VF et al (2017) Organic transistor platform with integrated microfluidics for in-line multi-parametric in vitro cell monitoring. Microsyst Nanoeng 3(1):17028
79. Beebe DJ, Mensing GA, Walker GM (2002) Physics and applications of microfluidics in biology. Annu Rev Biomed Eng 4(1): 261–286
80. Sweet RG (1965) High frequency recording with electrostatically deflected ink jets. Rev Sci Instrum 36(2):131–136
81. Rayleigh L (1879) On the capillary phenomena of jets. Proc R Soc Lond 29(196–199): 71–97
82. Duncombe TA, Tentori AM, Herr AE (2015) Microfluidics: reframing biological enquiry. Nat Rev Mol Cell Biol 16(9):554–567
83. Oh KW et al (2012) Design of pressure-driven microfluidic networks using electric circuit analogy. Lab Chip 12(3):515–545
84. Adamatzky A (2019) A brief history of liquid computers. Philos Trans R Soc B Biol Sci 374(1774):20180372
85. Simon SI, Goldsmith HL (2002) Leukocyte adhesion dynamics in shear flow. Ann Biomed Eng 30(3):315–332
86. Nobili M et al (2008) Platelet activation due to hemodynamic shear stresses: damage accumulation model and comparison to in vitro measurements. ASAIO J 54(1):64–72
87. Rana A et al (2019) Shear-dependent platelet aggregation: mechanisms and therapeutic opportunities. Front Cardiovasc Med 6:141
88. DeStefano JG et al (2018) Benchmarking in vitro tissue-engineered blood–brain barrier models. Fluids Barriers CNS 15(1):32
89. Dolan JM, Kolega J, Meng H (2013) High wall shear stress and spatial gradients in vascular pathology: a review. Ann Biomed Eng 41(7):1411–1427
90. Malek AM, Alper SL, Izumo S (1999) Hemodynamic shear stress and its role in atherosclerosis. JAMA 282(21):2035–2042
91. Koutsiaris AG et al (2007) Volume flow and wall shear stress quantification in the human conjunctival capillaries and post-capillary venules in vivo. Biorheology 44(5–6): 375–386
92. Zarrinkoob L et al (2016) Aging alters the dampening of pulsatile blood flow in cerebral arteries. J Cereb Blood Flow Metab 36(9): 1519–1527
93. Sohrabi S, Kassir N, Keshavarz Moraveji M (2020) Droplet microfluidics: fundamentals and its advanced applications. RSC Adv 10(46):27560–27574
94. Walsh EJ et al (2017) Microfluidics with fluid walls. Nat Commun 8(1):816
95. Wang X et al (2016) An on-chip microfluidic pressure regulator that facilitates reproducible loading of cells and hydrogels into microphysiological system platforms. Lab Chip 16(5): 868–876
96. Sobrino A et al (2016) 3D microtumors in vitro supported by perfused vascular networks. Sci Rep 6(1):31589
97. Moya ML et al (2013) In vitro perfused human capillary networks. Tissue Eng Part C Methods 19(9):730–737

98. Resto PJ et al (2010) An automated micro-droplet passive pumping platform for high-speed and packeted microfluidic flow applications. Lab Chip 10(1):23–26
99. Chen IJ, Eckstein EC, Lindner E (2009) Computation of transient flow rates in passive pumping micro-fluidic systems. Lab Chip 9(1):107–114
100. Bischel LL, Lee S-H, Beebe DJ (2012) A practical method for patterning lumens through ECM hydrogels via viscous finger patterning. J Lab Autom 17(2):96–103
101. Herland A et al (2016) Distinct contributions of astrocytes and pericytes to neuroinflammation identified in a 3D human blood-brain barrier on a chip. PLoS One 11:e0150360
102. Mao K et al (2020) Paper-based microfluidics for rapid diagnostics and drug delivery. J Control Release 322:187–199
103. Bhalerao A et al (2020) In vitro modeling of the neurovascular unit: advances in the field. Fluids Barriers CNS 17(1):22–22
104. Stamatovic SM et al (2016) Junctional proteins of the blood–brain barrier: new insights into function and dysfunction. Tissue Barriers 4(1):e1154641
105. Huh D et al (2012) A human disease model of drug toxicity-induced pulmonary edema in a lung-on-a-chip microdevice. Sci Transl Med 4(159):159ra147
106. Kasendra M et al (2020) Duodenum intestine-chip for preclinical drug assessment in a human relevant model. eLife 9:e50135
107. Jang KJ et al (2019) Reproducing human and cross-species drug toxicities using a Liver-Chip. Sci Transl Med 11(517):eaax5516
108. Jang KJ et al (2013) Human kidney proximal tubule-on-a-chip for drug transport and nephrotoxicity assessment. Integr Biol (Camb) 5(9):1119–1129
109. Sances S et al (2018) Human iPSC-derived endothelial cells and microengineered organ-chip enhance neuronal development. Stem Cell Reports 10(4):1222–1236
110. Vatine GD et al (2019) Human iPSC-derived blood–brain barrier chips enable disease modeling and personalized medicine applications. Cell Stem Cell 24(6):995–1005.e6
111. Maoz BM et al (2018) A linked organ-on-chip model of the human neurovascular unit reveals the metabolic coupling of endothelial and neuronal cells. Nat Biotechnol 36(9): 865–874
112. Vulto P et al (2011) Phaseguides: a paradigm shift in microfluidic priming and emptying. Lab Chip 11(9):1596–1602
113. Brown TD et al (2019) A microfluidic model of human brain (μHuB) for assessment of blood brain barrier. Bioeng Transl Med 4(2): e10126
114. Loessberg-Zahl J et al (2020) Patterning biological gels for 3D cell culture inside microfluidic devices by local surface modification through laminar flow patterning. Micromachines (Basel) 11(12):1112
115. Adriani G et al (2017) A 3D neurovascular microfluidic model consisting of neurons, astrocytes and cerebral endothelial cells as a blood–brain barrier. Lab Chip 17(3): 448–459
116. Bouhrira N et al (2020) Disturbed flow disrupts the blood–brain barrier in a 3D bifurcation model. Biofabrication 12(2):025020
117. Gelber MK, Bhargava R (2015) Monolithic multilayer microfluidics via sacrificial molding of 3D-printed isomalt. Lab Chip 15(7): 1736–1741
118. Miller JS et al (2012) Rapid casting of patterned vascular networks for perfusable engineered three-dimensional tissues. Nat Mater 11(9):768–774
119. Hasan A et al (2014) Microfluidic techniques for development of 3D vascularized tissue. Biomaterials 35(26):7308–7325
120. Lee E et al (2018) A 3D in vitro pericyte-supported microvessel model: visualisation and quantitative characterisation of multistep angiogenesis. J Mater Chem B 6(7): 1085–1094
121. Buzhdygan TP et al (2020) The SARS-CoV-2 spike protein alters barrier function in 2D static and 3D microfluidic in vitro models of the human blood–brain barrier. Neurobiol Dis 146:105131
122. de Graaf MNS et al (2019) Scalable microphysiological system to model three-dimensional blood vessels. APL Bioeng 3(2): 026105
123. Cassot F et al (2006) A novel three-dimensional computer-assisted method for a quantitative study of microvascular networks of the human cerebral cortex. Microcirculation 13(1):1–18
124. Smith AF et al (2019) Brain capillary networks across species: a few simple organizational requirements are sufficient to

reproduce both structure and function. Front Physiol 10:233
125. Heidari H, Taylor H (2018) A multi-layer, self-aligning hydrogel micro-molding process offering a fabrication route to perfusable 3D in-vitro microvasculature. bioRxiv:242156. https://doi.org/10.1101/242156
126. Zheng Y et al (2012) In vitro microvessels for the study of angiogenesis and thrombosis. Proc Natl Acad Sci U S A 109(24): 9342–9347
127. Kim S et al (2013) Engineering of functional, perfusable 3D microvascular networks on a chip. Lab Chip 13(8):1489–1500
128. Phan DTT et al (2017) A vascularized and perfused organ-on-a-chip platform for large-scale drug screening applications. Lab Chip 17(3):511–520
129. Whisler JA, Chen MB, Kamm RD (2014) Control of perfusable microvascular network morphology using a multiculture microfluidic system. Tissue Eng Part C Methods 20(7): 543–552
130. Bang S et al (2017) A low permeability microfluidic blood-brain barrier platform with direct contact between perfusable vascular network and astrocytes. Sci Rep 7(1):8083
131. Osaki T, Sivathanu V, Kamm RD (2018) Engineered 3D vascular and neuronal networks in a microfluidic platform. Sci Rep 8(1):5168
132. Usenovic M et al (2015) Internalized Tau oligomers cause neurodegeneration by inducing accumulation of pathogenic tau in human neurons derived from induced pluripotent stem cells. J Neurosci 35(42):14234–14250
133. Wu JW et al (2013) Small misfolded Tau species are internalized via bulk endocytosis and anterogradely and retrogradely transported in neurons. J Biol Chem 288(3):1856–1870
134. Song H-L et al (2014) β-Amyloid is transmitted via neuronal connections along axonal membranes. Ann Neurol 75(1):88–97
135. Brahic M et al (2016) Axonal transport and secretion of fibrillar forms of α-synuclein, Aβ42 peptide and HTTExon 1. Acta Neuropathol 131(4):539–548
136. Freundt EC et al (2012) Neuron-to-neuron transmission of α-synuclein fibrils through axonal transport. Ann Neurol 72(4):517–524
137. Park J et al (2018) A 3D human triculture system modeling neurodegeneration and neuroinflammation in Alzheimer's disease. Nat Neurosci 21(7):941–951
138. Achyuta AKH et al (2013) A modular approach to create a neurovascular unit-on-a-chip. Lab Chip 13(4):542–553
139. Cho H et al (2015) Three-dimensional blood-brain barrier model for in vitro studies of neurovascular pathology. Sci Rep 5: 15222–15222
140. Koo Y, Hawkins BT, Yun Y (2018) Three-dimensional (3D) tetra-culture brain on chip platform for organophosphate toxicity screening. Sci Rep 8(1):2841–2841
141. Shin Y et al (2019) Blood–brain barrier dysfunction in a 3D in vitro model of Alzheimer's disease. Adv Sci 6(20):1900962
142. Bersini S et al (2020) Transcriptional and functional changes of the human microvasculature during physiological aging and Alzheimer disease. Adv Biosyst 4(5):2000044
143. Bonakdar M, Graybill PM, Davalos RV (2017) A microfluidic model of the blood--brain barrier to study permeabilization by pulsed electric fields. RSC Adv 7(68): 42811–42818
144. Weller M et al (2015) Glioma. Nat Rev Dis Primers 1(1):15017
145. Xu H et al (2016) A dynamic in vivo-like organotypic blood–brain barrier model to probe metastatic brain tumors. Sci Rep 6: 36670
146. Valiente M et al (2020) Brain metastasis cell lines panel: a public resource of organotropic cell lines. Cancer Res 80(20):4314–4323
147. Raimondi MT, Albani D, Giordano C (2019) An organ-on-a-chip engineered platform to study the microbiota–gut–brain axis in neurodegeneration. Trends Mol Med 25(9): 737–740
148. Sung JH et al (2019) Recent advances in body-on-a-chip systems. Anal Chem 91(1): 330–351
149. Paşca SP (2018) The rise of three-dimensional human brain cultures. Nature 553(7689):437–445
150. Horvath P et al (2016) Screening out irrelevant cell-based models of disease. Nat Rev Drug Discov 15:751
151. Nestor CE et al (2015) Rapid reprogramming of epigenetic and transcriptional profiles in mammalian culture systems. Genome Biol 16:11
152. Yang N et al (2017) Generation of pure GABAergic neurons by transcription factor programming. Nat Methods 14(6):621–628
153. Cao S-Y et al (2017) Enhanced derivation of human pluripotent stem cell-derived cortical glutamatergic neurons by a small molecule. Sci Rep 7(1):3282
154. Mahajani S et al (2019) Homogenous generation of dopaminergic neurons from multiple

hiPSC lines by transient expression of transcription factors. Cell Death Dis 10(12):898
155. Lundin A et al (2018) Human iPS-derived astroglia from a stable neural precursor state show improved functionality compared with conventional astrocytic models. Stem Cell Reports 10(3):1030–1045
156. Hasselmann J, Blurton-Jones M (2020) Human iPSC-derived microglia: a growing toolset to study the brain's innate immune cells. Glia 68(4):721–739
157. Faal T et al (2019) Induction of mesoderm and neural crest-derived pericytes from human pluripotent stem cells to study blood-brain barrier interactions. Stem Cell Reports 12(3):451–460
158. Lu M et al (2021) Pluripotent stem cell-derived epithelium misidentified as brain microvascular endothelium requires ETS factors to acquire vascular fate. Proc Natl Acad Sci, 118(8):e2016950118. https://doi.org/10.1073/pnas.2016950118
159. Boström KI et al (2018) Endothelial cells may have tissue-specific origins. J Cell Biol Histol 1(1):104
160. van Meer BJ et al (2017) Small molecule absorption by PDMS in the context of drug response bioassays. Biochem Biophys Res Commun 482(2):323–328
161. Zhou XC et al (2017) Thiol–ene–epoxy thermoset for low-temperature bonding to biofunctionalized microarray surfaces. Lab Chip 17(21):3672–3681
162. Ren K, Zhou J, Wu H (2013) Materials for microfluidic chip fabrication. Acc Chem Res 46(11):2396–2406
163. Park T-E et al (2019) Hypoxia-enhanced blood–brain barrier chip recapitulates human barrier function and shuttling of drugs and antibodies. Nat Commun 10(1):2621

Chapter 12

Integrating Primary Astrocytes in a Microfluidic Model of the Blood–Brain Barrier

Eliana Lauranzano, Marco Rasile, and Michela Matteoli

Abstract

An in vitro blood–brain barrier (BBB) model must be highly reproducible and imitate as much as possible the properties of the in vivo environment, from both the functional and anatomical point of view. In our latest work, a BBB prototype was implemented through the use of human primary brain cells and then integrated in a microfluidic platform (Lauranzano et al., Adv Biosyst 3:e1800335, 2019). Here we describe, step by step, the setting of a customized bio-mimetic platform, which uses human brain endothelial cells and primary astrocytic cells to allow the study of the complex interactions between the immune system and the brain in healthy and neuroinflammatory conditions. The model can be exploited to investigate the neuroimmune communication at the blood–brain interface and to examine the transmigration of patient-derived lymphocytes in order to envisage cutting-edge strategies to restore barrier integrity and block the immune cell influx into the CNS.

Key words Primary human astrocyte isolation, Primary human astrocyte cultures, Blood–brain barrier in vitro model, Neurovascular unit, Microfluidic, T cell transmigration

1 Introduction

The managing interface between blood and brain parenchyma is the neurovascular unit (NVU), a highly specialized barrier composed by dynamic and interactive cellular and acellular elements. Besides endothelial cells (ECs) connected by tight junctions and lying on the basal lamina, which form the so-called blood–brain barrier (BBB), pericytes, astrocytes, and brain parenchymal cells, including neurons and microglia, are part of this anatomical and functional complex.

The NVU has the specialized role of preserving homeostasis within the central nervous system (CNS) environment. The contribution of non-fenestrated BBB capillaries, low pinocytic activity, and the presence of tight junctions between the ECs, in addition to the broad coverage of the BBB by pericytes and astrocytes [1, 2], not only restrict paracellular diffusion of water-soluble substances

Nicole Stone (ed.), *The Blood-Brain Barrier: Methods and Protocols*,
Methods in Molecular Biology, vol. 2492, https://doi.org/10.1007/978-1-0716-2289-6_12,

but also prevent CNS invasion by microbes and regulate the trafficking of blood-borne cells into the brain parenchyma. The NVU is indeed central to the neuroimmune communication that physiologically occurs between the immune and nervous systems and which is deeply altered in pathological conditions such as encephalomyelitis [3], brain tumors [4], autoimmune diseases like multiple sclerosis [5, 6] and narcolepsy [7], neurodegenerative processes such as Parkinson's [8] and Alzheimer's [9, 10] diseases, and ischemic stroke [11], all characterized by the significant elevation of pro-inflammatory molecules. Inflammatory conditions can have a profound impact on the barrier function: besides influencing vascular permeability and reducing the integrity of the BBB, inflammation can lead to the activation of resident glial cells and macrophages, which in turn modulate the infiltration of non-resident immune cells, thus representing by fact a positive feedback for the ongoing immune response in the CNS. Also, the inflammatory signals that are initiated in the periphery can propagate to the brain and influence resident cells, ultimately resulting in demyelination and neuronal damage [12]. Factors released from activated CNS resident glial cells and BBB endothelial cells may amplify the excessive immune response.

Astrocytes, which actively orchestrate key physiological aspects of brain development and functioning [13–18] and constitute by number the largest cellular component in the human brain [19], are nowadays recognized as necessary and nonredundant actors for NVU maintenance and repair [20, 21]. Besides providing a necessary component for the integrity of the BBB through the induced expression of polarized tight junction proteins, astrocytes play a critical role in regulating ionic homeostasis and guaranteeing the clearance of interstitial solutes. Further, astrocytes are primarily involved in the metabolic blood–brain communications, sensing synaptic activity, and regulating the supply of oxygen and glucose to the nervous tissue [22] and are emerging as gatekeepers for the neuroimmune communication, also reducing BBB damage and promoting its recovery [23].

The NVU acts as the main guardian and traffic regulator of the brain. For these reasons, the scientific community is in need of reliable tools for the study of the NVU and BBB transport mechanisms, which realistically assess alterations of the neuroimmune communication and rescue strategies. In this respect, several in vitro models of the human BBB (only EC)/NVU have been developed to investigate how neuroimmune communication orchestrates the passage of lymphocytes in healthy individuals and how pathological alterations enhance the anomalous extravasation of immune cells. These models can be classified into static (i.e., mere physical separations of the brain and blood environment usually achieved by Transwell™ systems) or dynamic, which include a physiological-like intraluminal flow. We sought to develop a human NVU model by the implementation of primary human astrocytes to a BBB model, suitable for being integrated with microfluidics. This model may overcome the limitations of several currently used models, including the absence of fluidic shear stress and/or the lack of a thin dual cell layer interface between astrocytes and ECs.

In order to overcome the important technical limitation represented by the isolation of live cells from human brain tissue, which is practically challenging and results in the loss of many vital cells during the process, we describe here a panning strategy, analogous to the one reported by Zhang et al. [24]. This method allows to isolate human primary astrocytes for functional studies (for a comparison between primary cells and commercially available cell lines, *see* ref. [25]). The biological entity of the hereby proposed NVU model consists of a direct-contact co-culture formed by ECs and astrocytes. Brain ECs are grown to form a homogeneous monolayer above a permeable porous membrane (luminal side) covered on both sides by a thin layer of extracellular matrix, thus mimicking a BBB structure. Primary astrocytic cells are grown on the opposite side of the membrane (abluminal side), reproducing an essential part of the physiological structure and function of the human NVU. The entire platform is compatible with immunochemistry and high-resolution microscopy and can be integrated on a microfluidic setting for human leukocyte transmigration studies [25].

2 Materials

Prepare all solutions under sterile hood, keep reagents sterile, and work rapidly. Prepare all solutions using ultrapure sterile water.

2.1 Preparation of Panning Dishes

1. Dulbecco's phosphate-buffered saline (dPBS), no calcium, no magnesium, no phenol red indicator, pH range 7.0–7.3.
2. Tris–HCl 50 mM, pH 9.5: weight 3.0285 g of Tris base (MW 121.14 g/mol) to make 500 mL. Add 350 mL of deionized water, and stir until completely dissolved. Slowly add HCl 37% under a chemical hood, while monitoring the pH, and stop when the pH of the solution reaches 9.5. Add enough deionized water to obtain a final volume of 500 mL. Under a sterile hood, filter the solution using a 0.22 μm filter membrane, and store at 4 °C.
3. Five 10-cm Ø Petri dishes, non-treated, sterile.
4. Antibodies for panning dishes coating (*see* **Note 1**):
   (a) Anti-rat IgG.
   (b) Anti-mouse IgG.
   (c) Anti-CD45 (BD, #550539).

(d) Anti-galactocerebroside (clone GalC, #MAB342).

(e) Anti-HepaCAM (clone 419305, #MAB4108).

(f) BSL-1 (*Bandeiraea simplicifolia* lectin-1).

5. Bovine serum albumin (BSA) stock solution: prepare 10% stock solution diluting 1 g in 10 mL dPBS (*see* **Note 2**). Filter using 0.22 μm membrane filters, and store at 4 °C.
6. Deoxyribonuclease I (DNase I) stock solution: prepare 0.4% stock solution (12,500 U/mL) solubilizing it with dPBS (*see* **Note 3**).
7. Antibody diluent: prepare 0.2% BSA/0.0004% DNase solution diluting BSA and DNase stock solutions with dPBS. Filter using 0.22 μm membrane filters, and store at 4 °C.
8. Cell incubator at 37 °C, 5% CO_2/saturated humidity.

2.2 Cell Culture Plate Coating

1. Sterile cell culture plates, glass coverslips.
2. Human fibronectin 1 mg/mL.
3. Geltrex™ (*see* **Note 4**).
4. SATO 100X. Final concentration of the 1X solution: 100 μg/mL transferrin, 100 μg/mL BSA, 16 μg/mL putrescine, 60 ng/mL progesterone, 40 ng/mL sodium selenite (*see* **Note 5**):

(a) Prepare a 79.5 mM progesterone stock solution: weight 2.5 mg, and add 100 μL of absolute ethanol; gently swirl to dissolve.

(b) Prepare a 2.3 mM sodium selenite stock solution: weight 4 mg, and add 10 mL neurobasal (NB) medium + 10 μL NaOH; gently swirl to dissolve.

(c) Dilute 100 mg apo-transferrin human in 4 mL NB to obtain a 2.5% solution.

(d) Weight 100 mg BSA, and dilute in 4 mL NB to obtain a 2.5% solution.

(e) Weight 16 mg putrescine dihydrochloride, and dilute in 1 mL NB to obtain a 99 mM solution.

(f) Mix 4 mL transferrin 2.5% with 4 mL BSA 2.5%, 1 mL putrescine 99 mM, 2.5 μL progesterone 79.5 mM, 100 μL sodium selenite 2.3 mM. Make up to 10 mL with NB. Mix well and filter through pre-rinsed 0.22 μm filter. Make 500 μL aliquots and store at −20 °C.

5. Astrocyte growth medium: 48% NB, 48% Dulbecco's Modified Eagle Medium (DMEM, Lonza), 100 U/mL penicillin, 100 μg/mL streptomycin, 1 mM sodium pyruvate, 2 mM L-glutamine, 5 μg/mL N-acetylcysteine, 5 ng/mL heparin-binding epidermal growth factor-like growth factor (HB-EGF), 1% SATO 100X. Vacuum filter in tissue culture hood and keep sterile. Store at 4 °C.
6. Endothelial basal medium (EBM™-2).

2.3 Purification of Primary Human Brain Cells

1. RPMI 1640 medium at 4 °C for the transport of brain tissue to the lab after resection.
2. Scalpel blade No. 10, 70 μm cell strainer gasket filter, Accutase™.
3. Enzyme stock buffer: mix 10% Earle's Balanced Salt Solution (EBSS) 10X, 0.46% D-(+)-glucose, 25 mM $NaHCO_3$. Bring to the desired volume with water. Filter using 0.22 μm filter membrane, and store at 4 °C. Equilibrate with sterile 95% O_2:5% CO_2 before each use.
4. Papain stock solution: dilute 1 mM L-cysteine and 0.5 mM EDTA in enzyme stock buffer. Check the pH to be ~7 before adding 400 units papain (*see* **Note 6**).
5. Ovomucoid stock solution: solubilize 6 mg/mL ovomucoid protease inhibitor and 6 mg/mL BSA with enzyme stock solution. Mix well to dissolve. Check pH, and, if necessary, use 1N NaOH to adjust pH to 7.4. Filter using 0.22 μm filter membrane, and store at 4 °C.
6. Papain inhibitor solution: prepare 1 mg/mL ovomucoid solution diluting ovomucoid stock solution with enzyme stock buffer. Add 0.005% DNase (*see* **Note 7**).

2.4 Immunopanning

1. Resuspension solution: prepare 0.02% BSA/0.0004% DNase solution diluting BSA and DNase stock with enzyme stock buffer. Filter using 0.22 μm membrane filters, and store at 4 °C.
2. Endothelial growth medium: endothelial basal medium (EBM™-2) supplemented with 5% heat-inactivated fetal bovine serum (FBS), 100 U/mL penicillin, 100 μg/mL streptomycin, 1.4 μM hydrocortisone, 5 μg/mL ascorbic acid, 1% chemically defined lipid concentrate, 10 mM 4-(2-hydroxyethyl)-1-piperazineethanesulfonic acid (HEPES), 1% long R insulin-like growth factor-1 (R^3-IGF-1), 1% vascular endothelial growth factor (VEGF), 1% epidermal growth factor (hEGF), and 1 ng/mL basic fibroblast growth factor (FGF-B). Vacuum filter in tissue culture hood, and store at 4 °C.

2.5 NVU Construction

1. Six-well tissue culture plates, 24 mm Ø glass slides.
2. Isopore™ Membrane Filters, 8.0 μm pore size, hydrophilic polycarbonate membrane, 25 mm Ø (*see* **Note 8**).
3. Trypsin–EDTA (0.25%).
4. Endothelial coating solution: dilute 0.1 mg/mL collagen type I (low viscosity) and 2 μg/mL human fibronectin in serum-free ice-cold EBM-2 medium. Mix thoroughly, and store at 4 °C.

3 Methods

Perform all procedures in a sterile environment, using aseptic techniques such as working under a cell culture hood to prevent contamination. All procedures should be performed at room temperature unless otherwise specified.

3.1 Preparation of Panning Dishes

The day before starting human brain cell purification, prepare panning dishes.

1. Use five Petri dishes for the coating with secondary antibodies. Select the secondary antibodies based on the species of primary antibodies host:
   (a) One Petri dish for CD45: dilute 1:400 anti-rat IgG in 50 mM Tris–HCl, pH 9.5.
   (b) Two Petri dishes for GalC: dilute 1:400 anti-mouse IgG in 50 mM Tris–HCl, pH 9.5.
   (c) One Petri dish for HepaCAM: dilute 1:400 anti-mouse IgG in 50 mM Tris–HCl, pH 9.5.
   (d) One Petri dish for BSL-1: dilute 1:400 BSL-1 in 50 mM Tris–HCl, pH 9.5.
2. Incubate overnight at 37 °C.
3. The next day, wash the plates three times with sterile dPBS. Do not dry.
4. Add primary antibody diluted in antibody diluent solution to the correspondent panning dish:
   (a) Anti-CD45 1:600.
   (b) Anti-GalC 1:2000.
   (c) Anti-HepaCAM 1:800.
5. Incubate at room temperature for the time of tissue processing.
6. Immediately before use, wash each panning dish three times with sterile dPBS. Do not dry.

3.2 Cell Culture Plate Coating

The day of the dissociation, before starting human brain cell purification, prepare sterilized coverslips and/or culture plates.

1. Coating for astrocytes cultures

   Dilute Geltrex™ 1:200 in ice-cold serum-free medium. Add Geltrex™ as a drop. Make sure to cover the entire surface. Incubate at 37 °C for 90 min. Wash twice with serum-free medium. Do not dry.

2. Coating for endothelial cultures

   Dilute 5 μg/mL fibronectin in serum-free EBM-2 medium. Add fibronectin as a drop covering the entire surface. Incubate at RT for 30–60 min. Wash twice with serum-free EBM-2 medium. Do not dry.

3.3 Purification of Primary Human Brain Cells

Primary human brain cells can be purified from cortical brain tissue obtained from surgical procedures for treating tumors or epilepsy (*see* Fig. 1). Surgical samples must be collected exclusively from consenting patients (*see* **Note 9**).

1. Immediately following surgery, immerse the cerebral tissue resection specimen in ice-cold RPMI medium, and transfer it to the lab for tissue dissociation within 1 h maximum.

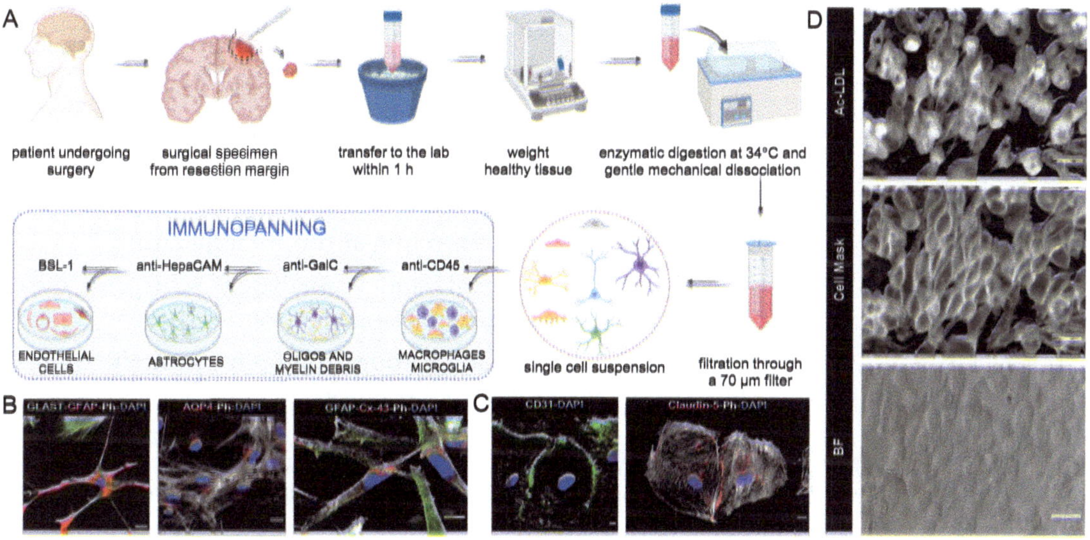

Fig. 1 Primary human brain cell purification. (**a**) Schematic of endothelial cell and astrocyte purification protocol, depicting major steps and immunopanning strategy. (**b**) Characterization by confocal immunofluorescence of HepaCAM-purified astrocytes stained for: left, GLAST (green), GFAP (red), Actin (Ph, gray); center, Aquaporin-4 (AQP4, red), Actin (Ph, gray); right, GFAP (green), Connexin-43 (Cx-43, red), Actin (gray). Nuclei of all cells (DAPI, blue). Scale bars: 10 μm. (**c**) Characterization by confocal immunofluorescence of BSL-1-purified endothelial cells stained for: left, CD31 (green); right, Claudin-5 (red), Actin (Ph, gray). Nuclei of all cells (DAPI, blue). Scale bars: 10 μm. (**d**) Identification of endothelial cells based on (top) acetylated-LDL uptake (DiI), (central) CellMask Green Plasma Membrane Stain, (bottom) bright field (BF)

2. Use a sterile 10 cm Ø Petri dish to accommodate the surgical specimen removed from the resection margin (*see* **Note 10**). Use a sterile n. 10 scalpel blade to dissect out gray matter from human brain specimen. Quickly weigh the healthy tissue that is going to be processed.
3. Put ~0.5 g of tissue into each 10-cm Ø Petri dish. Finely chop the brain specimen into <1 mm^3 pieces (*see* **Note 11**).
4. Collect minced tissue in a 15 mL sterile centrifuge tube. Add 20 units/mL papain stock solution and 0.005% DNase. Incubate at 34 °C in a water bath for 60–90 min with constant agitation (*see* **Note 12**).
5. After digestion, gently triturate the mixture with a 5 mL pipette (*see* **Note 13**).
6. If any piece of undissociated tissue remains after trituration, allow larger clumps to settle to the bottom of the tube for 1–2 min.
7. Collect the cloudy cell suspension carefully in a sterile centrifuge tube. Avoid including any piece of undissociated tissue.
8. Centrifuge at 300 × g for 5 min at room temperature. While centrifuging, prepare papain inhibitor working solution.
9. Discard the supernatant, and immediately resuspend the cell pellet in 4 mL papain inhibitor solution to fully interrupt the digestion phase.
10. Use a clean 15 mL centrifuge tube to prepare a discontinuous density gradient of protease inhibitor solution. Add 7 mL of ovomucoid stock solution to the bottom of the tube, and carefully overlay the papain inhibitor solution with cell suspension (*see* **Note 14**).
11. Centrifuge at 70 × g for 6 min at room temperature.
12. Discard the supernatant, and immediately resuspend dissociated cells pellet in resuspension solution.
13. Filter the cell suspension through a 70 μm cell strainer gasket filter to obtain a single-cell solution.

3.4 Immunopanning

Pass the single-cell suspension of dissociated tissue sequentially over a series of precoated cell-type-specific Petri dishes (prepared in Subheading 3.1).

1. Add the cell suspension to the first panning dish (CD45) as a negative selection for microglia/macrophages. Incubate at RT for 15 min. Shake the CD45 plate, and collect the cell suspension; use 1 mL resuspension solution to wash the CD45 plate, and collect the solution from the plate together with the unattached cells.

2. Transfer the cells to the second panning dish (GalC) as a negative selection of oligodendrocyte-lineage cells. Incubate at RT for 10 min, shake the GalC plate, and collect the cell suspension; use 1 mL resuspension solution to wash the CD45 plate, and collect the solution from the plate together with the unattached cells.
3. Repeat **step 2** with a second GalC plate.
4. Transfer unbound cells to the fourth panning dish (HepaCAM) as a positive selection for astrocytes. Incubate at RT for 15 min, shake the HepaCAM plate, and collect the cell suspension; use 1 mL resuspension solution to wash the HepaCAM plate, and collect the solution from the plate together with the unattached cells.
5. Transfer the cells to the fifth panning dish (BSL-1) as a positive selection of endothelial cells. Incubate at RT for 30 min.
6. Wash the panning plates from **steps 4** (HepaCAM/astrocytes) and **5** (BSL-1/endothelial cells) five times with dPBS to remove loosely bound contaminating cells.
7. Detach the cells from antibody-coated dishes for cell culture and in vitro experiments, adding pre-warmed Accutase™, and incubate at 37 °C for 5–10 min (*see* **Note 15**).
8. Stop the enzymatic digestion, adding astrocyte growth medium to the HepaCAM plate and endothelial growth medium to the BSL-1 dish.
9. Harvest the cells squirting gently the medium around the plate, going through every part of the Petri dish. Collect the cell suspension in a 50 mL centrifuge tube, and add DNase. Use additional growth medium to wash the plates, and collect each solution from the plates pooling with the corresponding cell suspension.
10. Pellet cells at 150 × g for 10 min at RT. Discard supernatant, and resuspend cells in 2 mL growth medium.
11. Count viable cells diluting the cell suspension 1:10 with trypan blue (*see* **Note 16**).
12. Pre-plate cells in a small volume, and add as a drop at the center of the precoated slides (prepared in Subheading 3.2). Gently transfer to the incubator for 30 min.
    (a) Purified primary human astrocytes resuspended in astrocyte growth medium are cultured on Geltrex® precoated slides at a cell density of 0.8–1 × 10^5 cells/cm^2.
    (b) Human primary brain endothelial cells are grown in endothelial growth medium at a concentration of 27,000 cells/cm^2 in tissue culture supports precoated with fibronectin.

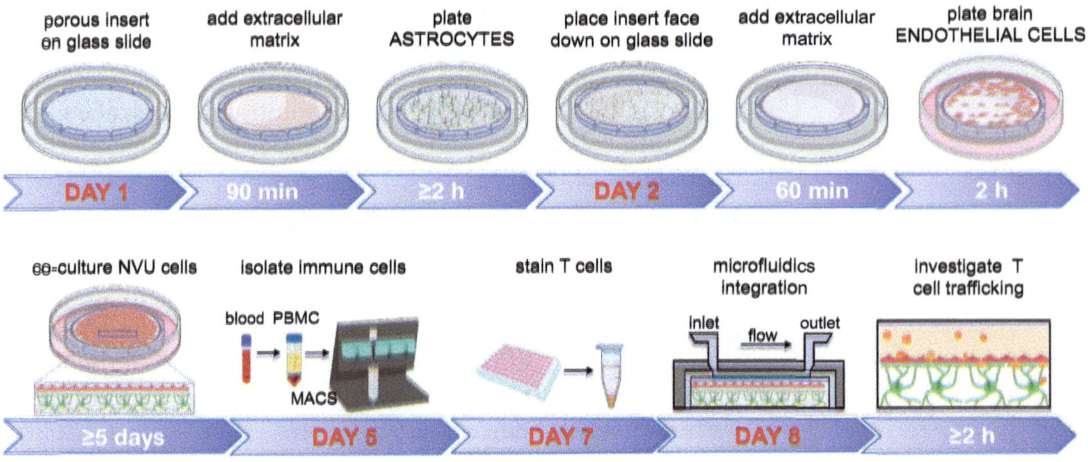

Fig. 2 NVU bio-mimetic model integration into a microfluidic platform. Schematic of the NVU construction protocol, depicting major steps and NVU in vitro model microfluidic device integration exploited to investigate human T lymphocyte recruitment and transmigration with real-time imaging

13. Remove the drop of cell suspension, and carefully add growth medium to adhering cells.
14. Cells are cultured at 37 °C, 5% CO_2/saturated humidity (*see* **Note 17**).

3.5 NVU Construction

The Transwell® culture system can be used as permeable support providing independent access to both sides of the insert for static experiments, while the Isopore™ membranes can be successfully employed for integration in a customized platform for in vitro dynamic assays (*see* Fig. 2).

1. Use sterile tweezers to place a sterile 24 mm Ø glass slide in each well of a six-well tissue culture plate. Lay down one Isopore™ membrane over each glass slide, with the basolateral (opaque) side facing upwards.
2. Dilute Geltrex® 1:10 in ice-cold dPBS, and mix thoroughly.
3. Add 333 μL of diluted Geltrex® on each Isopore™ membrane as a drop (*see* **Note 18**).
4. Carefully place the plate in the incubator at 37 °C for 90 min.
5. In the meantime, gently wash primary human astrocyte culture (prepared in Subheading 3.4) with warmed dPBS for three times. Add trypsin–EDTA, and incubate at 37 °C for 5–10 min (*see* **Note 19**).
6. Add astrocyte growth medium to stop the digestion. Squirt the medium gently around the plate. Pay attention not to introduce bubbles. Collect the cell suspension in a centrifuge tube. Use 1 mL growth medium to wash the plate, and collect the solution from the plate together with the detached cells.

7. Centrifuge at 350 × g for 5 min at RT. Discard the supernatant. Resuspend astrocytes pellet in 1 mL astrocyte growth medium.
8. Count viable cells diluting the cell suspension 1:10 with trypan blue (*see* **Note 15**).
9. Dilute 1 × 10^6 human astrocytes in 2 mL astrocyte growth medium.
10. Carefully take the plate with the Isopore™ membranes out of the incubator. Wash twice with warm dPBS. Do not dry.
11. Add 400 μL cell suspension as a drop at the center of each precoated Isopore™ membrane. 2 × 10^5 human astrocytes are plated on the abluminal side of matrix-coated inserts. Gently transfer to the incubator for at least 2 h at 37 °C (*see* **Note 20**).
12. Use sterile tweezers to place a sterile 24 mm Ø glass slide in each well of a new six-well tissue culture plate.
13. Carefully take out of the incubator the plate with the human astrocytes seeded on the Isopore™ membrane. Use a p1000 micropipette to remove the drop of cell suspension. Do not touch the surface of the filter with the pipette tip. Do not dry.
14. With a rapid and precise movement, lift the Isopore™ membrane using sterile tweezers, and lay it down on top of a new glass slide (prepared in **step 12**) overlapping the edges. Add 3 mL astrocyte growth medium, and incubate at 37 °C for 1 day.
15. The next day use sterile tweezers to place a sterile 24 mm Ø glass slide in each well of a new six-well tissue culture plate. Take out of the incubator the plate with the human astrocytes (prepared in **step 13**), and visually inspect it under the microscope.
16. With a rapid and precise movement, gently lift the Isopore™ membrane using sterile tweezers, and flip it, laying it facedown on top of a new glass slide, with the apical (translucid) side facing upwards. Do not dry the membrane.
17. Gently add 250 μL of endothelial coating solution on each Isopore™ membrane as a drop (*see* **Note 18**). Incubate at 37 °C for 60 min. Wash twice with serum-free EBM-2 medium. Do not dry.
18. In the meantime, gently wash human brain endothelial cells with warmed dPBS for three times. Add trypsin–EDTA, and incubate at 37 °C for 5–10 min (*see* **Note 19**).
19. Add endothelial growth medium to stop the digestion. Squirt the medium gently around the plate. Pay attention not to introduce bubbles. Collect the cell suspension in a centrifuge

tube. Use 1 mL growth medium to wash the plate, and collect the solution from the plate together with the detached cells.

20. Centrifuge at 350 × g for 5 min at RT. Discard the supernatant. Resuspend endothelial cells pellet in 1 mL endothelial growth medium.
21. Count viable cells diluting the cell suspension 1:10 with trypan blue (*see* **Note 15**).
22. Dilute 0.8×10^6 human endothelial cells in 1 mL endothelial growth medium.
23. Carefully take the plate with the Isopore™ membranes out of the incubator. Gently wash twice with warmed EBM-2 medium. Do not dry.
24. Add 200 μL human brain endothelial cell suspension as a drop at the center of each precoated Isopore™ membrane. Seed 3×10^4 cells/cm^2 on the luminal side of matrix-coated inserts. Gently transfer the plate to the incubator for at least 20 min at 37 °C (*see* **Note 20**).
25. Carefully take the plate out of the incubator. Use a p200 micropipette to remove the drop of cell suspension. Do not touch the surface of the filter with the pipette tip. Do not dry.
26. With a rapid and precise movement, lift the Isopore™ membrane using sterile tweezers, and lay it down on top of a new glass slide (prepared in **step 12**) overlapping the edges. Add 1.25 mL astrocyte growth medium and 1.5 mL endothelial growth medium. Incubate at 37 °C. Replace half medium with fresh growth medium every other day. Co-culture cells for 5–10 days (*see* **Note 21**).

The fully differentiated human BBB model featuring cellular elements of the NVU (i.e., astrocytes) is integrable in a customized microfluidic platform endowed with a flow-based system, where vascular endothelial cells can be exposed to shear stress. The dynamic interactions at the interface between astrocytes and EC, combined with the optical transparency and high permeability of the model, make this NVU platform an ideal scaffold for the assembly of an in vitro microfluidic model designed to meet specific requirements for different endpoints (e.g., co-cultures, functional studies, live cell imaging). Moreover, the NVU bio-mimetic model is adaptable for 3D construction with the different cellular and acellular components of the NVU and can provide a valuable model for the integration with patient-derived primary cells (for T cells study, *see* ref. [25]), thus representing a promising tool to investigate neuroimmune crosstalk.

4 Notes

1. Primary antibodies for panning dish coating should be aliquoted in ready-to-use volume aliquots and stored at −20 °C, to avoid repeated freeze and thaw.
2. Use BSA globulin free. BSA can be dissolved quickly if the solution is cooled to about 4 °C.
3. DNase is sensitive to shear denaturation. Avoid vigorous mixing. Check pH: it should be comprised in the range between 7 and 7.5; if not, adjust by adding a small amount of NaOH or HCl solutions. Filter under sterile hood using a 0.22 μm filter membrane. Store at 4 °C between uses. Prepare it fresh every time.
4. Thaw Geltrex™ Basement Membrane Matrix at 2–8 °C overnight. Mix Geltrex™ matrix solution by slowly pipetting up and down, being careful not to introduce air bubbles. Use ice-cold serum-free medium to dilute. Keep on ice until use.
5. Exposure to progesterone and sodium selenite powders is dangerous: potentially hazardous reagents should be handled and stored properly, according to institutional and governmental guidelines and regulations. Proper personal protective equipment, like gloves or goggles and facial masks, should always be used. To avoid co-worker exposure, weight powder under a dedicated fume hood or biosafety cabinet directly in the resealable tube where it will be diluted. Prepare fresh progesterone and sodium selenite stocks; do not re-use stock solutions and dispose of reagents following safety as stated by guidelines for waste management.
6. Briefly incubate papain stock solution at 34 °C to ensure full solubility and activity of the enzyme immediately before use. The solution should appear clear. Filter using 0.22 μm filter membrane, and promptly use the solution. It can be kept at RT during tissue dissociation. Papain dissociation protocol is based on instructions provided by the manufacturer (http://www.worthington-biochem.com/pds/default.html) and optimizations from a previous work [25].
7. Prepare immediately before use, and keep on ice until use. Prepare fresh every time.
8. Isopore™ Membrane Filters are not sterile. Use tweezers (disinfect using 70% ethanol) under a cell hood to put a single membrane in a 3 cm Ø Petri dish, and seal it with Parafilm M. To sterilize, expose the top and bottom surfaces of the membrane to ultraviolet (UV) light for 20–30 min. Sterilized filters can be stored at RT for several months.

9. Research on human brain tissue should always be conducted in accordance with governmental and institutional guidelines and regulations. Human brain specimens must be obtained with informed consent and under the approval of the Independent Ethical Committee of the provider hospital. Cortical brain tissue can be obtained from surgeries performed to treat tumors or epilepsy.
10. Before starting the procedure, prepare a 15 cm Ø Petri dish with ice inside, and seal it with Parafilm M. Use the 15 cm Ø Petri dish as a base for the 10 cm Ø Petri dish in order to keep the brain tissue cold. If a precision balance is not available under a cell culture hood, weight the empty 10 cm Ø Petri dish, and annotate the weight, so that it can be subtracted when weighting brain tissue. These precautions are important to keep the tissue in a sterile environment.
11. Brain tissue mincing is performed on ice for no more than 2 min.
12. The amount of time must be determined empirically.
13. Quickly suck up and release brain homogenate, repeating for ~20 times. Stop trituration when the solution appears homogeneous. Keep the pipette into the solution, and work carefully to avoid the introduction of bubbles.
14. During this relatively slow centrifugation, cells separate according to size: cells sediment at the bottom of the tube. Membrane fragments being smaller remain at the interface. The interface between the two layers of the gradient should be clearly visible. Be careful when handling the tube, minimizing intermixing of the solutions. However, minimal mixing at this boundary does not affect the result.
15. Incubation time with Accutase™ is determined empirically by visual examination of the plates under a differential interference contrast (DIC) microscope. Incubation time can differ from the suggested time in this protocol due to lot-dependent variation in enzyme activity. Gently tap the side of the Petri dish, and stop the digestion when visual examination of the plate indicates that about half of the cells are dislodged. Increase incubation time if there are lots of cells stuck. Enzymatic digestion should generally not exceed 15 min.
16. The cells that exhibit a blue cytosol are dead; viable cells do not allow trypan blue to enter inside the cell cytosol.
17. Half of the cell culture medium should be removed and replaced with freshly prepared growth medium every 2–3 days (depending on cell density). Upon isolation, primary cells should be tested for mycoplasma contamination. Primary human brain cell can be maintained in vitro at least 3 weeks or until they expand adequately for plating or passaging. The

phenotype of these cells can be confirmed by immunostaining for cell-type-specific marker expression and according to previously published protocols [24–26].

18. Use a p1000 micropipette to distribute the solution uniformly on the whole surface of the Isopore™ membrane. Pay attention to how deep the pipette tip is immersed into the solution: do not allow the pipette tip to touch the filter. Do not introduce bubbles.
19. The duration of trypsin–EDTA digestion should be determined empirically. After 3 min, take out the plate from the incubator and evaluate under the microscope when most cells are dislodged. Tap the side of the plate before visual examination. Enzymatic digestion should generally not exceed 15–20 min.
20. Seed the same number of human cells on a precoated glass slide without the Isopore™ membrane. This will allow the visual inspection of adhered cells under the microscope and, in the following days, the examination of cell growth.
21. If needed, after 5 days of co-culture, endothelial cells cultured to confluence on the luminal side of the filter can be stained with CellMask™ Green Plasma Membrane Stain, whereas astrocytes, grown on the abluminal side, can be stained with CellMask™ Orange Plasma Membrane Stain, in order to mark cellular boundaries and test confluency by direct observation at inverted fluorescence microscope.

Acknowledgments

We would like to thank Humanitas Neuro Center neurosurgeons for providing brain tissues and Pardi's Lab (San Raffaele Hospital) for the NVU microfluidic integration. This work was supported by the Ministero della Salute (Grant No. GR-2018-12367117 and GR-2019-12370776) to E.L and by Fondazione Italiana Sclerosi Multipla (grant FISM 2019/R-Single/032) to M.M. Illustrations were created with BioRender (©BioRender: biorender.com).

References

1. Kacem K, Lacombe P, Seylaz J, Bonvento G (1998) Structural organization of the perivascular astrocyte endfeet and their relationship with the endothelial glucose transporter: a confocal microscopy study. Glia 23:1–10. https://doi.org/10.1002/(SICI)1098-1136(199805)23:1<1::AID-GLIA1>3.0.CO;2-B
2. Mathiisen TM, Lehre KP, Danbolt NC, Ottersen OP (2010) The perivascular astroglial sheath provides a complete covering of the brain microvessels: an electron microscopic 3D reconstruction. Glia 58:1094–1103. https://doi.org/10.1002/glia.20990
3. Furtado GC, Piña B, Tacke F et al (2006) A novel model of demyelinating encephalomyelitis induced by monocytes and dendritic cells. J Immunol 177:6871–6879. https://doi.org/10.4049/jimmunol.177.10.6871

4. Herting CJ, Chen Z, Maximov V et al (2019) Tumour-associated macrophage-derived interleukin-1 mediates glioblastoma-associated cerebral oedema. Brain 142:3834–3851. https://doi.org/10.1093/brain/awz331
5. Codarri L, Gyülvészi G, Tosevski V et al (2011) RORγt drives production of the cytokine GM-CSF in helper T cells, which is essential for the effector phase of autoimmune neuroinflammation. Nat Immunol 12:560–567. https://doi.org/10.1038/ni.2027
6. Holman DW, Klein RS, Ransohoff RM (2011) The blood–brain barrier, chemokines and multiple sclerosis. Biochim Biophys Acta Mol basis Dis 1812:220–230. https://doi.org/10.1016/j.bbadis.2010.07.019
7. Latorre D, Kallweit U, Armentani E et al (2018) T cells in patients with narcolepsy target self-antigens of hypocretin neurons. Nature 562:63–68. https://doi.org/10.1038/s41586-018-0540-1
8. Brochard V, Combadière B, Prigent A et al (2009) Infiltration of CD4+ lymphocytes into the brain contributes to neurodegeneration in a mouse model of Parkinson disease. J Clin Invest 119:182–192. https://doi.org/10.1172/JCI36470
9. Baruch K, Rosenzweig N, Kertser A et al (2015) Breaking immune tolerance by targeting Foxp3+ regulatory T cells mitigates Alzheimer's disease pathology. Nat Commun 6:7967. https://doi.org/10.1038/ncomms8967
10. Dansokho C, Ait Ahmed D, Aid S et al (2016) Regulatory T cells delay disease progression in Alzheimer-like pathology. Brain 139: 1237–1251. https://doi.org/10.1093/brain/awv408
11. Ito M, Komai K, Mise-Omata S et al (2019) Brain regulatory T cells suppress astrogliosis and potentiate neurological recovery. Nature 565:246–250. https://doi.org/10.1038/s41586-018-0824-5
12. Obermeier B, Daneman R, Ransohoff RM (2013) Development, maintenance and disruption of the blood-brain-barrier. Nat Methods 19:1584–1596. https://doi.org/10.1038/nm.3407.Development
13. Allen NJ, Barres BA (2009) Glia—more than just brain glue. Nature 457:675–677. https://doi.org/10.1038/457675a
14. Clarke LE, Barres BA (2013) Emerging roles of astrocytes in neural circuit development. Nat Rev Neurosci 14:311–321. https://doi.org/10.1038/nrn3484
15. Molofsky AV, Krenick R, Ullian E et al (2012) Astrocytes and disease: a neurodevelopmental perspective. Genes Dev 26:891–907. https://doi.org/10.1101/gad.188326.112
16. Khakh BS, McCarthy KD (2015) Astrocyte calcium signaling: from observations to functions and the challenges therein. Cold Spring Harb Perspect Biol 7:a020404. https://doi.org/10.1101/cshperspect.a020404
17. Lee J-H, Kim J, Noh S et al (2021) Astrocytes phagocytose adult hippocampal synapses for circuit homeostasis. Nature 590:612–617. https://doi.org/10.1038/s41586-020-03060-3
18. Fossati G, Matteoli M, Menna E (2020) Astrocytic factors controlling synaptogenesis: a team play. Cell 9:2173. https://doi.org/10.3390/cells9102173
19. Liddelow SA, Barres BA (2017) Reactive astrocytes: production, function, and therapeutic potential. Immunity 46:957–967. https://doi.org/10.1016/j.immuni.2017.06.006
20. Abbott NJ, Rönnbäck L, Hansson E (2006) Astrocyte-endothelial interactions at the blood-brain barrier. Nat Rev Neurosci 7: 41–53. https://doi.org/10.1038/nrn1824
21. Heithoff BP, George KK, Phares AN et al (2021) Astrocytes are necessary for blood--brain barrier maintenance in the adult mouse brain. Glia 69:436–472. https://doi.org/10.1002/glia.23908
22. Liu C-Y, Yang Y, Ju W-N et al (2018) Emerging roles of astrocytes in neuro-vascular unit and the tripartite synapse with emphasis on reactive gliosis in the context of Alzheimer's disease. Front Cell Neurosci 12:193
23. Michinaga S, Koyama Y (2019) Dual roles of astrocyte-derived factors in regulation of blood-brain barrier function after brain damage. Int J Mol Sci 20:1–22. https://doi.org/10.3390/ijms20030571
24. Zhang Y, Sloan SA, Clarke LE et al (2017) Purification and characterization of progenitor and mature human astrocytes reveals transcriptional and functional differences with mouse. Neuron 89:37–53. https://doi.org/10.1016/j.neuron.2015.11.013
25. Lauranzano E, Campo E, Rasile M et al (2019) A microfluidic human model of blood–brain barrier employing primary human astrocytes. Adv Biosyst 3:e1800335. https://doi.org/10.1002/adbi.201800335
26. Guttenplan KA, Liddelow SA (2018) Astrocytes and microglia: models and tools. J Exp Med 216:71–83. https://doi.org/10.1084/jem.20180200

Chapter 13

Fabrication of Microtube-Embedded Chip to Mimic Blood–Brain Barrier Capillary Vessels

Dilshan Sooriyaarachchi, Shahrima Maharubin, and George Z. Tan

Abstract

Capillary vessels of the blood–brain barrier (BBB) regulate the transportation of solutes into the brain and provide defense against the disease-causing pathogens and neurotoxins present in the blood. Paradoxically, this regulation also prevents drug transportation into the brain. These unique characteristics of the BBB cause impediment in the treatment of neurological diseases. The development of preclinical models that mimic the BBB capillary vessel is crucial to investigate the complex transport mechanism. Microfluidics-based in vitro models are now extensively investigated for therapeutic applications due to the ability to create a tunable dynamic extracellular microenvironment. One of the main challenges of creating a BBB-on-a-chip is to recapitulate the tubular capillary structure. This chapter presents two novel fabrication methods for microfluidic devices embedded with tubular micro-channels that resemble the diameter and morphology of capillary vessels. These microfluidic devices can be seeded with cells for physiological and pathological studies to support future drug development.

Key words Electrospinning, Multi-jet printing, Fibrous scaffolds, Hybrid bioprinting

1 Introduction

The brain is protected through layers of defense including a thick skull (7 mm), a protective fluid (cerebrospinal), and a protective membrane (meninges) from any possible physical damage or injury. An additional distinct protective element in the brain is the blood–brain barrier (BBB), a barricade that offers a shield against the undesired elements in the blood. The key role of this barrier is to offer protection against disease-causing pathogens and neurotoxins present in the blood [1]. Another function of this barrier is to preserve the homeostasis of the brain through regulating the transportation of necessary ions, hormones, nutrients, and water between the blood and the brain [2]. The barrier itself consists of specialized brain microvascular endothelial cells, which are similar to other vertebrates [3], but the biochemistry and function of these cells are different from other endothelium cells in the body. Inside

Nicole Stone (ed.), *The Blood-Brain Barrier: Methods and Protocols*,
Methods in Molecular Biology, vol. 2492, https://doi.org/10.1007/978-1-0716-2289-6_13,

the capillaries, these highly specialized brain microvascular endothelial cells are compressed extremely close together, forming tight junctions. These specialized junctions regulate paracellular and transcellular permeability and are essential for BBB integrity. A loss of BBB integrity can affect neurological diseases such as brain cancer, stroke, autism, and Alzheimer's disease [4].

Under homeostatic conditions, the BBB prevents undesirable elements from entering the brain. However, this same feature of BBB regulates the essential drug transportation into the brain, posing an obstacle to drug delivery in the treatment of neurological diseases and brain tumors [5, 6]. Insufficient penetration of therapeutic drugs across the BBB remains a pivotal reason for failure in drug development and disease control [4]. Hence, the development of preclinical models of the BBB is crucial to evaluate drug permeability and analyze the impairment of the BBB. For years, in vivo models have been the most widely used and effective approach to study drug delivery. However, the use of animals in both pathology and drug development is costly, labor-intensive, and under constant scrutiny for ethical reasons. Correspondingly, the reliability of these in vivo techniques also remains in question. More than 80% of successful drug candidates in animal models were unsuccessful later in clinical trials [7, 8]. There has been a growing research interest in developing reliable, robust, and cost-effective in vitro models that closely mimic the human BBB microenvironment. These models will pave the road for effective disease diagnosis and drug delivery. Microfluidics-based in vitro models, also known as tissue/organ-on-chip, have emerged as a popular therapeutic tool due to the ability to create a tunable dynamic extracellular microenvironment [9–11]. The key limitations of widely used in vitro platforms for BBB is to recapitulate the three-dimensional tubular structure [12] and achieve the required diameter (7–10 μm) that resembles the diameter of a capillary vessel [13].

This chapter presents two novel fabrication methods for microtube-embedded microfluidic devices that mimic the geometry of capillary vessels.

1. Core-sheath electrospun microfiber-embedded microfluidic device for BBB model: In this method, polycaprolactone (PCL) microtubes (with a diameter < 10 μm) were fabricated by core-sheath electrospinning and embedded in a bridge between two reservoirs in a PDMS chip.
2. Polymer-coated sugar microfiber-embedded microfluidic device for BBB model: In this method, polycaprolactone (PCL)-coated sugar microfibers were embedded in the PDMS compartments. The sugar cores were then dissolved to form microtunnels with a diameter < 20 μm.

2 Materials

Store all materials at room temperature. Diligently follow all waste disposal regulations when disposing waste materials.

2.1 Electrospinning

1. Polycaprolactone (PCL, molecular weight = 80,000).
2. Chloroform (anhydrous, ≥99%, contains 0.5–1.0% ethanol as stabilizer).
3. Polyethylene glycol (PEG, molecular weight = 300,000).
4. Magnetic stirrer.
5. Glass beaker (250 mL).
6. Parafilm film (4-inch width, clear).

2.2 Sugar Fiber Coating

1. Pure cane sugar.
2. Ceramic stirring hot plate (at least 7 × 7 inches).
3. Glass beaker (250 mL).
4. Glass stir rod.

2.3 Polydimethylsiloxane Chip

1. Fused deposition modelling 3D printer.
2. Polylactic acid (PLA) filament that fits your 3D printer.
3. Polydimethylsiloxane, (PDMS, SYLGARD 184 Elastomer, DOW Corning Co., Midland, MI).
4. SYLGARD 184 cure agent.
5. Glass beaker (250 mL).
6. Glass stir rod.
7. Fibronectin solution from human fibroblasts (0.5 mg/mL).
8. Hanks' balanced salt solution (HBSS).
9. Deionized water.

3 Methods

All solutions are prepared at ambient temperature unless indicated otherwise. Please carefully follow all waste disposal regulations when disposing of any material.

3.1 PCL Solution (9.3% w/v) for Electrospinning

1. Add 9.3 g PCL pellets in a glass beaker.
2. Add 100 mL chloroform into the beaker. [Note 1]
3. Place a 2 cm ceramic stirrer bar in the beaker, and cover the beaker with a piece of paraffin film.
4. Place the beaker on a magnetic stirring, and stir at 500–1000 revolution per minute (RPM) for at least 4 h at room temperature.

3.2 PEG Solution (30% w/v) for Electrospinning

1. Add 15 g PCL pellets in a glass beaker.
2. Add 50 mL chloroform into the beaker.
3. Place a 2 cm ceramic stirrer bar in the beaker, and cover the beaker with a piece of paraffin film.
4. Place the beaker on a magnetic stirring, and stir at 500–1000 RPM for at least 4 h at room temperature.

3.3 Fibronectin Solution (10 μg/mL) for Coating

1. Prepare 5 mL HBSS in a sterile 15 mL centrifuge tube.
2. Add 100 μL fibronectin solution (0.5 mg/mL) to the HBSS.

3.4 Fabrication of PDMS Chip

1. Design a mold (1 cm × 1 cm × 0.4 mm) for the chip by a computer-aided design (CAD) software. The mold is the reverse of the chip with two reservoirs. The size of the mold can vary according to the need.
2. 3D print two molds by fused deposition modelling using PLA filament. Other materials such as PCL and ABS are also acceptable.
3. Mix 50 mL SYLGARD 184 elastomer with the curing agent at a 10:1 ratio. [Note 2] Stir well until the curing agent is homogeneously dispersed.
4. Place the molds on the bottom of a petri dish. Keep a distance of at least 1 cm between the two molds.
5. Pour the mixed PDMS onto the molds to fully cover them.
6. Let the PDMS cure for 36 h. [Note 3]
7. Peel the PDMS off carefully. [Note 4]
8. Cut the PDMS into a rectangular strip (4 cm × 2 cm) as shown in Fig. 3a and two separate squares (2 cm × 2 cm) as shown in Fig. 3b. The rectangular chip is for core-sheath electrospinning. The square chips are for sugar fiber-incorporated chip.

3.5 Fabrication of Microfluidic Chip Embedded with PCL Microtubes

The fabrication of porous microtubes is shown in Fig. 1. A core-sheath spinneret is needed for the electrospinning process.

1. Fill the two-channel syringe pump with PEG (30% w/v) and PCL solutions (9.3% w/v).
2. Connect the PEG syringe to the core channel of the spinneret and the PCL syringe to the sheath channel of the spinneret. The nozzle sizes are 25 gauge for the core and 18 gauge for the sheath, respectively.
3. Place two parallel metal bars under the spinneret. The distance between the spinneret and the chip is 16 cm.

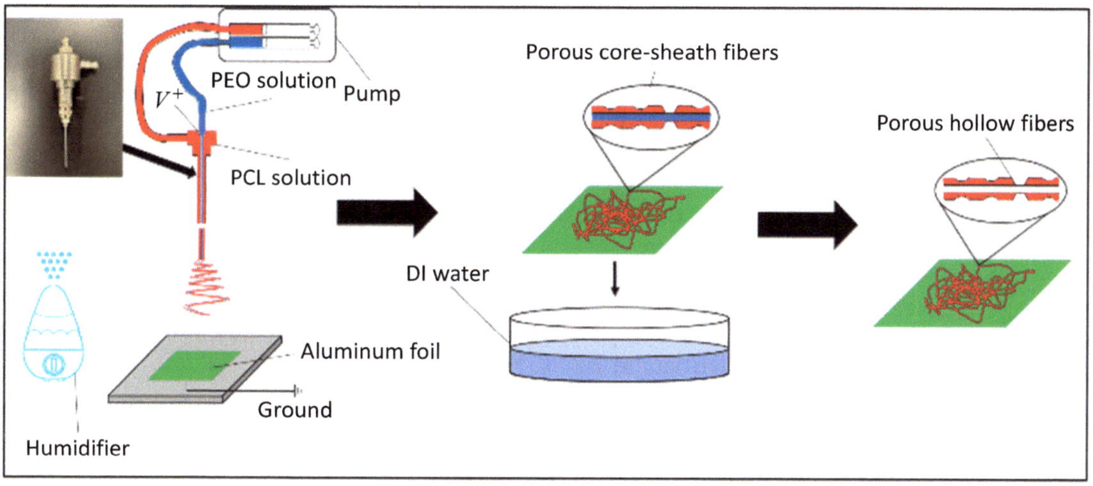

Fig. 1 Fabrication of PCL microtubes

4. Connect the two parallel bars to the grounding wire of the electrospinning device.
5. Place the rectangular PDMS chip under the spinneret and between the two parallel bars. A stand may be needed to support the chip to make sure the top surface of the chip is 1–5 mm above the top surface of the metal bars.
6. Place a portable humidifier in the electrospinning chamber to control the humidity. Turn on the humidifier, and maintain the humidity at 50–60%.
7. Set the electrospinning process parameters as follows, and start the electrospinning:
   - Pump rate: 0.5 mL/h
   - Spinneret voltage: 8.5 kV
   - Electrospinning time: 3–5 min
8. After the electrospinning, a layer of aligned porous microfibers will be deposited onto the chip. Carefully trim off excessive microfibers around the chip by a scissor.
9. Trim off the microfibers so that the two ends of the microfibers are in two reservoirs of the chip respectively. The aligned microtubes should be laid on the ridge between the two reservoirs of the chip with the two ends connecting the two reservoirs.
10. Add 2 mL of deionized water onto the microfibers using a pipette to dissolve the PEG core. Carefully remove the water. Repeat this rinsing process for three times. The microfibers will turn to microtubes.
11. Sterilize the chip with microtubes by ultraviolet exposure for 30 min.

12. Air-dry the chip in a sterile environment for 24 h.
13. Apply a thin layer of mixed PDMS around the reservoir. Make sure the microtubes are sealed in the PDMS at the ridge between the two reservoirs. [Note 5]
14. Let the PDMS cure for 36 h.
15. Seal the chip on a glass slide by plasma bonding or PDMS bonding.
16. Create the inlets at the two reservoirs of the chip.
17. Fill the chip with human fibronectin solution (10 μg/mL), and incubate at 37 °C for 2 h.
18. Drain the fibronectin solution, and flush the chip with HBSS.

3.6 Fabrication of Microfluidic Chip with Micro-channels

1. Add 10 g cane sugar in a glass beaker, and heat the sugar at 160 °C until a semi-solid texture is obtained.
2. Insert a glass rod into the sugar melt, and draw a sugar fiber from it.
3. Draw numerous fibers with different diameters.
4. Examine the fiber diameter under a microscope. Select ten fibers with diameter less than 50 μm.
5. Cut the selected fibers into 2 cm.
6. Place fibers onto the square microfluidic chip bridging the two reservoirs. Make sure the fibers are laid on the ridge between the two reservoirs of the chip with the two ends connecting the two reservoirs. Place a total of five to ten fibers in parallel.
7. Sterilize the chip with sugar fibers by ultraviolet exposure for 30 min.
8. Apply a thin layer of mixed PDMS around the reservoir. Make sure the fibers are sealed in the PDMS at the ridge between the two reservoirs.
9. Let the PDMS cure for 36 h.
10. Seal the chip on a glass slide by plasma bonding or PDMS bonding.
11. Create the inlets/outlets at the two reservoirs of the chip.
12. Feed hot deionized water (75 °C) into the chip through the inlet to dissolve the sugar fibers and form the micro-channels between the two reservoirs.
13. Fill the chip with human fibronectin solution (10 μg/mL), and incubate at 37 °C for 2 h.
14. Drain the fibronectin solution, and flush the chip with HBSS.

The two types of microfluidic chips are illustrated in Fig. 2. A complete device is shown in Fig. 3.

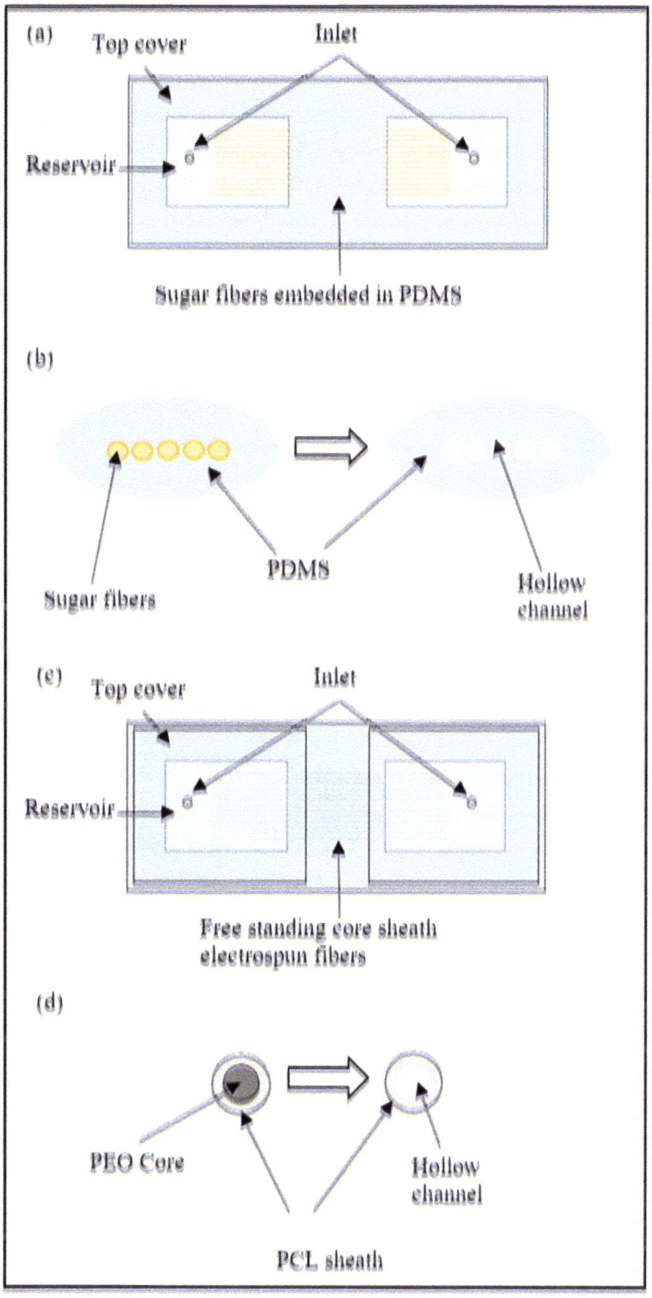

Fig. 2 (**a**) Top view of the BBB-on-a-chip model incorporated with sugar fibers. (**b**) Cross-section view of the micro-channels of the BBB-on-a-chip model before and after dissolving the sugar fibers. (**c**) Top view of the BBB-on-a-chip model incorporated with PCL/PEO fibers. (**d**) Cross section of a PCL/PEO fiber before and after dissolving the PEO core [13]

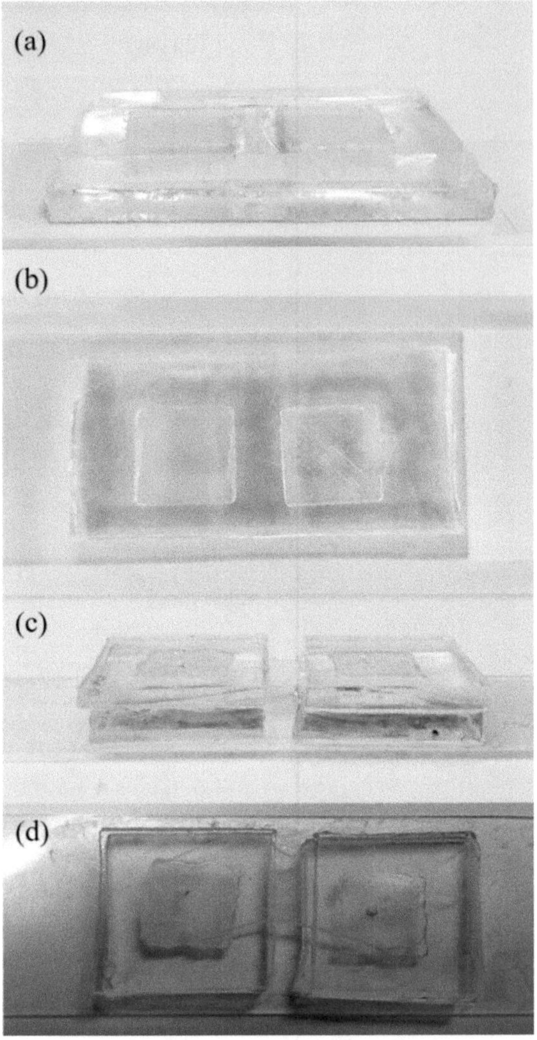

Fig. 3 (**a**) Front view of a BBB-on-a-chip with micro-channels created by sugar fibers. (**b**) Top view of a BBB-on-a-chip with micro-channels created by sugar fibers. (**c**) Front view of a BBB-on-a-chip incorporated with PCL micro-channels. (**d**) Top view of a BBB-on-a-chip model incorporated with PCL micro-channels [13]

4 Notes

1. Prepare the organic solvents in a laboratory fume hood.
2. Always prepare the PDMS by adding the curing agent to the SYLGARD 184, not the other way.
3. Curing the PDMS in an oven at 37 °C will help reduce the risk of shape deformation or formation of air bubbles.

4. Using a plastic petri dish instead of a glass one will help you peel the PDMS easily.
5. Once the thin layer of PDMS has been applied, let it dry for 15–30 min before placing the top layer. Place a heavy object on the top to ensure proper sealing. Apply superglue around the boundary can ensure proper sealing.

References

1. Abbott NJ (2005) Dynamics of CNS barriers: evolution, differentiation, and modulation. Cell Mol Neurobiol 25(1):5–23
2. Abbott NJ, Patabendige AA, Dolman DE et al (2010) Structure and function of the blood-brain barrier. Neurobiol Dis 37(1):13–25
3. Booth R, Kim H (2012) Characterization of a microfluidic in vitro model of the blood-brain barrier (μBBB). Lab Chip 12(10):1784–1792
4. Cardoso FL, Brites D, Brito MA (2010) Looking at the blood–brain barrier: molecular anatomy and possible investigation approaches. Brain Res Rev 64(2):328–363
5. Cucullo L, Aumayr B, Rapp E et al (2005) Drug delivery and in vitro models of the blood-brain barrier. Curr Opin Drug Discov Devel 8(1):89–99
6. Hatherell K, Couraud P-O, Romero IA et al (2011) Development of a three-dimensional, all-human in vitro model of the blood–brain barrier using mono-, co-, and tri-cultivation Transwell models. J Neurosci Methods 199(2):223–229
7. Pamies D, Hartung T, Hogberg HT (2014) Biological and medical applications of a brain-on-a-chip. Exp Biol Med 239(9):1096–1107
8. Pardridge WM (2002) Why is the global CNS pharmaceutical market so under-penetrated? Drug Discov Today 7(1):5–7. Elsevier
9. Pardridge WM (2003) Blood-brain barrier drug targeting: the future of brain drug development. Mol Interv 3(2):90
10. Pardridge WM (2005) The blood-brain barrier: bottleneck in brain drug development. NeuroRx 2(1):3–14
11. Perrin S (2014) Preclinical research: make mouse studies work. Nature 507(7493):423
12. Risau W, Wolburg H (1990) Development of the blood-brain barrier. Trends Neurosci 13(5):174–178
13. Sooriyaarachchi D, Zhou Y, Maharubin S et al (2020) Microtube-embedded microfluidic devices for potential applications in blood brain barrier research. Procedia Manuf 48: 294–301

Chapter 14

BBB-on-a-Chip: Modeling Functional Human Blood-Brain Barrier by Mimicking 3D Brain Angiogenesis Using Microfluidic Chip

Somin Lee, Minhwan Chung, and Noo Li Jeon

Abstract

Organ-on-a-chip enables human cell-based 3D tissue culture, which recapitulates the physiological structure and function of the tissue. In terms of the blood–brain barrier (BBB) modeling, the 3D structure of the vessel is essential for studying the cellular interactions among BBB composing cells and investigating the barrier function. Here, we describe a BBB-on-a-chip model with 3D perfusable human vasculature tri-cultured with pericytes and astrocytes. The culture method is based on mimicking angiogenic sprouting since the barrier formation is parallel with angiogenesis during the developmental process. This microfluidic-based 3D tri-culture system enables the comparative study on how surrounding BBB-related cells affect brain angiogenic sprouting. Moreover, the engineered perfusable vasculature is eligible for quantitative analysis on barrier function such as efflux transport system. We expect the BBB-on-a-chip could be used to enhance understanding BBB-related pathologies as well as the drug modulating barrier function of BBB.

Key words Human blood–brain barrier, Efflux transport system, CNS angiogenesis, Organ-on-a-chip, Microfluidic chip, 3D in vitro vessel model

1 Introduction

The blood–brain barrier (BBB) is an important research topic in the central nervous system (CNS) disease research and its drug development [1, 2]. However, traditional in vitro models have been unable to recapitulate the structural and pathophysiological characteristics of BBB, especially the barrier function and cellular interactions among BBB components [3–5]. The advent of microfluidic-based organ-on-a-chip has contributed to the physiological modeling of human BBB, and numerous BBB-on-a-chip models with various culture methods were introduced previously [6, 7].

Nicole Stone (ed.), *The Blood-Brain Barrier: Methods and Protocols*,
Methods in Molecular Biology, vol. 2492, https://doi.org/10.1007/978-1-0716-2289-6_14,

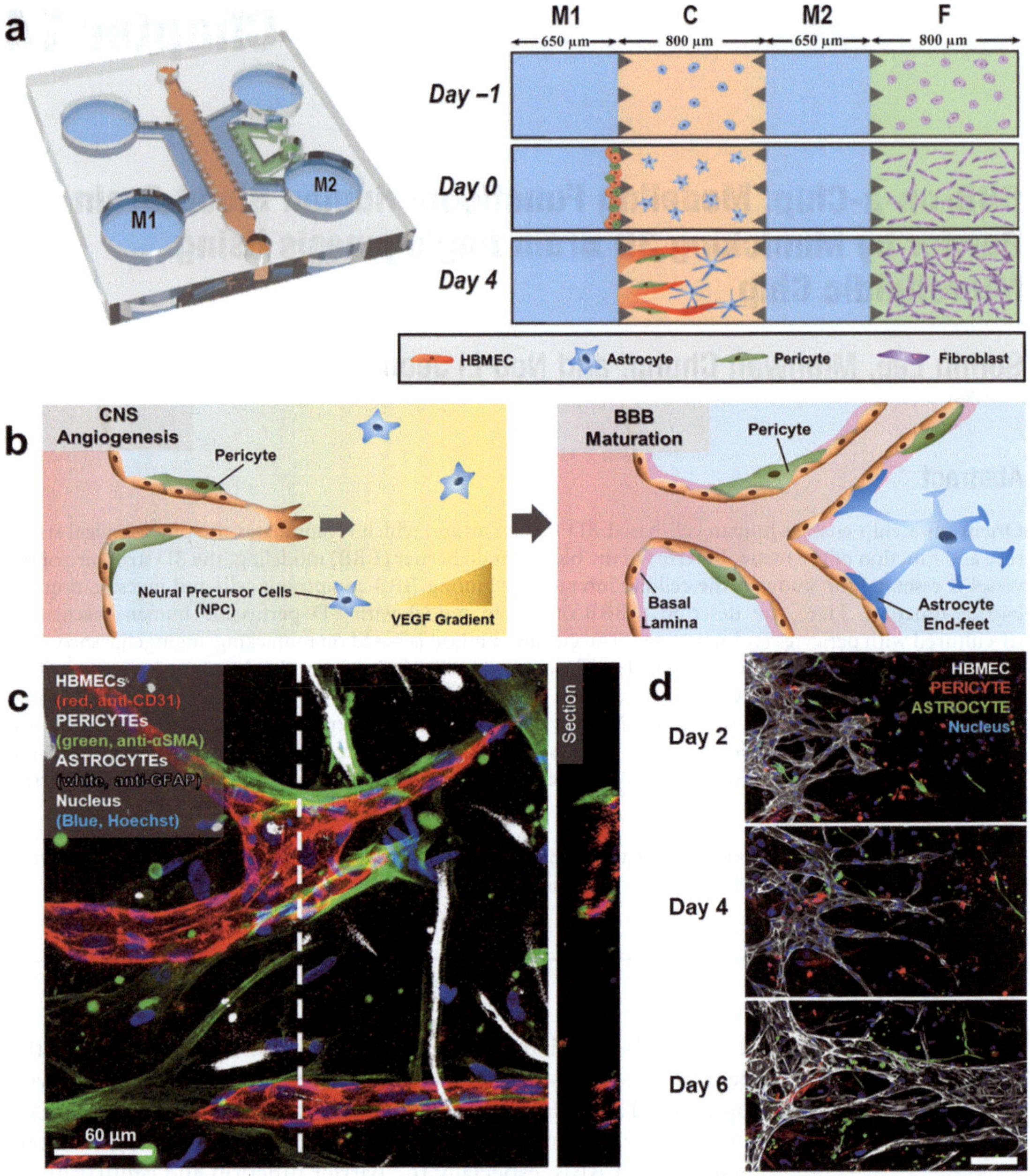

Fig. 1 Design of microfluidic chip and culture method of 3D BBB-on-a-chip. (**a**) Schematic view of the microfluidic chip and section view of microchannels. Dimensions of each microchannel and sequential cell loading to reconstitute 3D brain angiogenesis. (**b**) Conceptual description of BBB modeling though mimicking CNS angiogenesis. (**c**) Representative z-projected confocal image of BBB tri-culture in day 6. Scale bar = 60 μm. (**d**) z-projected confocal image of day-by-day sprouting of brain microvessel. Scale bar = 100 μm

Our BBB-on-a-chip is based on mimicking human CNS angiogenesis [8] (Fig. 1b), which is a unique feature in the development of CNS vasculature and also correlates with BBB barriergenesis

[9]. During this process, it is important for angiogenic sprouting to interact with surrounding astrocytes and pericytes. We adopted a microfluidic 3D angiogenic sprouting model initially developed by Kim et al. [10] and optimized the tri-culture method for mimicking the CNS microenvironment (Fig. 1). Physiological characteristics of BBB were verified by examining distinct morphological phenotypes of the vessel when tri-cultured with BBB composing cells. The 3D co-culture platform enabled quantitative analysis of the morphological characteristics of vasculature in different co-culture conditions.

The barrier function of our BBB-on-a-chip was verified in diverse aspects, such as measuring the expression level of tight junctions or vessel permeability. Among previous BBB-on-a-chip models, our model is specialized in the examination of the efflux transport system (Fig. 3a). This system mainly consists of a multidrug resistance (MDR) protein family, which is a critical reason why the drug penetration is so restricted in CNS [11, 12]. To investigate the functionality of the efflux transport system within the reconstructed model, we tried to assess the efficacy of the inhibitors of this system through the calcein-AM assay. This assay is a well-established method to easily visualize the efflux rate of fluorescent molecules intracellularly untaken by cells [13, 14]. By applying this method, we could measure the difference in the efflux rate of the fluorescent molecules depending on the presence of efflux transport inhibitors. The platform is one of the few in vitro models eligible for studying the function of the efflux transport system and testing the efficacy of transport inhibitors, which are now being considered an effective strategy to improve drug delivery success rate [15].

This chapter covers how the 3D human BBB can be reconstructed using microfluidic organ-on-a-chip platform, starting from chip fabrication. Moreover, we will introduce how the morphological phenotype of BBB vascular could be quantitatively analyzed. Specifically, we will also inform how to apply this model to study the drug transport system within BBB. This functional 3D human model will be a novel alternative to study deep biology of BBB and will serve as a useful tool to test therapeutics to improve drug delivery.

2 Materials

2.1 Microfluidic Chip Preparation

The microfluidic chips are prepared in advance. All cells and biochemical reagents are aliquoted as stock, and the necessary amount for each assay is prepared right before the experiment.

1. Microfluidic chip: manufactured by soft-lithography replica molding process using polydimethylsiloxane.

2. The master mold for soft lithography: drawing of the chip design was conducted using computer-aided designing (CAD) software. Dimension of the microfluidic channel is described in Fig. 1a (*see* **Note 1**). The master mold was fabricated through photo-lithography. SU-8 100 from MicroChem was used to pattern the microstructure with a height of 150–200 μm on a silicon wafer.
3. PDMS pre-polymer and curing agent from Dow Corning (Sylgard 184).
4. Biopsy puncher for punching the inlets, outlets, and reservoirs on PDMS chip.
5. Cover glass as the bottom substrate of the chip.

2.2 Cell Culture and Immunofluorescent Reagents

1. Fetal bovine serum (FBS).
2. Dimethyl sulfoxide (DMSO).
3. Phosphate-buffered saline (PBS).
4. Trypsin–EDTA.
5. Primary human brain endothelial cells (HBMECs) from Cell-Systems were passaged up to passage number 4 and preserved in liquid nitrogen until chip culture (*see* **Note 2**).
6. HBMEC culture medium: endothelial basal medium-2 (EBM-2) supplemented with EGM-2 MV BulletKit (EGM-2 MV) from Lonza.
7. Primary normal human astrocytes from Lonza were passaged up to passage number 4 and preserved in liquid nitrogen until chip culture.
8. Astrocyte culture medium: astrocyte growth medium (AGM) supplemented with AGM Bullet Kit from Lonza.
9. Primary human pericytes from PromoCell were passage up to passage number 6 and preserved in liquid nitrogen until chip culture.
10. Pericyte culture medium: pericyte growth medium 2 (PGM2) from PromoCell.
11. Primary normal human lung fibroblasts from Lonza were passaged up to passage number 6 and preserved in liquid nitrogen until chip culture.
12. Fibroblast culture medium: fibroblast growth medium (FGM) from Lonza.
13. Media for cell culture in microfluidic chip: EGM-2 MV and AGM are mixed in 1:1 ratio.
14. 10 mg/mL fibrin hydrogel solution: 20 mg of fibrinogen powder (from bovine plasma) diluted in PBS (*see* **Notes 3** and **4**).

15. 4-TIU/mL aprotinin stock solution (filtered through 0.22 μm-pore-size syringe filter).
16. 50-unit/mL thrombin stock solution (filtered through 0.22 μm-pore-size syringe filter).
17. 4% paraformaldehyde (PFA).
18. 0.2% Triton X-100.
19. 3% bovine serum albumin (BSA).
20. Antibodies to label endothelial vessel network: Alexa Fluor-conjugated mouse anti-human cluster of differentiation 31 (anti-CD31), Alex Fluor-conjugated mouse anti-human zonula occludens-1 (anti-ZO-1), Alex Fluor-conjugated mouse anti-human claudin-5, Alexa Fluor-conjugated mouse anti-human occluding, and Alexa Fluor-conjugated mouse anti-human VE-cadherin.
21. Antibody to label human astrocyte: Alex Fluor-conjugated anti-human glial fibrillary acidic protein (anti-GFAP).
22. Antibody to label human pericyte: Alex Fluor-conjugated anti-human alpha-smooth muscle actin (anti-α-SMA).
23. Hoechst 33342.
24. Calcein-AM solution.
25. Efflux transporter inhibitors: Valspodar (PSC-833, Adooq) for inhibiting p-glycoprotein (p-gp) and Elacridar (GF120918, Tocris) for inhibiting p-gp and breast cancer resistance protein (BCRP).
26. Dimethyl sulfoxide (DMSO).

3 Methods

3.1 Primary Cell Culture

All primary cells were cultured in a humidified incubator with condition of CO2 5% and 37 °C. Culture solution for cryopreservation was fetal bovine serum (FBS) with 5% dimethyl sulfoxide (DMSO). Phosphate-buffered saline (PBS) and trypsin with EDTA were used in cell passaging.

3.2 Preparing Microfluidic Chip

The master mold with microstructure patterned on silicon wafer is firmly attached on the bottom of plastic container, ready to be used for replica molding.

1. Liquid PDMS mixture (pre-polymer and curing agent is completely mixed in 10:1 weight ratio) was poured and degassed to remove all bubbles from PDMS mixture.
2. PDMS is polymerized on a hot plate for 30 min until it becomes solid state.

3. The cross-linked PDMS polymer, which is solid but flexible, is peeled off from the master, and the holes for hydrogel injection and media reservoirs are punched using biopsy puncher.
4. Prepared PDMS piece is covalently bonded to clean coverslip right after oxygen plasma treatment.
5. The chip is stored in dry oven for at least 2 days before hydrogel patterning to restore hydrophobicity of the microchannel surface.

3.3 Cellular Hydrogel Loading and 3D Cell Culturing

1. PDMS chips for the experiment are sterilized using UV lamp on cell bench for 20–30 min.
2. 10 mg/mL fibrin hydrogel solution is made by adding PBS to fibrinogen powder. After waiting 30 min at room temperature to make sure the powder is dissolved enough, the solution is filtered through 0.22 μm-pore-size syringe filter. Then, 250 μL of 10 mg/mL fibrin hydrogel solutions is mixed with 40 μL of 4-TIU/mL aprotinin solution (or in volume ratio of 25:4), and the mixture is referred to as "hydrogel solution" from now on.
3. Astrocytes and fibroblasts are detached from cell culture dish using trypsin. The volume of trypsin depends on the total area of culture dish or culture flask. (1–2 mL of trypsin is enough for detaching cells cultured on 100 mm-diameter culture dish.) The cells were incubated in incubator for 2 min and taken out. The dish or flask is hand-tapped gently to apply physical force for cells to be detached. Cells were collected in the tube and are centrifuged and resuspended using media mixture for chip culture.
4. Prepare astrocyte and fibroblast suspension in cell concentration of 6×10^6 cells/mL of 8×10^6 cells/mL, respectively. Then, the cell suspension in desired concentration is aliquoted independently in the Eppendorf tube at a volume of 37.5 μL for each tube (the tube with astrocyte suspension is referred to as *tube A*, and the tube with fibroblast suspension is referred to as *tube F*).
5. 1 μL of thrombin solution is dispensed on each well of 96-well plate.
6. 12.5 μL of hydrogel solution is added in a single tube of fibroblast suspension (*tube F*). A total of 50 μL of cellular hydrogel solution is gently mixed to make the cell distribution even.
7. The cellular hydrogel solution is added on the single well of 96-well plate where thrombin is prepared in advance. Quickly mix the solution by pipetting once, and collect them without producing bubbles.

8. Load the mixture on the inlet of fibroblast channel (Channel F) gently until the solution appears on the other injection port. Wait 5 min until the cellular hydrogel gets cross-linked (*see* **Note 5**). Final concentration of fibroblasts loaded on the microfluidic chip is 6×10^6 cells/mL.
9. 12.5 μL of hydrogel solution is added in a single tube of astrocyte suspension (*tube A*). A total of 50 μL of cellular hydrogel solution is gently mixed to make cell distribution even.
10. The cellular hydrogel solution is added on the single well of 96-well plate where thrombin is prepared in advance. Quickly mix the solution by pipetting once, and collect them without producing bubbles.
11. Load the mixture on the inlet of the central channel (Channel C) gently until the solution appears on the other injection port. Wait 5 min until the cellular hydrogel gets cross-linked. Final concentration of astrocytes loaded on the microfluidic chip is 6×10^6 cells/mL.
12. Chip cell culture medium (mixture of EGM-2 MV and AGM 1: 1) is loaded only on the two top reservoirs among the total four reservoirs of each chip. Gentle vacuum aspiration is done on both channel ports of the bottom reservoirs to make the medium fill inside the media channels M1 and M2 (*see* **Note 6**).
13. Medium is filled on all four reservoirs of each chip, and the chip is cultured inside the cell culture incubator overnight.
14. The following day, HBMECs and pericytes are prepared as the previous **step 3** in Subheading 3.3.
15. Prepare HBMEC and pericyte suspension in both cell concentration of 5×10^6 cells/mL. Then, HBMEC and pericyte suspension is mixed in 5:1 volume ratio.
16. Medium in all four reservoirs in each chip is aspirated, while a small amount of medium is left inside the media channel.
17. 5 μL of the cell mixture is loaded on the one side of reservoir connected to medium channel M1, and the chip is tilted at 90° as the media channel M1 is positioned top and M2 is positioned bottom. Remain for 30 min as tilted inside the incubator.
18. Medium is filled again on all four reservoirs.
19. Take 5 μL of medium in one of the reservoirs connected to channel M1, and put on one of the reservoirs connected to channel M2. This generates a difference in medium level of the left and right side of the chip, which leads to hydrostatic pressure and interstitial flow inside the central channel filled with cellular hydrogel. The flow is generated toward the right side to the left side, as opposed to the sprouting direction. This interstitial flow encourages the angiogenic sprouting (*see* **Note 7**).

20. The culture is continued for 4–5 days to investigate the difference in sprouting rate depending on co-culture condition. Otherwise, the culture is continued over a week until the sprout reaches the other end of the central channel and both ends of the lumen are opened to generate perfusable network, eligible for permeability measurement or measurement of efflux transporter function. Media is changed every 2–3 days during culture by aspiration of the old media in reservoirs and filling them again.

3.4 Immuno-fluorescent Imaging for Quantitative Analysis on Vessel Morphology

1. After 4–5 days of culture, the angiogenic sprouting shows phenotypic difference in length or width depending on culture condition (Fig. 2a). The sample could be fixed anytime for a specific purpose of the experiment.
2. Medium in all four reservoirs is removed by aspiration, with a small portion of media remaining inside the media channels M1 and M2.
3. 4% PFA is added on two top reservoirs in each chip, enough to wet the bottom surfaces; incubate for 20 min at room temperature.
4. PFA is removed in all four reservoirs by aspiration. 0.2% Triton X-100 is added on two top reservoirs in each chip, enough to wet the bottom surfaces. Incubate for 20 min at room temperature.
5. Triton X-100 is removed in all four reservoirs by aspiration. Two top reservoirs in each chip are filled with 3% BSA, enough to wet the bottom surface. Incubate for 40 min at room temperature.
6. 3% BSA is removed in all four reservoirs by aspiration. The two top reservoirs in the chip are filled with 20 and 10 μL of immunofluorescent staining solution (reconstituted with 3% BSA at a concentration according to the manufacturer's recommendation; we did a1:200 dilution for anti-CD31 staining), respectively (*see* **Note 8**). The samples are stored at 4 °C for 2–3 days.
7. Immunostaining solution in all four reservoirs is removed by aspiration, and PBS is fully filled in the reservoirs, ready for fluorescence imaging or long-term storage in 4 °C.
8. To quantitatively analyze the morphological phenotype of angiogenic sprouting, anti-CD31 immunostained or lectin-stained vessel sprouting was imaged using 3D confocal microscopy.
9. Using Fiji, open-access software (http://fiji.sc.), the 3D confocal images were z-projected and converted to binary mask by applying threshold for pixel fluorescent intensity (Fig. 2b).

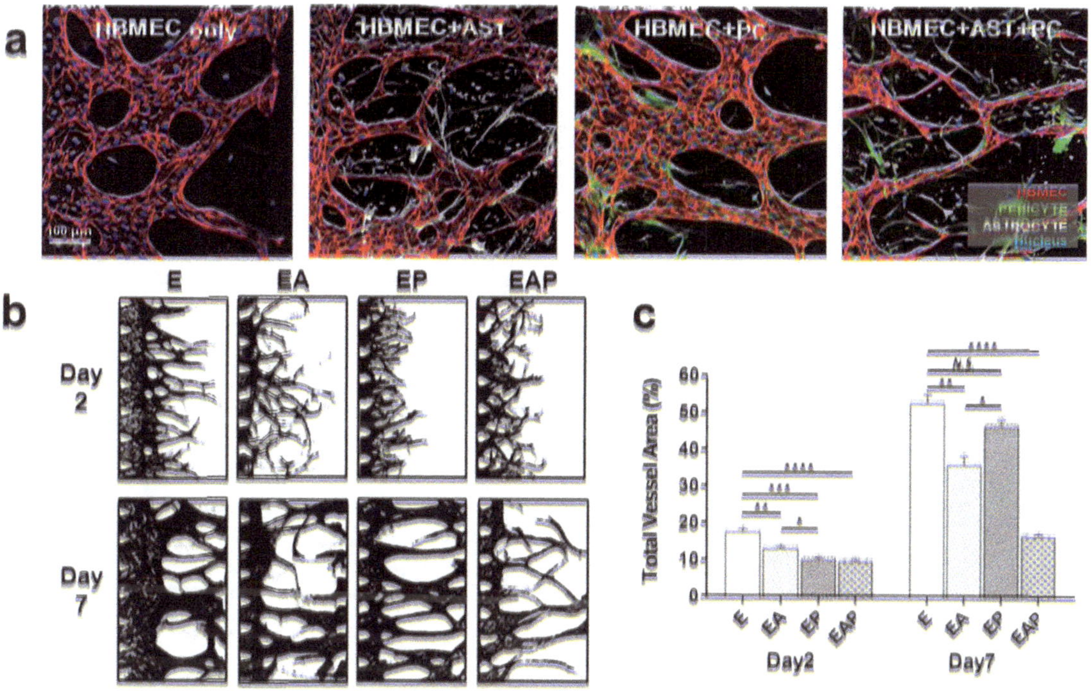

Fig. 2 Quantitative analysis on the morphological phenotype of human brain microvessel sprouting. (**a**) Representative z-projected confocal image of human brain microvasculature (red) in different co-culture conditions with BBB composing cells (astrocyte in white and pericytes in green). Scale bar = 100 μm. (**b**) The confocal image is z-projected and masked by setting manual threshold to the fluorescence intensity. Example of masked culture images in each co-culture condition in days 2 and 7. (**c**) Masked images are quantitatively analyzed in diverse parameters of vessel morphology. Example data of quantification of total vessel area among culture conditions. Unpaired two-tailed Student's t test was performed to obtain statistical comparisons of analyzed values, with the p value threshold for statistical significance set at *$p < 0.05$; **$p < 0.005$; ***$p < 0.0005$; ****$p < 0.00001$

10. Masked image was applied to AngioTool, free plug-in software in Fiji, which enables semi-automatic measurement of diverse vessel parameters including average sprout length or the number of vascular junctions (Fig. 2c) (*see* **Note 9**) [16].

3.5 Calcein-AM Efflux Assay for Testing the Efflux Transporter Function

The sequential process of the assay is displayed on Fig. 3b. Also, schematic description of the basic rationale of this assay is introduced on Fig. 3a.

1. After 6–7 days of culture, angiogenic sprouting starts to reach at the end of the opposite side of the central channel (Channel C), and both sides of the lumen are opened as perfusable vessel network (*see* **Note 10**).
2. Culture until day 9–10 to mature vasculature. At that time, samples were treated with Valspodar or Elacridar at a concentration of 10 μM for 10 h. Control groups are also treated with DMSO.

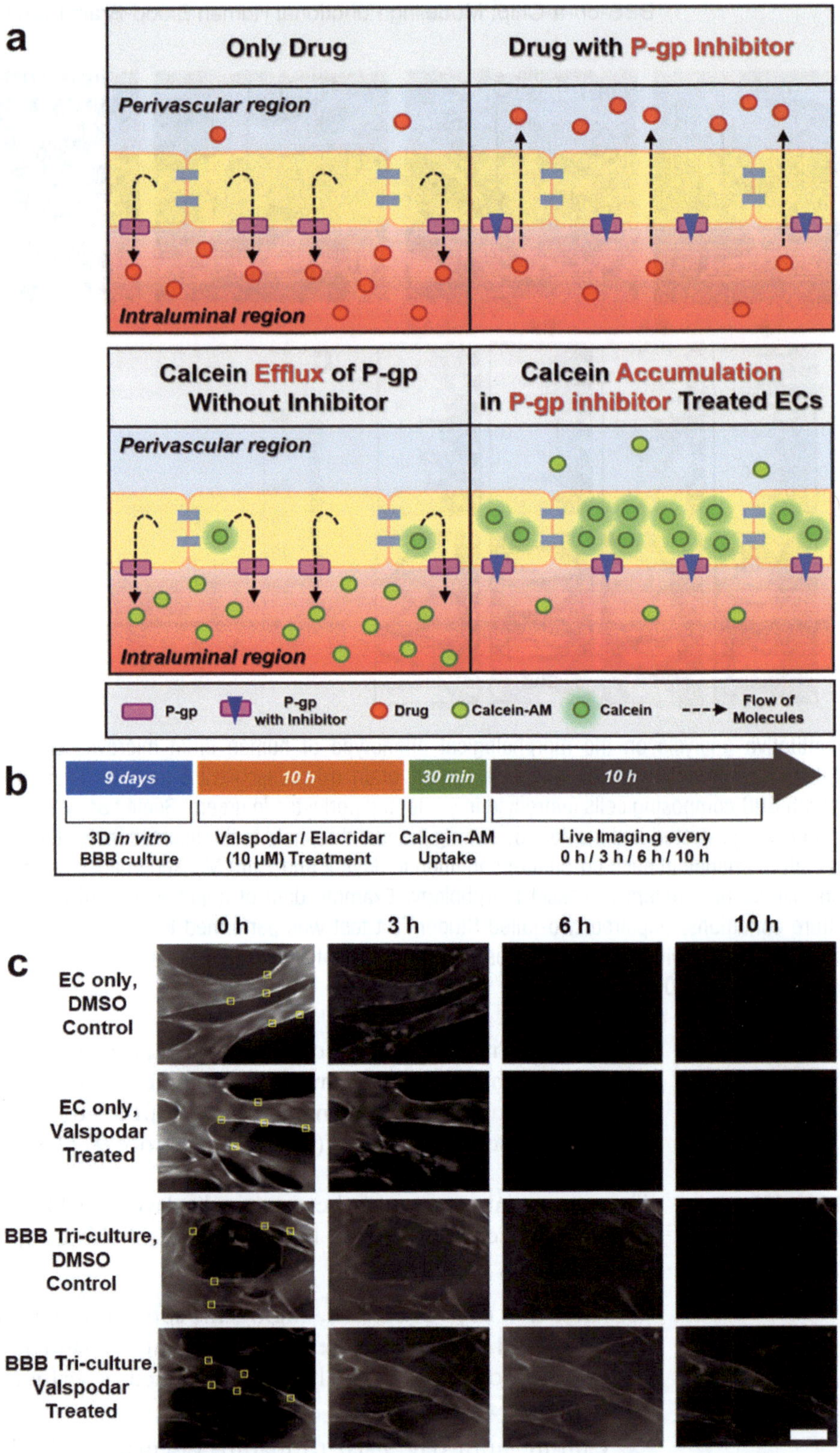

Fig. 3 Application of the BBB model to investigate efflux transport function using calcein-AM efflux assay. (**a**) Schematic description of adopting calcein-AM assay to visually analyze the effect of efflux transport inhibitor which enhances drug transport across BBB. (**b**) Timeline of calcein-AM assay on BBB-on-a-chip with treatment of efflux transporter inhibitors. (**c**) Example data of live imaging to measure efflux rate of calcein-AM in different co-culture conditions or drug treatment conditions. Scale bar = 100 μm

3. 10 h after treatment, the medium with drug was removed by aspiration. 2 μM calcein-AM solution containing the same concentration of the drug was treated on the two top reservoirs, 100 μL individually.
4. After 30 min, the calcein-AM solution was removed, and the medium with the drug is filled on all four reservoirs again, ready for live imaging.
5. The central channel (Channel C) was imaged by confocal microscopy every 0, 3, 6, and 10 h after washing with calcein-AM solution.
6. All images in different timepoints were z-projected. In each sample, five to ten regions of interest (ROIs) are designated at the boundary of the endothelium of vasculature, and all designated ROIs have the same size of square.
7. By fixing the positions of ROI in each sample, the change of the mean fluorescent intensity of ROI was measured along the timepoints (*see* **Note 11**).

4 Notes

1. The specific dimensions of the microfluidic chip design could be altered depending on the purpose of the experiment. However, several key structural dimensions are recommended for successful formation of angiogenic sprouting: the distance between the post and the post length (100 μm), width of the central channel (800–1000 μm), and the height of microchannel (100–200 μm).
2. It is important to passage endothelial cells before the confluency on the petri dish is no more than 90%. Also, it is recommended to passage the cells into a new petri dish with an initial density of no less than 50%. Endothelial cells are best to use in passage number 4, and it is not recommended to exceed passage number 6.
3. It is important to check the lot information (Certificate of Analysis) before purchasing fibrinogen powder. Buy the lot with % of clottable protein more than 89%.
4. Weigh the powder precisely as possible. Centrifuge enough to avoid powder being attached to the wall of the tube. Make sure to gather the powder entirely at the bottom of the tube. When adding warm PBS in calculated amount, never let the pipette tip directly touch the powder at the bottom. Leave it stable to be dissolved in 20–30 min, without voltexing.
5. For 50 μL of cellular hydrogel mixture, a maximum of eight microfluidic channels could be loaded at once with fast skillful

hands. (Approximately 5 μL is loaded per channel.) However, as the hydrogel cross-links quickly, it is recommended to load three channels at a time. Two experimenters could load simultaneously at a time to double the sample numbers.

6. As the microfluidic chip is initially hydrophobic before cell loading, media cannot be filled in the media channels M1 and M2 by itself which requires manual pressure by aspiration. However, heading the aspiration tip directly to the port of the media channel should be avoided unless it will disrupt the hydrogel pattern.
7. Interstitial flow generated by the difference in media level could activate or inhibit the angiogenic sprouting depending on the direction of the flow. To avoid the flow with same direction with sprouting, which inhibits the sprouting, generating a 10 μL difference in the left and right reservoir media level is recommended at the first day of EC culture. More details about the effect of the flow on angiogenic sprouting are described on the work by Kim et al. [17].
8. It is important to perfuse the staining solution entirely through the 3D tissue inside the microfluidic chip. Putting the device on horizontally shaking rocker (in slow speed) will help better perfusion of the solution.
9. As the ROI for quantitative analysis is defined by individual researchers, comparative description of the data normalized by control groups is recommended.
10. The perfusability of vessel network could be verified by flowing FITC-Dextran solution from one side of the media channel, which flows out to the other side of the media channel immediately in a couple of seconds through the vessel lumen.
11. Measured mean fluorescent intensity of each ROI could be interpreted in diverse ways depending on the purpose of the experiment. For example, the initial mean intensity (intensity at the point of starting live imaging) could be compared between the efflux transporter inhibited group and control group. Moreover, the portion of remaining mean intensity over initial mean intensity could be compared to infer the efflux rate between the experimental groups.

Acknowledgments

This work was supported by Basic Science Research Program through the National Research Foundation of Korea (NRF) funded by the Ministry of Science and Technology (2019R1A4A2001651). In addition, it was supported by the Global Ph.D. Fellowship Program also funded by the Ministry of Science and Technology (2016H1A2A1907378).

References

1. Banks WA (2016) From blood–brain barrier to blood–brain interface: new opportunities for CNS drug delivery. Nat Rev Drug Discov 15:275
2. Obermeier B, Daneman R, Ransohoff RM (2013) Development, maintenance and disruption of the blood-brain barrier. Nat Med 19:1584
3. Hatherell K, Couraud PO, Romero IA et al (2011) Development of a three-dimensional, all-human in vitro model of the blood–brain barrier using mono-, co-, and tri-cultivation Transwell models. J Neurosci Methods 199:223–229
4. Nakagawa S, Deli MA, Kawaguchi H et al (2009) A new blood–brain barrier model using primary rat brain endothelial cells, pericytes and astrocytes. Neurochem Int 54:253–263
5. Nakagawa S, Deli MA, Nakao S et al (2007) Pericytes from brain microvessels strengthen the barrier integrity in primary cultures of rat brain endothelial cells. Cell Mol Neurobiol 27:687–694
6. Lee CS, Leong KW (2020) Advances in microphysiological blood-brain barrier (BBB) models towards drug delivery. Curr Opin Biotechnol 66:78–87
7. Oddo A, Peng B, Tong Z et al (2019) Advances in microfluidic blood–brain barrier (BBB) models. Trends Biotechnol 37:1295–1314
8. Lee S, Chung M, Lee SR et al (2020) 3D brain angiogenesis model to reconstitute functional human blood–brain barrier in vitro. Biotechnol Bioeng 117:748–762
9. Umans RA, Henson HE, Mu F et al (2017) CNS angiogenesis and barriergenesis occur simultaneously. Dev Biol 425:101–108
10. Kim S, Lee H, Chung M et al (2013) Engineering of functional, perfusable 3D microvascular networks on a chip. Lab Chip 13:1489–1500
11. Kusuhara H, Suzuki H, Terasaki T et al (1997) P-Glycoprotein mediates the efflux of quinidine across the blood-brain barrier. J Pharmacol Exp Ther 283:574–580
12. Schinkel AH (1999) P-Glycoprotein, a gatekeeper in the blood–brain barrier. Adv Drug Deliv Rev 36:179–194
13. Holló Z, Homolya L, Davis CW et al (1994) Calcein accumulation as a fluorometric functional assay of the multidrug transporter. Biochim Biophys Acta 1191:384–388
14. Tiberghien F, Loor F (1996) Ranking of P-glycoprotein substrates and inhibitors by a calcein-AM fluorometry screening assay. Anti-Cancer Drugs 7:568–578
15. Miller DS, Bauer B, Hartz AM (2008) Modulation of P-glycoprotein at the blood-brain barrier: opportunities to improve central nervous system pharmacotherapy. Pharmacol Rev 60:196–209
16. Zudaire E, Gambardella L, Kurcz C et al (2011) A computational tool for quantitative analysis of vascular networks. PLoS One 6:e27385
17. Kim S, Chung M, Ahn J et al (2016) Interstitial flow regulates the angiogenic response and phenotype of endothelial cells in a 3D culture model. Lab Chip 16:4189–4199

Part V

Models to Study Specific Pathologies at the BBB: Cancer, Neurodegeneration, and More

Chapter 15

Cell Interplay Model to Assess the Impact of Glioma Cells on Blood–Brain Barrier Permeability

Cláudia Martins and Bruno Sarmento

Abstract

The blood–brain barrier (BBB) is the most selective protecting layer of the central nervous system (CNS) with unique neurovascular features. The BBB is known to undergo a process of molecular alterations during disease state, such as in the case of glioma. This results in a non-uniform permeability along the BBB layer, which retains intact regions but develops focal sites of higher leakiness, especially in the surrounds of the tumor core. Although essential to guarantee brain homeostasis, the BBB has been the Achilles heel of drug delivery to the brain since the early times of the first classification as "barrier," more than a century ago. Due to the presence of the BBB, the transport of drug molecules from the bloodstream to the brain parenchyma is highly restricted, and, therefore, clinically relevant therapeutic concentrations cannot be achieved. Research efforts have focused on the development of novel tools to ameliorate drug permeability across the BBB, including drug formulation into non-invasive delivery systems with brain targeting properties and techniques that allow a temporary disruption of the BBB. To strengthen the advancement of potential drug candidates, in vitro models that recapitulate the main in vivo features of BBB are required to perform a preliminary screening of permeability, both in health and disease conditions. Herein, a protocol to assemble a BBB in vitro model to screen drug permeability in a glioma disease state is detailed. The model consists of a BBB and glioma cell co-culture and aims at exploiting the effect of the interplay between the cell constituents on the permeability of drug molecules. Although simple and straightforward, the herein in vitro model presents a high reproducibility, cost-effectiveness, and a favorable time–benefit balance.

Key words Blood–brain barrier, Cell interplay, Co-culture, Drug permeability, Glioma

1 Introduction

The blood–brain barrier (BBB) acts as the most selective gatekeeper of the central nervous system (CNS). The "barrier" classification was first termed by Stern and Gautier in 1918 [1]. More than a century of research has allowed to unravel the structural composition of the specialized capillary beds of the BBB. They are mainly constituted by endothelial cells interconnected by tight junctions and surrounded by a basal lamina that is in intimate contact with astrocytic end feet, pericytes, and, sparsely, microglia and neuronal

Nicole Stone (ed.), *The Blood-Brain Barrier: Methods and Protocols*,
Methods in Molecular Biology, vol. 2492, https://doi.org/10.1007/978-1-0716-2289-6_15,

projections [2]. Moreover, the BBB presents a complex network of transport systems that orchestrate the efflux of toxic byproducts and influx of systemic molecules pivotal for a proper function of the CNS. Among other biological key roles, the BBB regulates molecular traffic, maintains ionic gradients, controls the central–peripheral neurotransmitter pool crosstalk, and modulates the immune surveillance [3].

Although endothelial cells are the major structural constituent of the BBB, they do not behave independently of the surrounding cells, and the BBB permeability is, thus, also affected by a molecular crosstalk between astrocytes, pericytes, microglia, neurons, and even circulating immune cells [3]. The same crosstalk is observed in brain disease state, such as in the case of glioma, between the BBB cellular components and tumor cells, ultimately leading to alterations in the BBB permeability [4]. Gliomas are characterized by a heterogeneous BBB, in which localized sites of higher leakiness are observed along the tumor core compared to the peritumoral region and the surrounding brain tissue [2]. The mechanisms behind this higher leakiness are associated with displacement or loss of astrocytic end feet, abnormal distribution of pericytes, and decrease of expression of endothelial cell junction proteins [5–7]. Notwithstanding, it is also clear that even high-grade gliomas, such as glioblastomas, present a clinically significant mass of tumor protected by an intact BBB, which emphasizes the importance of investigating drug distribution across an intact barrier as a critical first step in the development of novel therapies for brain diseases [8].

Although vital to assure CNS homeostasis, the BBB is also a major bottleneck in the field of drug delivery to the brain. The BBB has been reported to allow the brain influx of only 2% of all small-molecule drugs and exclude virtually all large-molecule therapeutics [9]. This is attributed to the tightly packed BBB endothelium characterized by low rates of transcytosis, as well as the presence of robust junction mechanisms and efflux pumps [10]. Therefore, efforts have focused on the exploitation of non-invasive drug delivery systems featuring brain targeting properties, minimally or medium invasive technologies such as microbubble-mediated focused ultrasound or convection-enhanced delivery, and also alternatives to the systemic route such as intranasal or intrathecal administration [11].

Advances in the field of drug discovery and development require translational in vitro models with human-relevant features to screen the permeability of drug candidates across the BBB, both in health and disease states.

Despite the need to approach in vivo conditions as much as possible, the 3R principle—Replacement, Reduction, and Refinement—has aroused to regulate animal experimentation and energize the creation of novel in vitro cell culture technologies [12]. The hCMEC/D3 cell line is a simple but widely used model to conduct preliminary studies of drug permeability across the BBB, presenting advantageous features compared to other primary culture models, such as a relatively easy cell source, low-complexity characterization, high reproducibility, and good time- and cost-effectiveness [13, 14]. Moreover, hCMEC/D3 cells express most of the transporters and receptors involved in BBB permeability in vivo [15]. Therefore, several authors have explored hCMEC/D3 BBB in vitro models for transmigration studies, either as monoculture [16–18] or co-culture [19–21].

Gathering all this information, the herein protocol describes the assembly of a BBB in vitro model to screen drug permeability in a glioma disease state. The model exploits the interplay between co-cultured hCMEC/D3 BBB endothelial and glioma cell lines of human source and the impact of this cell interplay on the BBB permeability of drug candidates. This protocol presents a straightforward methodology to obtain a BBB-glioma in vitro model of high reproducibility and relatively low cost and with an excellent time–benefit balance.

2 Materials

2.1 Cell Culture Lines, Reagents, and Disposables

1. hCMEC/D3 cell line, up to the 35th passage (*see* **Note 1**).
2. U87 cell line, up to the 35th passage (*see* **Note 1**).
3. hCMEC/D3 working medium: endothelial basal medium 2 (EBM-2) supplemented with 5% (v/v) fetal bovine serum (FBS), 1% (v/v) penicillin–streptomycin, 1.4 μM hydrocortisone, 5 μg/mL ascorbic acid, 1% (v/v) chemically defined lipid concentrate, 10 mM HEPES, and 1 ng/mL basic fibroblast growth factor (bFGF) (*see* **Note 2**).
4. U87 working medium: high glucose with ultraglutamine Dulbecco's Modified Eagle Medium (DMEM) supplemented with 10% (v/v) FBS and 1% (v/v) penicillin–streptomycin.
5. Rat tail collagen type I: 150 μg/mL in 0.02 M acetic acid (*see* **Note 3**).
6. Acetic acid: 0.02 M.
7. Phosphate-buffered saline (PBS; Sigma-Aldrich, St. Louis, MO, USA) at 0.01 M, pH = 7.4.
8. Trypsin–EDTA.
9. Trypan blue.
10. Transparent Transwell® cell culture inserts of polyester and 0.4 μm pore size.
11. Tissue culture-treated microplates.
12. Tissue culture flasks.
13. Centrifuge conical tubes.

2.2 Equipment

1. Laminar flow hood.
2. Cell culture incubator set at 37° and 5% CO_2 under a water saturated atmosphere.
3. Water bath set at 37°.
4. Inverted light microscope.
5. Hemocytometer.
6. Conical tube centrifuge.
7. EVOM^2 voltohmmeter equipped with STX2 chopstick electrodes (World Precision Instruments, Inc., Sarasota, FL, USA).

3 Methods

3.1 Separate Culture of Each Cell Line

3.1.1 Defrosting Process and Subsequent Maintenance

This section describes a step-by-step procedure from the defrosting process until cells reach around 80% confluency.

All reagents should be pre-warmed in the water bath at least 10 min prior to use on cells.

1. Defrost the cell cryovial by placing it in the water bath for about 5 min.
2. Seed the cryovial cell content in 0.2 mL medium per cm^2 tissue culture flask.
3. Keep the culture flask in the cell culture incubator.
4. Renew the medium after 24 h of culture and afterwards every 2 days (*see* **Note 4**).

3.1.2 Cell Detachment After Confluency

This section describes a step-by-step cell detachment procedure to be followed when cells reach around 80% cell confluency, allowing cell counting for routine subculture or experiment initiation.

All reagents should be pre-warmed in the water bath at least 10 min prior to use on cells.

1. Remove the cell culture medium.
2. Rinse the cell layer with 0.08 mL PBS per cm^2 tissue culture flask. Gently rock the flask back and forth to wash the residual medium from the cells.
3. Add 0.04 mL trypsin–EDTA per cm^2 tissue culture flask. Gently rock the flask back and forth, side to side, to assure a uniform distribution of trypsin. Place the cell culture back in the cell incubator for 3–5 min or until cells start detaching from the flask surface (*see* **Note 5**). If necessary, tap the flask gently to help detach the cells.
4. After confirming complete cell detachment under the inverted light microscope, add at least two times more culture medium than the trypsin–EDTA volume.
5. Collect the cell suspension in a conical tube, and centrifuge it for 5 min at 300 g and room temperature.

6. Remove the supernatant, and resuspend the cell pellet in the desired amount of culture medium (*see* **Note 6**).
7. Calculate the cell density by diluting an aliquot of cell suspension into trypan blue and counting the number of cells using a hemocytometer (*see* **Note 7**).
8. Determine the volume of cell suspension corresponding to the number of cells required to proceed with routine subculture or experiment initiation (*see* **Notes 8** and **9**).

3.2 Establishment of the In Vitro Cell Model

The in vitro model is composed of Transwell® cell culture inserts fitting the size of the respective microplate. The Transwell® cell culture inserts are composed of a microporous membrane that separates each microplate well into an upper (apical) and lower (basolateral) compartment representing the bloodstream and the brain parenchyma, respectively. To screen drug permeability across the BBB, from the bloodstream to the brain, drug transport is tested from the apical to the basolateral compartment.

All reagents should be pre-warmed in the water bath at least 10 min prior to use on cells.

1. Coat the Transwell® insert membrane with 100 μL of 150 μg/mL collagen coating solution per cm^2 insert cell growth area (*see* **Note 10**). Place the Transwell® microplate in the cell incubator for 1 h.
2. Remove the coating solution from the Transwell® insert, and wash it twice with approximately 500 μL PBS per cm^2 insert cell growth area.
3. Seed hCMEC/D3 cells on the apical compartment of the model by filling the Transwell® insert with a hCMEC/D3 cell suspension at a density of 4.6×10^4 cells per cm^2 insert cell growth area (around 500 μL per 0.9 cm^2 cell growth area) (*see* **Notes 11, 12,** and **13**). Fill the basolateral compartment with hCMEC/D3 working medium. Keep the Transwell® microplate in the cell incubator.
4. After 24 h, remove the hCMEC/D3 working medium from the basolateral compartment, and wash it twice with PBS. Seed U87 cells on the basolateral compartment by filling the microplate well with a U87 cell suspension at a density of 4.6×10^4 cells per cm^2 well cell growth area (around 1.5 mL per 3.8 cm^2 cell growth area) (*see* **Note 14**). Keep the Transwell® microplate in the cell incubator.
5. Maintain the co-culture for an additional 5 days, after which the model is ready to be used at day 7. Keep the Transwell® microplate in the cell incubator, and renew both the apical and basolateral medium in days 2, 4, and 6. Before renewing the medium, perform TEER measurements to monitor the integrity of the model (*see* **Note 15**).

6. At day 7, if conducting permeability experiments in cell culture medium-free conditions such as in Hanks' Balanced Salt Solution (HBSS), wash twice and fill both apical and basolateral compartments with HBSS. Place the Transwell® microplate back in the cell incubator, and allow the model to equilibrate for 30 min before initiating the experiment (*see* **Note 16**). A resume scheme of the model is presented in Fig. 1.

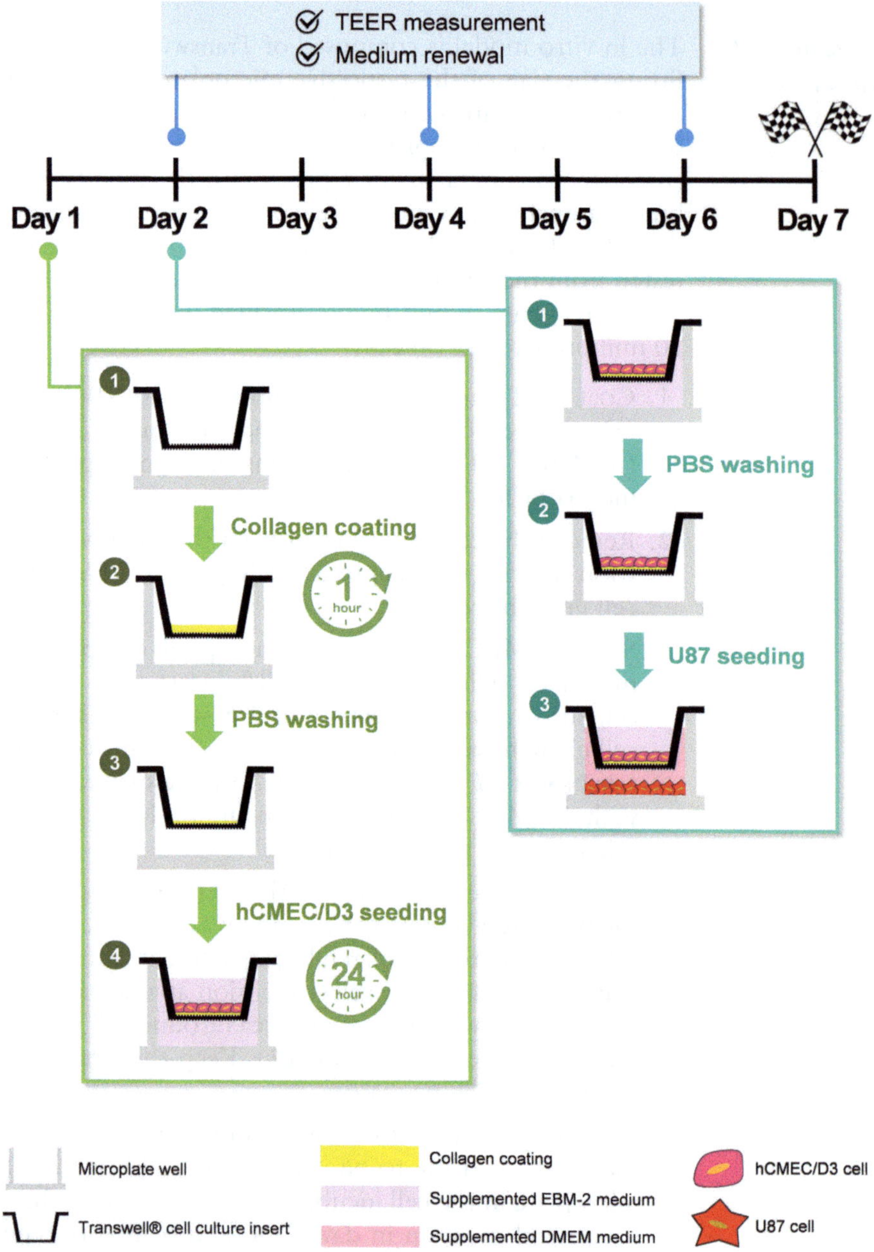

Fig. 1 Graphical scheme of the 7-day in vitro model assembly

4 Notes

1. Cell line features, such as morphology, growth rate, protein expression, and stimuli response, may suffer significant alterations over passage number. Thus, to ensure reliable and reproducible results, it is of importance to conduct experiments using a similar range of cell passage number. It is reported that hCMEC/D3 cells are capable of maintaining a stable growth and expression of characteristic endothelial markers (e.g., CD34, CD31, CD40, CD105, CD144) at least up to the 35th passage [22], whereas U87 cells are described to present variable tumorigenicity over increasing cell passage number, including lower invasion properties and a more epithelial phenotype with decreased PI3K/Akt and TGF-β pathway expression [23].
2. If some of the supplements or supplement solutions cannot be kept sterile, sterilize them by filtration using 0.22 μm syringe filters prior to medium preparation.
3. Prepare a 0.02 M acetic acid solution, and sterilize it by filtration using 0.22 μm syringe filters. Keep the commercially available concentrated solution of rat tail collagen type I sterile, and dilute it into the filtered 0.02 M acetic acid solution.
4. Cell morphology should be monitored daily using the inverted light microscope to allow an early detection of possible phenotypic alterations or contaminations. Moreover, it is recommended to perform mycoplasma testing frequently, since mycoplasma contaminations over the period of cell culture may result in non-reproducible data due to alterations in the cell growth profile, inhibition of cell metabolism, and chromosomal aberrations, among others.
5. Prolonged exposure of cells to trypsin may cause cytotoxicity. Therefore, it is recommended to shorten the trypsinization time as much as possible and neutralize trypsin with serum-containing medium as soon as the majority of cells detach from the surface of the tissue culture flask.
6. Resuspend the pellet vigorously to avoid the presence of cell aggregates in the following step of cell counting. This is particularly important for U87 cells, which tend to spontaneously cluster into spheres when close to confluency. Moreover, adjust the volume of resuspending medium to a reasonable level in order to allow sufficient pellet dilution for a clear physical separation of each cell in the following step of cell counting. Herein, it is worthy of note that, at confluency, the average number of both hCMEC/D3 and U87 cells per cm^2 tissue culture flask ranges from approximately 9×10^4 to 11×10^4.

7. The percentage of dead cells should not exceed 5%.
8. For routine subculture, it is recommended to split cells no less than at a ratio of 1:10, whereas for experiment initiation, it is recommended to adjust the volume of resuspending medium mentioned in **Note 5** in order to directly take an aliquot corresponding to the required number of cells. Alternatively, centrifuge the volume of resuspending medium corresponding to the required number of cells, and resuspend the cell pellet in a suitable volume of fresh medium. It is of highlight that, although residual, each centrifugation cycle is necessarily associated with a loss in the total number of cells.
9. After the defrosting process, cells should be cultured for at least two passages before use in experiments and undergo mycoplasma testing.
10. Pipette tips should never touch the Transwell® insert membrane to avoid damage. The coating solution should be poured without the formation of air bubbles since this leads to "air gaps" that compromise the distribution of a uniform coating throughout the Transwell® membrane. If necessary, a pipette tip may be used to establish superficial electrostatic interactions with the drop of coating solution and help spread it throughout the Transwell® membrane.
11. Use the respective hCMEC/D3 working medium.
12. To confirm a proper cell seeding, assess cell density similarities between wells using the inverted light microscope.
13. If using 96-well plates, reduce the so-called edge effect, which leads to higher levels of evaporation at the edge wells of the microplate. To do so, create a moisture barrier by filling the outer edge wells with PBS, and assemble the in vitro cell model only in the inner wells of the microplate.
14. Use the respective U87 working medium.
15. TEER is calculated by subtracting the average resistance of collagen-coated Transwell® cell culture inserts without cells growing on their surface and considering the surface area of the Transwell® cell culture inserts. From our experience, typical TEER values for this in vitro co-culture model at day 7 should average 20 Ω/cm^2 [24]. However, several factors, such as pH, temperature, and cell passage, are able to deviate TEER from the expected value.
16. TEER fluctuations should be monitored throughout the permeability experiment to assure that the integrity of the model is kept. At the end of the permeability experiment, both hCMEC/D3 and U87 cell layers may be disrupted with either 1% Triton X-100 or an organic solvent (e.g., DMSO), for quantification of the percentage of drug internalized by cells.

Acknowledgments

This work was financed by Portuguese funds through Fundação para a Ciência e a Tecnologia (FCT)/Ministério da Ciência, Tecnologia e Ensino Superior, in the framework of the project "Institute for Research and Innovation in Health Sciences" UID/BIM/04293/2019. CM gratefully acknowledges Fundação para a Ciência e a Tecnologia (FCT), Portugal, for financial support (grant SFRH/BD/137946/2018).

References

1. Saunders NR, Dreifuss JJ, Dziegielewska KM, Johansson PA, Habgood MD, Møllgård K, Bauer HC (2014) The rights and wrongs of blood-brain barrier permeability studies: a walk through 100 years of history. Front Neurosci 8:404. https://doi.org/10.3389/fnins.2014.00404
2. Arvanitis CD, Ferraro GB, Jain RK (2020) The blood–brain barrier and blood–tumour barrier in brain tumours and metastases. Nat Rev Cancer 20:26–41. https://doi.org/10.1038/s41568-019-0205-x
3. Pandit R, Chen L, Götz J (2020) The blood-brain barrier: physiology and strategies for drug delivery. Adv Drug Deliv Rev 165–166: 1–14. https://doi.org/10.1016/j.addr.2019.11.009
4. Liebner S, Dijkhuizen RM, Reiss Y, Plate KH, Agalliu D, Constantin G (2018) Functional morphology of the blood–brain barrier in health and disease. Acta Neuropathol 135: 311–336. https://doi.org/10.1007/s00401-018-1815-1
5. Achrol AS, Rennert RC, Anders C, Soffietti R, Ahluwalia MS, Nayak L, Peters S, Arvold ND, Harsh GR, Steeg PS, Chang SD (2019) Brain metastases. Nat Rev Dis Primers 5:5. https://doi.org/10.1038/s41572-018-0055-y
6. Dubois LG, Campanati L, Righy C, D'Andrea-Meira I, Spohr TC, Porto-Carreiro I, Pereira CM, Balça-Silva J, Kahn SA, DosSantos MF, Oliveira Mde A, Ximenes-da-Silva A, Lopes MC, Faveret E, Gasparetto EL, Moura-Neto V (2014) Gliomas and the vascular fragility of the blood brain barrier. Front Cell Neurosci 8: 418. https://doi.org/10.3389/fncel.2014.00418
7. Watkins S, Robel S, Kimbrough IF, Robert SM, Ellis-Davies G, Sontheimer H (2014) Disruption of astrocyte-vascular coupling and the blood-brain barrier by invading glioma cells. Nat Commun 5:4196. https://doi.org/10.1038/ncomms5196
8. Sarkaria JN, Hu LS, Parney IF, Pafundi DH, Brinkmann DH, Laack NN, Giannini C, Burns TC, Kizilbash SH, Laramy JK, Swanson KR, Kaufmann TJ, Brown PD, Agar NYR, Galanis E, Buckner JC, Elmquist WF (2018) Is the blood-brain barrier really disrupted in all glioblastomas? A critical assessment of existing clinical data. Neuro Oncol 20:184–191. https://doi.org/10.1093/neuonc/nox175
9. Pardridge WM (2005) The blood-brain barrier: bottleneck in brain drug development. Neurotherapeutics 2:3–14. https://doi.org/10.1602/neurorx.2.1.3
10. Lee CS, Leong KW (2020) Advances in microphysiological blood-brain barrier (BBB) models towards drug delivery. Curr Opin Biotechnol 66:78–87. https://doi.org/10.1016/j.copbio.2020.06.009
11. Haumann R, Videira JC, Kaspers GJL, van Vuurden DG, Hulleman E (2020) Overview ofcurrent drug delivery methods across the blood–brain barrier for the treatment of primary brain tumors. CNS Drugs 34(11): 1121–1131. https://doi.org/10.1007/s40263-020-00766-w
12. Graham ML, Prescott MJ (2015) The multifactorial role of the 3Rs in shifting the harm-benefit analysis in animal models of disease. Eur J Pharmacol 759:19–29. https://doi.org/10.1016/j.ejphar.2015.03.040
13. Buckley ST, Fischer SM, Fricker G, Brandl M (2012) In vitro models to evaluate the permeability of poorly soluble drug entities: challenges and perspectives. Eur J Pharm Sci 45: 235–250. https://doi.org/10.1016/j.ejps.2011.12.007
14. Jamieson JJ, Searson PC, Gerecht S (2017) Engineering the human blood-brain barrier in vitro. J Biol Eng 11:37. https://doi.org/10.1186/s13036-017-0076-1
15. Garberg P, Ball M, Borg N, Cecchelli R, Fenart L, Hurst RD, Lindmark T, Mabondzo A, Nilsson JE, Raub TJ,

Stanimirovic D, Terasaki T, Öberg JO, Österberg T (2005) In vitro models for the blood--brain barrier. Toxicol In Vitro 19:299–334. https://doi.org/10.1016/j.tiv.2004.06.011

16. Gomes MJ, Kennedy PJ, Martins S, Sarmento B (2017) Delivery of siRNA silencing P-gp in peptide-functionalized nanoparticles causes efflux modulation at the blood–brain barrier. Nanomedicine 12:1385–1399. https://doi.org/10.2217/nnm-2017-0023
17. Yamaguchi S, Ito S, Masuda T, Couraud P-O, Ohtsuki S (2020) Novel cyclic peptides facilitating transcellular blood-brain barrier transport of macromolecules in vitro and in vivo. J Control Release 321:744–755. https://doi.org/10.1016/j.jconrel.2020.03.001
18. Łukasiewicz S, Błasiak E, Szczepanowicz K, Guzik K, Bzowska M, Warszyński P, Dziedzicka-Wasylewska M (2017) The interaction of clozapine loaded nanocapsules with the hCMEC/D3 cells – in vitro model of blood brain barrier. Colloids Surf B Biointerfaces 159:200–210. https://doi.org/10.1016/j.colsurfb.2017.07.053
19. Yokoyama R, Taharabaru T, Nishida T, Ohno Y, Maeda Y, Sato M, Ishikura K, Yanagihara K, Takagi H, Nakamura T, Ito S, Ohtsuki S, Arima H, Onodera R, Higashi T, Motoyama K (2020) Lactose-appended β-cyclodextrin as an effective nanocarrier for brain delivery. J Control Release 328:722–735. https://doi.org/10.1016/j.jconrel.2020.09.043
20. Strazza M, Maubert ME, Pirrone V, Wigdahl B, Nonnemacher MR (2016) Co-culture model consisting of human brain microvascular endothelial and peripheral blood mononuclear cells. J Neurosci Methods 269:39–45. https://doi.org/10.1016/j.jneumeth.2016.05.016
21. Kulczar C, Lubin KE, Lefebvre S, Miller DW, Knipp GT (2017) Development of a direct contact astrocyte-human cerebral microvessel endothelial cells blood–brain barrier coculture model. J Pharm Pharmacol 69:1684–1696. https://doi.org/10.1111/jphp.12803
22. Weksler B, Romero IA, Couraud P-O (2013) The hCMEC/D3 cell line as a model of the human blood brain barrier. Fluids Barriers CNS 10:16–16. https://doi.org/10.1186/2045-8118-10-16
23. Zeng Y, Wang X, Wang J, Yi R, Long H, Zhou M, Luo Q, Zhai Z, Song Y, Qi S (2018) The tumorgenicity of glioblastoma cell line U87MG decreased during serial in vitro passage. Cell Mol Neurobiol 38:1245–1252. https://doi.org/10.1007/s10571-018-0592-7
24. Mendes B, Marques C, Carvalho I, Costa P, Martins S, Ferreira D, Sarmento B (2015) Influence of glioma cells on a new co-culture in vitro blood–brain barrier model for characterization and validation of permeability. Int J Pharm 490:94–101. https://doi.org/10.1016/j.ijpharm.2015.05.027

Chapter 16

An In Vitro Human Blood–Brain Barrier Model to Study Breast Cancer Brain Metastasis

Caroline Mysiorek, Lucie Dehouck, Fabien Gosselet, and Marie-Pierre Dehouck

Abstract

Studying the mechanisms of breast cancer cells in brain metastases is challenging, considering the high specificity of the blood–brain barrier (BBB) with whom breast cancer cells have to interact and cross in order to reach the brain parenchyma. While numerous in vitro BBB models are available, the setting of the model and phenotype of the endothelial cells (ECs) of the BBB model are essential to obtain relevant results.

In this chapter, we describe a method to establish a human in vitro BBB model to study adhesion of breast cancer cells and the adaptation of the method for trans-endothelial migration assay keeping the appropriate BBB phenotype of the ECs.

Key words Blood–brain barrier, Brain metastases, Adhesion assay, Trans-endothelial migration assay

1 Introduction

The development of breast cancer brain metastases is a long and sequential process including numerous steps that the cells have to achieve to develop metastasis in the delicate organ that is the brain [1].

At the metastatic stage, understanding the invasive properties of cancer cells is crucial to predict their dissemination. In vitro systems are widely developed to study these properties with both proliferation and invasion assays using Transwell filters. In these assays cancer cells have to migrate through the filter membrane, which is either empty or coated, with a predefined extracellular matrix [2]. However, studying extravasation through a vessel wall requires the use of endothelial cells (ECs) that should possess the phenotypic characteristics specific of the vasculature of the target organ. At the blood–brain barrier (BBB), this is more complex considering the highly restrictive and specific properties of the brain endothelial cells.

Nicole Stone (ed.), *The Blood-Brain Barrier: Methods and Protocols*,
Methods in Molecular Biology, vol. 2492, https://doi.org/10.1007/978-1-0716-2289-6_16,

Hence, the brain is protected by the BBB, a both physical and metabolic specific protection, located at the level of brain capillary ECs and whose properties spread along the microvasculature from pre-capillary arteriole to post-capillary venule [3]. Many reliable and well-characterized in vitro BBB models, developed with animal cells (rodents, bovine, porcine, monkey), are available and have proven valuable tools for studying the physiology and pathology of the BBB but also in central nervous system drug development and toxicology [4]. Nevertheless, deciphering at the molecular level the mechanisms of interaction of human breast cancer cells with the BBB requires a syngeneic approach using a human in vitro BBB model. However, the quality of the model used is often a limitation to extrapolate the results to the in vivo situation, especially when the BBB ECs don't display the required BBB properties as measured with HUVEC cells which are human ECs isolated from the umbilical vein [5].

Moreover, to avoid that the filter itself represents a barrier against cancer cells migration, the trans-endothelial migration assay requires the use of larger-pore-size filter which is an additional challenge because it is important that during model establishment, ECs are prevented from migrating through the filter before forming an endothelium monolayer.

Hence, the in vitro BBB model used should present, in addition to the restrictive and specific BBB properties, an EC monolayer, according to the capillary structure, when the ECs are cultivated on large-pore-size filters [6].

The model presented in this chapter exhibits low paracellular and transcellular permeability, criteria required for a well-differentiated BBB model. The adhesion differential of different breast cancer cell lines to the human ECs of this BBB model showed a correlation with the relative aggressiveness of these cell lines, and the trans-endothelial migration rate was in accordance to their adhesion potential. Moreover, the model proves valuable for the molecular studies highlighting the species specificities [7].

In this chapter we will describe how to develop a human in vitro BBB model, consisting of brain-like endothelial cells (BLECs) seeded on an insert filter co-cultivated with brain pericytes, to study the interactions of breast cancer cells during adhesion and trans-endothelial migration assays. The model presents the advantages of being developed with human ECs to have a syngeneic approach. Moreover, the adaptation of the culture protocol for the use of larger-pore-size filter allows to study the trans-endothelial migration assay keeping an appropriate monolayer of ECs.

2 Materials

All the procedures detailed below are performed under sterile conditions.

All the cell types in this protocol are cultivated in incubators at 37 °C, in a humidified atmosphere with 5% CO_2/95% air. Before being incorporated in the culture media, any serum is previously heat-inactivated by incubation at 58 °C for 30 min.

2.1 Material for Media Preparation

1. Dulbecco's Modified Eagle's Medium (DMEM) 25 mM glucose (glc, D5648, Sigma Aldrich) supplemented with 23 mM $NaHCO_3$.
2. Endothelial Cell Medium (ECM) (1001, ScienCell™).
3. Dulbecco's Modified Eagle's Medium (DMEM) 5 mM glc (D5523, Sigma-Aldrich).
4. Sera: fetal calf serum (FCS) (F7534, Sigma-Aldrich).
5. Amino acid: L-glutamine (1.00289, Merck Chemicals).
6. Penicillin–streptomycin (0503, ScienCell™).
7. Gentamicin (A2712, Biochrom AG).
8. Endothelial cell growth factor supplement (ECGS, 1052, ScienCell™).
9. Vascular endothelial growth factor (VEGF, 293-VE-050, Bio-Techne).
10. Basic fibroblast growth factor (bFGF, F0291; Sigma-Aldrich).

2.2 Material for Cell Isolation

1. Bovine brain isolated from a freshly slaughtered calf (for pericyte extraction).
2. Human cord blood (specific authorization for the use of human cells should be obtained from the local ethic committee. Moreover, written informed consent has to be signed by the patients).
3. MCF-7 cells are purchased from Sigma-Aldrich (N°86012803).
4. MDA MB 231 cells are purchased from ATCC (HTB-26).
5. Forceps and scalpel.
6. Dounce homogenizer (40 and 100 mL).
7. Glass pestles: 40 mL type A clearance 76–152 μm, 100 mL type B clearance 12–140 μm.
8. Nylon Sieve, pore size 60 and 180 μm, (Blutex®, Saati, France).
9. Phosphate Buffer Saline Calcium Magnesium-Free (PBS-CMF): 137 mM NaCl, 2.7 mM KCl, 1.8 mM

KH_2PO_4, 20 mM $Na_2HPO_4(12H2O)$, pH 7.4 in sterile deionized water.

10. Mini-MACS and midi-MACS immunomagnetic separation system, with LS and MS column (Miltenyi Biotec, Bergisch Gladbach).
11. CD34 microbeads kit (130-046-702, Miltenyi).
12. Ficoll, (Histopaque-1077 Hybri-Max; Sigma-Aldrich).
13. VEGF solution (ref 293-VE-050 Bio-Techne).

2.3 Material for Cell Culture and Dissociation

1. Petri dish 100 mm diameter (430293, Corning Incorporated).
2. 12-well cell culture plate (3512, Costar, Corning Incorporated).
3. Polycarbonate membrane Transwell® insert, diameter 12 mm, pore size 0.4 μm (3401, Costar, Corning Incorporated).
4. Polycarbonate membrane Transwell® insert, diameter 12 mm, pore size 3 μm (3402, Costar, Corning Incorporated).
5. Matrigel™ growth factors reduced (354230, BD BioCoat).
6. Gelatin type A (from pig skin) (G-2500, Sigma-Aldrich).
7. Trypsin/ethylenediaminetetraacetic acid (EDTA) 0.05/0.02% (w/v) in PBS-CMF (L2143, Biochrom AG).
8. Trypsin/EDTA (0103, ScienCell)) and trypsin neutralization solution (TNS) (0113, ScienCell).

2.4 Material for Adhesion and Trans-endothelial Migration Assays

1. CellTracker™ Green 5-chloromethyl fluorescein diacetate (CMFDA, C7025, Invitrogen).
2. EDTA solution 5 mM.
3. 4% paraformaldehyde solution (J61984, AK, Alfa Aesar, VWR).
4. Ready-to-use mounting solution containing 4′,6-diamidino-2-phénylindole (DAPI) for nuclei staining (4′,6-diamidino-2-phénylindole ProLong Diamond Antifade mounting medium with DAPI, P36962, Life Technologies).
5. Fluorescent microscope with ×400 magnification and appropriate filters to observe CellTracker™ (492/517 nm).

3 Methods

3.1 Isolation and Culture of Bovine Brain Pericytes

Pericytes are extracted from freshly collected bovine brain capillaries isolated from freshly slaughtered calves.

1. Remove meninges and white matter.
2. Cut the gray matter into 2 mm^3 pieces, and rinse it twice in PBS-CMF.

3. Transfer the brain tissue pieces with two volumes of PBS-CMF in the 40 mL Dounce homogenizer, and homogenize them by 15 up-and-down strokes using the type A glass pestle.
4. Pass the homogenate through a 180 μm nylon sieve to obtain a filtrate containing microvessels.
5. Homogenate a second time using the type B glass pestle.
6. Pass the homogenate through a 60 μm nylon sieve, and wash with PBS-CMF.
7. Collect the capillaries on the 60 μm nylon sieve, and resuspend in pericyte culture medium (DMEM 5 mM glc supplemented with 20% FCS, 2 mM L-glutamine, 50 μg/mL gentamycin, and 1 ng/mL bFGF).
8. Seed at a low density the capillaries onto Matrigel™-coated 60 mm petri dish containing 5 mL of pericytes growth culture medium.
9. Change the medium every 2 days.
10. Check the appearance of capillaries under microscope, and remove larger vessels when observed.
11. The pericytes migrated out of the capillaries and rapidly invaded the whole surface of the petri dish.

3.2 Isolation and Culture of Human CD34⁺-Derived Endothelial Cells

The protocol of differentiation of $CD34^+$ cells into ECs is described according to the protocol developed by Pedroso et al. (2011) [8].

Isolation of $CD34^+$ cells from cord blood

1. Isolate the buffy coat by mixing with Ficoll medium, and centrifuge to obtain a density gradient separation.
2. Select twice the $CD34^+$ cells using the midi- and mini-MACS immunomagnetic separation system.

Differentiation of $CD34^+$ cells into ECs

3. Plate the isolated $CD34^+$ cells at a density of 2×10^5 cells/well on 0.2% (w/v) gelatin-coated 24-well plates containing ECM medium supplemented with 20% of FCS, ECGS, 50 μg/mL gentamicin, and 50 ng/mL of VEGF.
4. Add fresh medium to the well every 2 days when the cells are still nonadherent and then every 2 days from the fifth day of culture, once the cells start to adhere.
5. After 15–20 days, the $CD34^+$ cells start their differentiation into ECs.
6. Treat the cells with trypsin/EDTA, and plate the $CD34^+$-derived ECs on 0.2% gelatin-coated 6-well plate in ECM medium containing 5% FCS, ECGS, and 50 μg/mL gentamycin and 50 ng/mL of VEGF. Change the medium every 2 days.

7. Once CD34$^+$-derived ECs are confluent, treat with trypsin/EDTA, and plate on a 0.2% gelatin-coated 100 mm-diameter petri dish in ECM medium containing 5% FCS, ECGS, and 50 μg/mL gentamicin.
8. At this stage, subculture of CD34$^+$-derived ECs can be done to obtain important quantity of cells. To do so, do again **step 7** when the cells reach confluence, and split the cell suspension into four 100 mm petri dishes. This step can be done maximum four times.

3.3 BBB Co-culture System

The protocol of co-culture between CD34$^+$-derived ECs and pericytes for the induction of BBB properties is described according to the protocol developed by Cecchelli et al. (2014) [9] (Figs. 1 and 2).

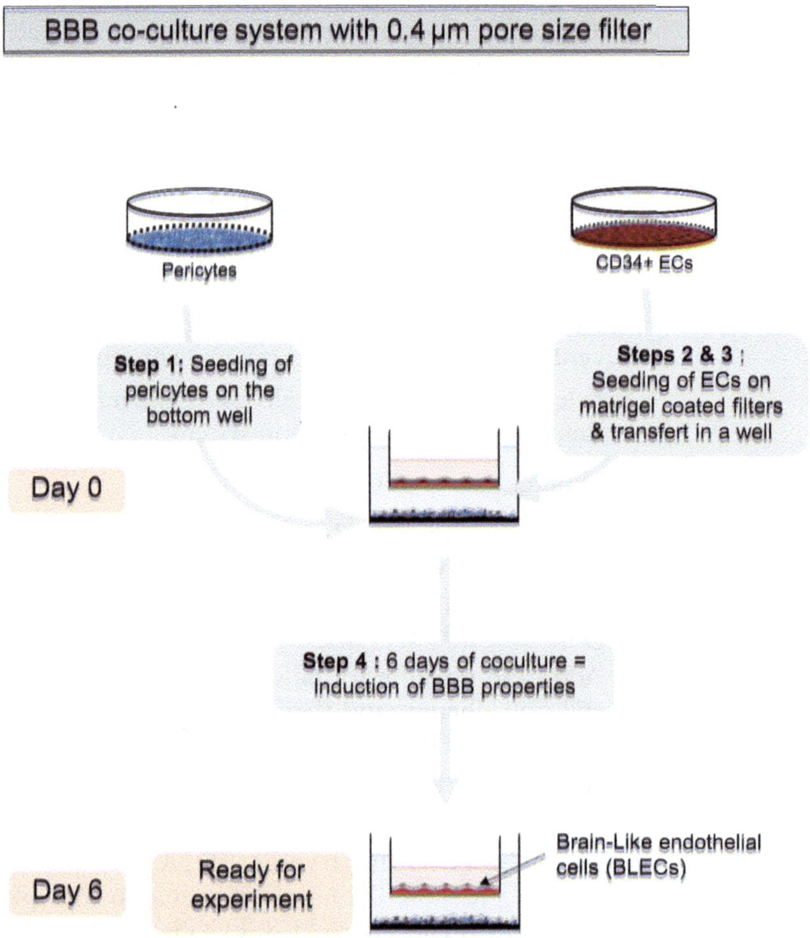

Fig. 1 Schematic representation of seeding protocol to set up co-culture for the human BBB model using 0.4 μm-pore-size filter. The step numbers indicated correspond to steps in the text in Subheading 3.3

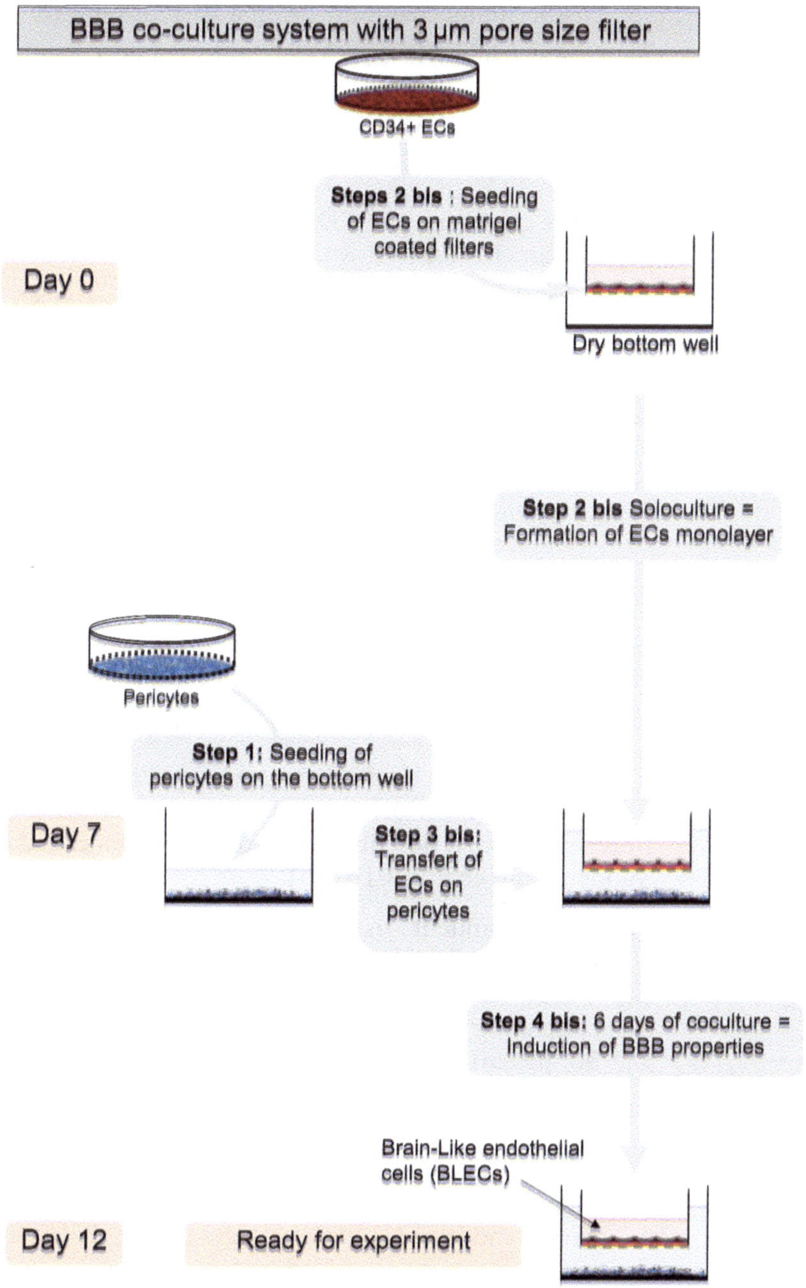

Fig. 2 Schematic representation of seeding protocol to set up co-culture for the human BBB model using 3 µm-pore-size filter. The step numbers indicated correspond to steps in the text. The step numbers indicated correspond to steps in the text in Subheading 3.3

3.3.1 Pericyte Seeding

1. Once the pericyte culture is confluent, treat the cells using trypsin/EDTA saline solution.
2. Once the cells start to detach, remove the trypsin/EDTA, and add complete medium.

3. Dissociate mechanically using a Pasteur pipette.
4. Count using Malassez cells, and seed at a density of 5 × 10^4 cells/wells on 12-well plates containing 1.5 mL ECM medium supplemented with 5% FCS, ECGS, and 50 μg/mL gentamicin.

3.3.2 Endothelial Cell Seeding

1. Once confluent, the $CD34^+$-derived ECs are treated with 5 mL trypsin/EDTA to dissociate the cells.
2. Once the cells start to detach, remove the trypsin/EDTA, and add 5 mL of complete medium.
3. Dissociate mechanically using a Pasteur pipette.
4. Count using Malassez cells, and seed at a density of 8 × 10^4 cells/insert on 1/50e diluted Matrigel™-coated filters containing 0.5 mL ECM supplemented with 5% FCS, ECGS, and 50 μg/mL gentamicin.

For the seeding on the 0.4 μm-pore-size filters

1. Transfer the filters containing ECs upon a well containing pericyte culture. The co-culture starts at day 0 and lasts 6 days.
2. The medium of the co-culture is renewed at days 2 and 5. After 6 days of co-culture, the model is ready for experiment.

For the seeding on the 3 μm-pore-size filters

The adaptation of the co-culture seeding protocol for larger-pore-size filters is described according to the method developed by Vandenhaute et al. (2016) [6].

3. Once confluent, treat the $CD34^+$-derived ECs with trypsin/EDTA to dissociate the cells, and then seed at a density of 8 × 10^4 cells/insert on 1/50e diluted Matrigel™-coated filters containing ECM supplemented with 5% FCS, ECGS, and 50 μg/mL gentamicin. Cultivate the $CD34^+$-derived ECs alone during 6 days without medium in the bottom compartment.
4. After 6 days, transfer the filter containing $CD34^+$-derived ECs in a well containing pericyte culture. The co-culture starts at days 7 and lasts 6 additional days before the experiment.
5. The medium of the co-culture is renewed at days 2 and 5. After 6 days of co-culture, the model is ready for experiment.

Six days of co-culture are necessary for an induction of BBB properties in the ECs by pericytes. The ECs are then renamed brain-like endothelial cells (BLECs).

3.4 Cancer Cell Culture

1. Thaw a vial of breast cancer cells rapidly, and seed in 100 mm petri dish containing 5 mL DMEM supplemented with 20% FCS, 2 mM L-glutamine, and 5μg/mL penicillin–streptomycin.

2. Cancer cells are cultivated 3 weeks before being used or experiment. During this 3 weeks, each time the cells reach confluence, dissociate cells to do subculture using 5 mL of trypsin/EDTA solution.
3. When the cells start to detach, stop the action of trypsin using 3 mL of TNS.
4. Transfer the suspension in 15 mL tube, and centrifuge 5 min at 190 g.
5. Resuspend the cell suspension in 5 mL of complete medium.
6. Share the cell suspension in two 100 mm petri dishes containing 10 mL of complete medium.
7. Following the last seeding, cancer cells are ready to be used for experiment from 80% of confluence.

3.5 Adhesion and Trans-endothelial Migration Assays (Fig. 3)

Staining of the cancer cells

1. On the day of the experiment, incubate the cancer cells with 5 mL of CellTracker™ Green CMFDA diluted at 0.5 μL/mL in DMEM 4.5 g/L glc for 45 min at 37 °C.
2. Wash once with DMEM 4.5 g/L glc, and replace by 5 mL of complete medium for a minimum of 45 min at 37 °C. The cells are then ready for adhesion/trans-endothelial migration assay.

Dissociation of the cancer cells

3. Wash twice the cells with PBS-CMF.
4. Add EDTA solution at 5 mM, and keep it for a few minutes at 37 °C until the start of cell detachment.
5. Stop the reaction using 5 mL of complete medium containing 1% FCS, and dissociate mechanically using Pasteur pipette.

Adhesion assay

6. Add the cancer cells on ECs at a density of 20×10^4 cells per filter containing 0.5 mL of ECs medium.
7. After 120 min remove gently the supernatant, and wash once with DMEM 5 mM glc.
8. Fix the filters 10 min with 4% paraformaldehyde solution.
9. Cut the filter, put on glass slides, and mount under coverslip with mounting solution containing DAPI.

Trans-endothelial migration assay

10. (bis) Seed the cancer cells on ECs at a density of 80×10^4 cells per filter containing 0.5 mL of EC medium.
11. (bis) After 16 h remove gently the supernatant, and wash once with DMEM at 37 °C.

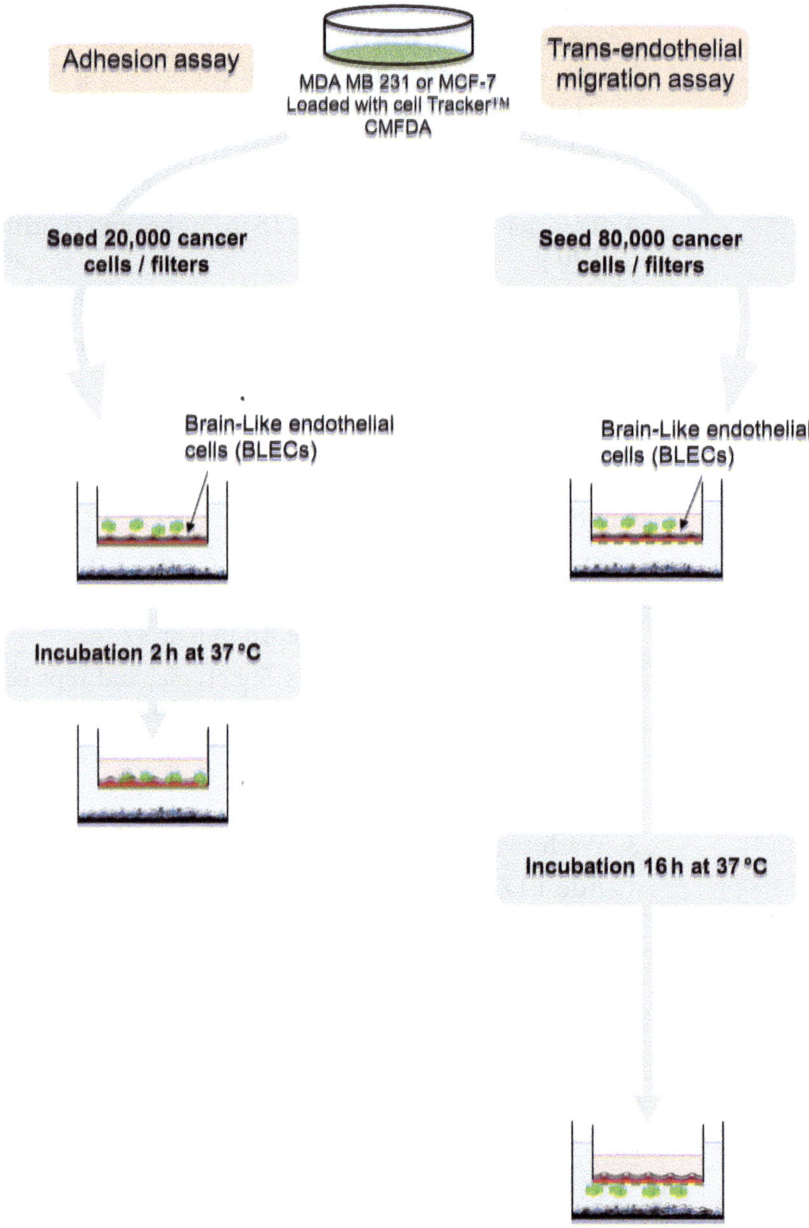

Fig. 3 Schematic representation of adhesion and trans-endothelial migration assays protocol

12. (bis) Fix the filters 10 min with 4% paraformaldehyde solution at room temperature.
13. (bis) Cut each filter, reverse upside down on glass slides, and mount on coverslip using mounting solution containing DAPI.

Quantification

14. Observe the filters using a fluorescent microscope using ×400 magnification (exc/em: 492/517 nm).
15. Quantify the adherent or transmigrated cells by counting the cells in each field in the total surface of each filter. Each experimental condition is performed using a triplicate.
16. Statistical analyses are performed using GraphPad Prism Software.

4 Notes

Contamination of cells by mycoplasma is a major problem in cell culture. This problem occurred in primary cells but more often with cancer cells which are cultivated for a longer period. Hence, for quality control, regularly check mycoplasma contamination in cancer cell culture, using commercial kit (MycoAlert detection kit; Ozyme Ref: LT07-318). This quality control is essential to obtain relevant results, considering that mycoplasma changes extensively the properties of cancer cells and could impact the proliferation and molecular interactions [10].

For amplification of cell, trypsin/EDTA can be used for the dissociation step. However, the enzymatic effect of trypsin can affect the cancer cell surface. For this reason, the cancer cells should not be treated with trypsin/EDTA 72 h before the adhesion trans-endothelial migration assays. The use of EDTA as a soft detaching method is required.

Acknowledgments

The authors greatly thank Dr Aurore Drolez and Dr Elodie Vandenhaute for their great work during their doctoral and post-doctoral positions respectively. This work was supported by "la Région Hauts-de-France" grants fellowship Aurore Drolez and Elodie Vandenhaute.

References

1. Witzel I, Oliveira-Ferrer L, Pantel K et al (2016) Breast cancer brain metastases: biology and new clinical perspectives. Breast Cancer Res 18(1):8
2. Katt ME, Placone AL, Wong AD et al (2016) In vitro tumor models: advantages, disadvantages, variables, and selecting the right platform. Front Bioeng Biotechnol 4:12
3. Francisco DMF, Marchetti L, Rodríguez-Lorenzo S et al (2020) Advancing brain barriers RNA sequencing: guidelines from experimental design to publication. Fluids Barriers CNS 17(1):51
4. Helms HC, Abbott NJ, Burek M et al (2015) In vitro models of the blood-brain barrier: an overview of commonly used brain endothelial

cell culture models and guidelines for their use. J Cereb Blood Flow Metab 36(5):862–890
5. Drolez A, Vandenhaute E, Julien S et al (2016) Selection of a relevant in vitro blood-brain barrier model to investigate pro-metastatic features of human breast cancer cell lines. PLoS One 11:e0151155
6. Vandenhaute E, Drolez A, Sevin E et al (2016) Adapting coculture in vitro models of the blood-brain barrier for use in cancer research: maintaining an appropriate endothelial monolayer for the assessment of transendothelial migration. Lab Invest 96:588–598
7. Drolez A, Vandenhaute E, Delannoy CP et al (2016) ST6GALNAC5 expression decreases the interactions between breast cancer cells and the human blood-brain barrier. Int J Mol Sci 17:1309
8. Pedroso DC, Tellechea A, Moura L et al (2011) Improved survival, vascular differentiation and wound healing potential of stem cells co-cultured with endothelial cells. PLoS One 6:e16114
9. Cecchelli R, Aday S, Sevin E et al (2014) A stable and reproducible human blood-brain barrier model derived from hematopoietic stem cells. PLoS One 9:e99733
10. Drexler HG, Uphoff CC (2002) Mycoplasma contamination of cell cultures: incidence, sources, effects, detection, elimination, prevention. Cytotechnology 39(2):75–90

Chapter 17

An In Vivo Mouse Model to Study Blood–Brain Barrier Destabilization in the Chronic Phase of Stroke

Svetlana M. Stamatovic, Chelsea M. Phillips, Richard F. Keep, and Anuska V. Andjelkovic

Abstract

Cerebral ischemic injury evokes a complex cascade of pathophysiological events at the blood–vascular–parenchymal interface. These evolve over time and space and result in progressive neurological damage. Emerging evidence suggests that blood–brain barrier (BBB) recovery and reestablishment of BBB impermeability are incomplete and that these could influence stroke injury recovery, increase the risk of new stroke occurrence, and be a solid substrate for developing vascular dementia. Recent work from the author's laboratory has established the existence of incomplete BBB recovery in chronic stroke conditions that was induced by structural alterations to brain endothelial junctional complexes and persistent BBB leakage. The experimental methodology presented here is focused on modelling chronic stroke injury using an in vivo thromboembolic mouse stroke model and how to evaluate the kinetics and magnitude of BBB hyperpermeability in chronic stroke conditions using a combination of magnetic resonance imaging, tracer studies, and immunohistochemistry.

Key words Blood–brain barrier, Permeability assays, Thromboembolic stroke, Magnetic resonance imaging, Tracers

1 Introduction

Stroke is an acute event caused by abrupt cessation of blood flow in a region of the brain (ischemic stroke) or rupture of a blood vessel (hemorrhagic stroke). Stroke injury evolves in four phases: hyperacute (~6 h from symptom onset), acute (6–24 h), subacute (24 h to 6 weeks), and chronic (>6 weeks) [1]. During this time, cerebral ischemia evokes a profound and deleterious upregulation in inflammation and triggers multiple cell death pathways. Concurrently, regenerative responses, including vascular remodelling, angiogenesis, and neurogenesis, are also induced [1]. Most attention on stroke injury has focused on the hyperacute, acute, and subacute phases due to a strategy of "early action, best prevention of stroke

Nicole Stone (ed.), *The Blood-Brain Barrier: Methods and Protocols*,
Methods in Molecular Biology, vol. 2492, https://doi.org/10.1007/978-1-0716-2289-6_17,

injury." In contrast, the chronic phase is generally understudied, and it is accepted as a time of post-stroke recovery. However, there are several pieces of evidence that pinpoint that there are pathological processes in the chronic phase which affect survival and functionality after stroke. Due to ongoing processes during the chronic phase after stroke, survivors are at increased risk of another stroke and for developing post-stroke cognitive impairment (vascular dementia) [2].

Clinical (magnetic resonance imaging; MRI) and experimental studies indicate chronic BBB dysfunction with increased permeability several weeks or months after the ischemic event, and this is closely associated with brain dysfunction and limited stroke recovery [3, 4]. The magnitude of this chronic BBB hyperpermeability is significantly less than in the acute phase of stroke (where there is uncontrolled and robust extravasation of plasma proteins, leucocyte extravasation, vasogenic edema). While the presence of small leaks during BBB recovery could be of short-term benefit, their persistence may lead to hemorrhagic transformation and cause excessive buildup of fluid leading to brain dysfunction and the progression to chronic inflammation and formation of microthrombi [5–7].

1.1 Models for the Chronic Stroke Condition

A first step in understanding BBB injury in chronic stroke conditions is the choice of an adequate model which will mimic the ongoing pathological process. Ischemic stroke is a complex and heterogeneous disorder. Adopted models to study stroke mostly cover some specific aspect of stroke pathogenesis and are based on the induction of focal cerebral ischemia. Models are generally divided into those inducing permanent and transient reductions in focal cerebral blood flow, although the majority of clinical stroke has some restoration of blood flow in the first 24 h. Frequently used models include using an intraluminal filament to induce middle cerebral artery occlusion (MCAO), cauterizing the MCA, photothrombotic stroke, thromboembolic (TE) stroke, injection of thrombin, and chemically induced stroke based on the injection of vasoconstrictor endothelin-1 [8, 9]. Characteristics of the stroke models are presented in Table 1.

Thromboembolic and MCAO models of stroke have a good potential to mimic chronic stroke conditions based on brain ischemic lesion composition (infarct core with penumbra), visibility of all processes (i.e., mechanisms of cell death, inflammation), and the presence of a full spectrum of stroke injury from acute to chronic phase (e.g., sensorimotor and behavioral deficits) [8–11]. MCAO models have been used in mice and rats as well primates to model the chronic phase of stroke, analyzing the post-stroke period from 7 days to several weeks [11–15]. A challenging issue for modelling chronic stroke is the duration of occlusion, which directly affects post-stroke survival time. In mouse models, 20–45-min MCA occlusion provides prolonged survival into the chronic phase

Table 1
Murine models of stroke

	Induction	Size of infarct	Reproducibility	Visibility of pathological process	Mimic phase of stroke	Other advantages/ disadvantages
Middle cerebral artery (MCA) occlusion	Mechanical occlusion (silicon-coated microfilament placed in MCA	Controllable size (regulated by time of occlusion)	Very good	Penumbra; all processes (e.g., apoptosis, inflammation, brain edema, neuronal depolarization))	Acute, subacute, chronic; mimic ischemia and reperfusion injury	Mimic large vessels in stroke/mechanical ischemic injury, absence of thrombus lodging and interaction with EC
Photothrombotic stroke	Intravascular photooxidation of photoactive dyes (i.e., Rose Bengal) by light	Controllable size, small infarct in specific area	High reproducibility	Absence of penumbra, brain edema, limited inflammation, and sensorimotor deficits	Acute and subacute; absence of reperfusion	Mimic stroke in cortex/no spontaneous restoration of blood flow, limited neurological deficit
Thromboembolic/ embolic stroke	Injection of mix platelets and fibrin (thromboemboli), fibrins, cholesterol droplets or spherule (emboli)	Variable and depend on the size of injected particles	Good/very good	Penumbra; visible apoptosis, inflammation, brain edema, neuronal depolarization	Acute, subacute, chronic; mimics "watershed infarct"	Stroke in small-caliber vessels; spontaneous recanalization; localization cortex and subcortical regions/ reproducibility, less controllable size
Thrombin-induced stroke	Microinjection of thrombin to induce clot in MCA	Ischemic stroke in territory of MCA	Good	Penumbra; visible apoptosis, inflammation, brain edema, neuronal depolarization	Acute, subacute chronic; mimic ischemia and reperfusion injury	Permanent and transient ischemia; occlusion and thrombus interaction/ reproducibility, needle effect
Chemically induced stroke (endothelin-1 injection)	Topical or ICV injection/ infusion of endothelin-1	Dose-dependent ischemia	Good	Ischemic-type injury; apoptosis, inflammation	Acute and subacute	Vasoconstriction of blood vessels/stability, brain needle damage

EC endothelial cell, *ICV* intracerebroventricular, *MCA* middle cerebral artery

[11, 15], while in rats or primates 60–90-min occlusion produces stroke injury with prolonged survival time [16–18]. Although the ischemic lesion size can be more variable in the TE stroke model, it has several advantages: (a) capturing the ongoing process of TE spontaneous resolution, (b) mirroring the contribution platelets to brain endothelial barrier injury/inflammation, and (c) direct effect on the small-caliber vessels and BBB dysfunction in stroke due to thromboemboli lodging. Other key elements for choosing an adequate model are the age of animals. Most published studies include only young animals (i.e., mice average age 10–12 weeks), and it is pivotal to modify the time of occlusion (MCAO) as well as the amount of injected thromboemboli in aged mice to model chronic stroke conditions. Our laboratory has developed both MCAO and TE models of chronic stroke [10, 11]. Comparing survival rate and variability of the size stroke injury, particularly in the light of mouse age, we have found that TE models have more advantages for studying chronic stroke conditions. This protocol will include the method of induction of mouse TE stroke for modelling the chronic phase of stroke.

1.2 Assessing BBB Injury in Chronic Stroke

The BBB is a highly complex and dynamic barrier, formed by an interdependent network of brain capillary endothelial cells, as well as perivascular cells, astrocytes, and pericytes, which are responsible for inducing and maintaining the barrier [19, 20]. The BBB strictly regulates paracellular permeability due to the presence of tight junctions (TJ) between endothelial cells. Any loosening of its adhesive interactions directly affects BBB integrity and increases paracellular permeability [19, 20].

BBB dysfunction in the chronic phase of stroke is a "driving force" for ongoing post-stroke pathology and is involved in the long-lasting consequences of stroke [5, 12, 21, 22]. Thus, to understand BBB injury in the chronic phase of stroke, a first step is to determine BBB integrity and the degree of BBB hyperpermeability. Assays to evaluate BBB dysfunction in vivo include tracer permeability, histological evaluation of changes in TJ protein expression (e.g., claudin-5), and the presence/accumulation of plasma proteins (e.g., fibrinogen) in the perivascular space.

Tracer studies can provide important semiquantitative and quantitative assessment of BBB function. Depending on the tracer and the method of examination, they can give spatial information on the BBB disruption and/or the time course of disruption (e.g., repeat measures in same animal). Fluorescence-based tracers (fluorescent-conjugated dextran, inulin, albumin, sodium fluorescein, cadaverine) or dyes (e.g., Evans blue) are widely used to assess paracellular and transcellular permeability. Quantitative assessment is based on determining the transfer coefficient (K_i) for tracer entry from blood to brain over time [11, 23]. Semiquantitative assessment is based on fluorescent angiography and analyzing the presence/absence of leaked tracer in brain tissue sections by fluorescent microscopy [24]. This provides additional visualization of the area of leakage. An advantage of this in vivo assay is utilizing tracers of different sizes, ranging from 376 Da to 70 kDa, to capture the degree of the barrier opening. It is important to highlight that tracer in vivo permeability assays with tissue sampling only provide a single "snapshot" of BBB leakage. Different experimental groups euthanized at different times are required to examine the temporal profile of BBB dysfunction.

Another approach for evaluating BBB injury in chronic stroke is MRI using conventional T1-weighted images and contrast-enhanced MRI using gadolinium as the contrast agent [10, 25, 26]. The quantitative, semiquantitative, and qualitative MRI is noninvasive real-time imaging for longitudinal assessment of BBB injury in chronic stroke conditions. The gadolinium-containing contrast agent injected *ip* or *iv* can leak through a disrupted BBB having effects on the water proton signal in the extravascular space. A qualitative assessment involves a comparison of T1-weighted images pre- and post-gadolinium injection, while quantitative assessment involves calculating T1 changes from pre- and post-contrast T1-maps or using dynamic contrast-enhanced MRI (DCE-MRI) to evaluate differences between alterations in vascular permeability and in the extravascular extracellular space.

The present protocol describes, stepwise, how to induce the TE chronic stroke model in mice and the methods to evaluate BBB status in chronic stroke conditions. The latter include using qualitative T1-weighted MRI (follow-up study); in vivo permeability assay calculating transfer coefficient for three different sized florescence tracers; and immunohistochemistry to analyze morphological alterations at the brain endothelial barrier by assessing TJ protein (claudin-5, ZO-1) status. The immunohistochemistry is performed in conjunction with a tracer assessment of BBB permeability.

2 Materials

2.1 Thromboembolic Stroke

1. 1 mL insulin syringe with 28 G gauge needle.
2. Polyethylene tube (I.D 0.32 mm; BD Diagnostic Systems).
3. PT-3100 Polytron homogenizer.
4. Mouse femoral vein catheter (Braintree Scientific, Inc.).
5. Zero Dead Volume 1 mL TB Syringe with 22G Catheter Connector (SAI Infusion Technologies).
6. Thrombin (from bovine plasma, Sigma-Aldrich).
7. Krebs–Ringer bicarbonate buffer (KRBB); 119 mM sodium chloride (NaCl), 4.7 mM potassium chloride (KCl), 1.2 mM

magnesium sulfate ($MgSO_4$), 1.3 mM $CaCl_2$, 25 mM $NaHCO_3$, and 1.2 mM KH_2PO_4. pH 7.4. For 500 mL KRBB: add 450 mL deionized H_20 in glass beaker. Weigh 3.4 g NaCl, 4.43 g KCl, 0.07 g $MgSO_4$, 0.072 g CaCl, 1.05 g $NaHCO_3$, and 0.08 g KH_2PO_4, and add to glass beaker. Mix and adjust pH to 7.4. Transfer KRBB to 500 mL glass cylinder, and add water to 500 mL. Filter KRBB with a 0.4 μm filter. Store KRBB at 4 °C. Maximal storage time 12 weeks.

8. Sterile nonpyrogenic isotonic 0.9% NaCl solution for injection (Mountainside Medical Equipment) (*see* **Note 1**).
9. Surgical microscope.

2.2 BBB Permeability Tracer Study

1. Mouse femoral vein catheter (Braintree Scientific, Inc.).
2. Polyethylene tube (I.D 0.32 mm; BD Diagnostic Systems).
3. 1 mL insulin syringe with 28 G gauge needle.
4. 2 mL syringe.
5. Microvette micro sample tube, serum 0.3 mL (Sarstedt).
6. Safe-Lock centrifuge tubes 0.5 and 1.5 mL.
7. Gadolinium-diethylenetriamine penta-acetic acid (Gd-DTPA) (BioPAL).
8. Fluorescein sodium salt (376.27 Da, Sigma-Aldrich).
9. Dextran, Cascade Blue™, 3000 MW, Anionic, Lysine Fixable (Thermo Fisher).
10. Dextran, Texas Red™, 10,000 MW, Anionic, Lysine Fixable (Thermo Fisher).
11. Sterile nonpyrogenic isotonic 0.9% NaCl solution for injection (Mountainside Medical Equipment).
12. Tris–HCl buffer, pH 7.6 (10×): 0.5 M Trizma base, HCl for adjusting pH. Add 800 mL distilled H_2O in glass beaker. Weigh 61 g Trizma base. Mix and adjust pH to 7.6 using 1 M HCl. Transfer buffer to glass cylinder and add H_2O to volume of 1 L. Store at room temperature. Dilute 1:10 prior to use (50 mM Tris–HCl), and adjust pH if necessary.
13. Methanol analytical grade reagent (Sigma-Aldrich).

2.3 Immunohistology with Tracer Injections

1. Anti-mouse claudin-5 Alexa Flour 488 (Thermo Fisher).
2. Anti-mouse ZO-1 Alexa Flour 594 (Thermo Fisher).
3. Phosphate-buffered saline (PBS) 0.01 M (1×) pH 7.4: 130 mM NaCl, 2.7 mM potassium chloride (KCl), 10 mM disodium hydrogen phosphate (Na_2HPO_4), 1.8 mM potassium dihydrogen phosphate (K_2HPO_4). Prepare 800 mL distilled H_2O in glass baker. Weigh and add 8 g NaCl, 200 mg KCl,

1.44 g N_2HPO_4, and 245 mg KH_2PO_4. Adjust pH on 7.4. Add H_2O until volume 1 L. Store at room temperature.

4. Normal goat serum (NGS, Vector Laboratories).
5. Triton X-100 (Sigma-Aldrich).
6. 4% paraformaldehyde (PFA) in 0.1 M phosphate buffer (Bioenno Tech).
7. Dextran, Cascade Blue, 3000 MW, Anionic, Lysine Fixable (Thermo Fisher).
8. 1 mL insulin syringe with 28 G gauge needle.

3 Methods

3.1 Mouse Thromboembolic Stroke

3.1.1 Thromboemboli Suspension

A thromboemboli suspension is prepared following an existing protocol with a few modifications [10, 27].

1. Fill 1 mL insulin syringe with a 28 G gauge needle with 50 mL of thrombin solution (60 U/mL thrombin in sterile 0.9% NaCl solution).
2. Collect arterial blood (200 μL) from the left ventricle of a donor mouse into another insulin syringe (*see* **Note 2**).
3. Connect two syringes from tip to tip with a 10 cm-long polyethylene tube (I.D 0.32 mm). The suspension containing blood and thrombin solution at a ratio of 4:1 is moved from one syringe to the other approximately 20 times.
4. Incubate the resultant mixture first at 37 °C for 1 h and then overnight at 4 °C.
5. Transfer suspension to a centrifuge tube containing 10 mL of Krebs–Ringer bicarbonate buffer (KRBB) and homogenize at 26,000 rpm for 1 min using a PT-3100 Polytron homogenizer. Wash the homogenate/pellet twice with KRBB by centrifugation at 3000 rpm for 10 min. After the last wash, resuspend the pellet in 1 mL of KRBB and centrifuge at 300 rpm for 5 min to collect large thromboemboli. Collect the resultant supernatant, and centrifuge at 6000 rpm for 10 min.
6. Resuspend the resulting pellet in 0.9% NaCl, and determine thromboemboli size using a ZM Coulter Counter (*see* **Note 3**).
7. Transfer thromboemboli suspension to a pre-weighed microcentrifuge tube, and centrifuge at 6000 rpm for 10 min at 4 °C. Discard supernatant and weigh remaining pellet. Resuspend collected pellet in 0.9% NaCl to a final concentration of 8 mg/100 μL. The composition of the thromboemboli suspension is shown in Fig. 1.

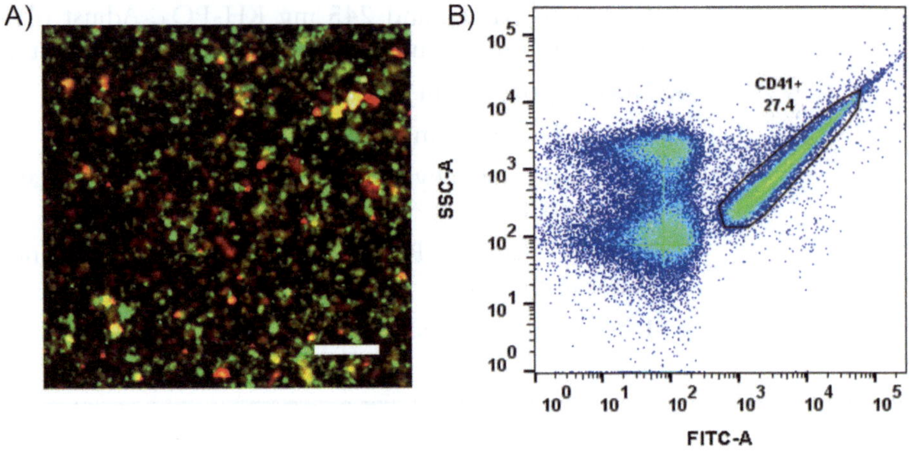

Fig. 1 (**a**) Representative cytospin smear of the thromboemboli suspension double labelled for calcein AM (green) and CD41 (platelets, red). Scale bar = 100 μm. (**b**) Platelet (CD41+ cells) content in thromboemboli suspension assessed by flow cytometry (FACS)

3.1.2 Thromboemboli Injection

1. Anesthetize mice (C57BL/6; 10–12 weeks of age; weight 28–30 g) by intraperitoneal injection of ketamine (100 mg/kg) and xylazine (10 mg/kg). Monitor and maintain body temperature and physiological parameters during surgical procedures using a temperature control system (Kent Scientific PhysioSuite Monitoring Device) (*see* **Note 4**).
2. Under a surgical microscope, perform a blunt dissection to expose the right common carotid artery (RCCA), the right external carotid artery (RECA), and the right internal carotid artery (RICA). Permanently ligate the RECA.
3. A mouse femoral vein catheter attached to a 1 mL syringe containing 100 μL of the thromboemboli suspension is inserted into the RCCA and guided to the bifurcation of the RCCA. Inject thromboemboli suspension through the catheter into the ICA. Withdraw catheter and permanently ligate cephalic RCC (*see* **Note 5**).

3.1.3 Infarct Assessment

Initially, infarct size is assessed 24 h after thromboemboli injection by MRI. It is recommended that infarct assessment be repeated over the first 5 days (1, 3, and 5 days after thromboemboli injection) to evaluate stroke injury development. An example of a TE-induced stroke monitored by T2 MRI is presented in Fig. 2.

1. Anesthetize mice with 2% isoflurane/air mixture prior to placement in a 7 T Agilent MR scanner, and maintain body temperature at 37 °C using forced heated air.
2. Acquire axial T2-weighted images using a fast spin-echo sequence with the following imaging parameters: repetition time/effective echo time, 4000/60 ms; average 4; echo

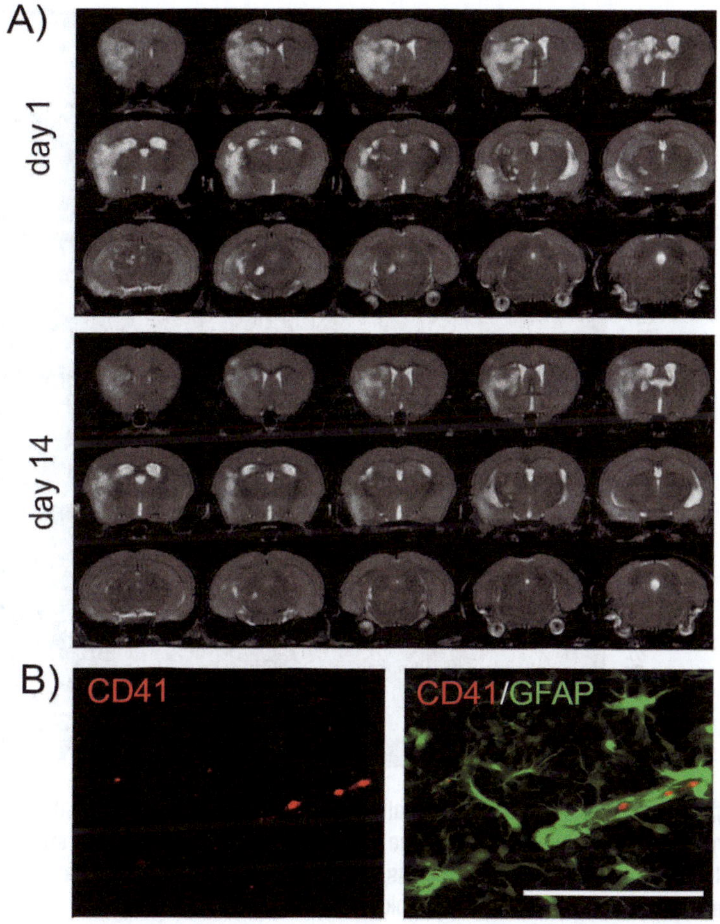

Fig. 2 (**a**) Representative images of brain infarction longitudinally monitored by T2 MRI in the same mouse on days 1 and 14 after intracarotid injection of thromboemboli suspension. (**b**) Double-label immunofluorescence of thromboembolic occlusion of cerebral microvessels in mice 1 day after injection. Thromboemboli are labelled with the platelet marker CD41, while astrocytes are labeled with GFAP. CD41-labelled thromboemboli can be seen within a blood vessel enveloped by astrocyte foot processes. Scale bar =50 μm

spacing, 15 ms; rare factor, 8; field of view, 20 × 20 mm; matrix, 256 × 128; slice thickness, 0.5 mm; number of slices, 25; and number of scans, 1 (t-scan time 4 min, 16 s).

3. Evaluate infarct volume measurement and location using T2-weighted coronal images and processed by ImageJ software (National Institutes of Health) by a blinded observer (*see* **Note 6**).

BBB leakage in chronic stroke conditions is evaluated by contrast-enhanced MRI using gadolinium-diethylenetriamine penta-acetic acid (Gd-DTPA) as a tracer at specific time points (7, 14,

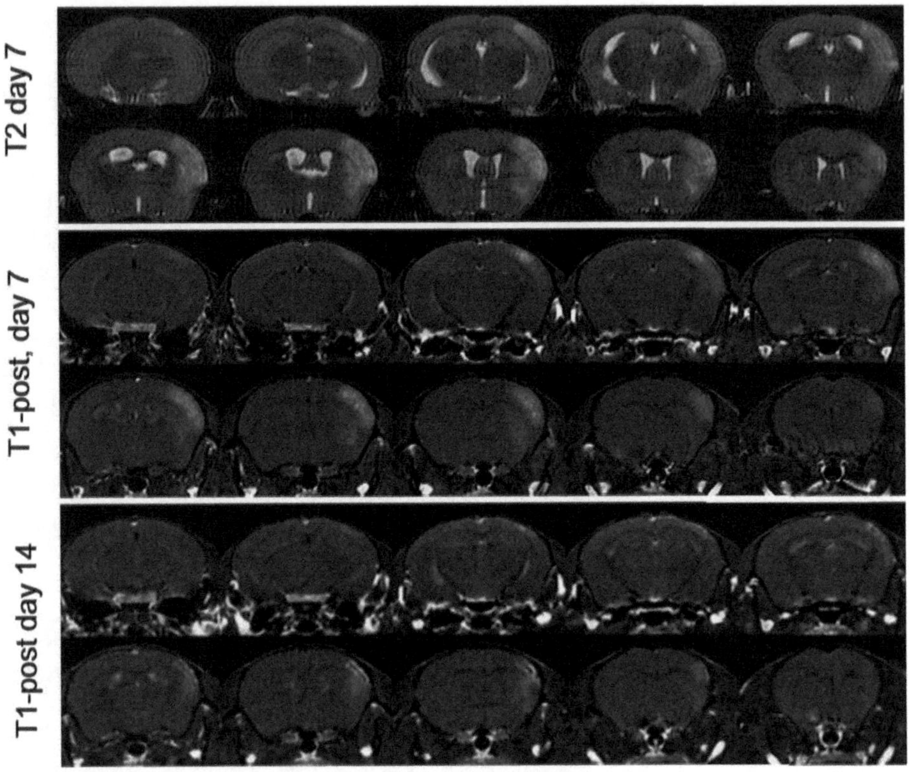

Fig. 3 Leakage of Gd-DTPA in brain infarct area 7 and 14 days after thromboemboli injection. The BBB permeability is followed in the same mouse in the chronic phase of stroke. T1-post represents data from T1 MRI monitoring at days 7 and 14. As part of monitoring, the mouse also underwent T2 MRI for evaluation the infarct size in chronic phase of stroke

3.2 BBB Permeability Assays

3.2.1 T1 MRI

21, 28 days, etc.). Figure 3 shows an example of longitudinal monitoring of BBB leakage during chronic TE stroke. The presented protocol for T1 MRI allows longitudinal monitoring of the BBB integrity over the course of chronic stroke condition, following the same group of mice.

1. Anesthetize mice with 2% isoflurane/air mixture prior to placement in a 7 T Agilent MR scanner, and body temperature is maintained at 37 °C using forced heated air.
2. Axial T1-weighted spin-echo images parameters: repetition time/effective echo time, 750/9 ms; average 4; echo spacing, 9 ms; rare factor, 2; field of view, 20 × 20 mm; matrix, 256 × 128; slice thickness, 0.5 mm; number of slices, 25; and number of scans, 1 (t-scan time 3 min, 12 s).
3. Obtain a T1 pre-contrast image. Inject mouse with pharmaceutical-grade tracer Gd-DTPA, via *ip* bolus injection (100 mL, 0.5 mM), and immediately image in five sessions over 10 min (T1 post-contrast images) (*see* **Note 7**).

4. MRI analysis of BBB permeability is evaluated from recording series of T1 pre- and post-contrast images obtained after administration of Gd-DTPA using ImageJ software. Analyze BBB permeability using subtracted images from the pre- and post-contrast T1-SE, by calculating the product of Gd-DTPA permeable BBB volume (PBV, in cm^3) and average pixel intensity (T1SI diff) from subtracted images [28].

3.2.2 Tracer Studies

An in vivo permeability assay is used to assess BBB integrity by determining the transfer coefficient (K_i) for fluorescence labelled tracers, sodium fluorescein (SF, 376 Da), Dextran-Cascade Blue (3 kDa), and Dextran-Texas Red (10 kDa) [11, 23]. The most frequent tracers used for in vivo BBB permeability in our laboratory are presented in Table 2 along with their specifics.

Table 2
Frequently used tracers in in vivo tracer and tracer+immunohistology studies

Tracers	Size (Da)	Fluorophore	Cell penetration/ barrier passage	Tracer study	Tracer study + immunohistology
Sodium fluorescein	376	Fluorescein sodium salt	No/ paracellular passage	Yes	Unstable after PFA fixation
Inulin	4000	Fluorescein isothiocyanate	No/ paracellular passage	Yes	Unstable after PFA fixation
Dextran	Range 3000–70,000	Cascade blue, fluorescein, Lucifer yellow, Oregon green, tetramethylrhodamine	No/ paracellular passage	Yes	Yes—only anionic, lysine fixable formula is stable after PFA fixation
Albumin	65,000	Fluorescein isothiocyanate	Transcellular and paracellular passage	Yes	Yes
Evans blue	960.8[a]	Colorimetric[b]	Paracellular and transcellular (binding to albumin)	No	No

PFA paraformaldehyde
[a]While Evans blue has a molecular weight of 960 Da, it is almost entirely bound to albumin in the bloodstream
[b]Evans blue is most commonly assessed colorimetrically. However, it can be examined by fluorescence (excitation 620 nm/emission 680 nm)

1. Prepare cocktail of three different sized tracers: SF, Dextran-Cascade Blue, and Dextran-Texas Red in 0.9% NaCl at a final concentration of 10 μg/mL. Protect tube from light and keep at 37 °C (water bath) until injection.
2. Anesthetize mice (C57BL/6) by intraperitoneal injection of ketamine (100 mg/kg) and xylazine (10 mg/kg).
3. Under a surgical microscope, perform a blunt dissection to expose the right and left femoral artery and veins, and insert catheters. Attach 1 mL syringe containing cocktail tracer solution to mouse femoral vein catheter. Inject 100 μL of the tracer cocktail over 20 sec. Replace syringe with 2 mL syringe filled with 0.9% NaCl.
4. The catheter placed in the left femoral artery is utilized for drawing blood samples (50 μL) at 0, 5, 10, 15, and 20 min. After drawing blood samples, inject 50 μL of 0.9% NaCl into the left femoral artery.
5. Place blood samples into micro sample tubes, keep at room temperature for 30 min, and centrifuge at 3000 rpm for 30 min. Collect serum and freeze at −80 °C. The serum samples are used to determine tracer concentration in blood over the 20 min of the experiment.
6. *Brain tissue preparation*: After collecting the last blood sample, perfuse mouse transcardially with 0.9% NaCl to remove tracer from the blood stream. Remove the brain, clean the meninges, and dissect into ischemic and contralateral hemisphere. Weigh brain samples, and homogenize in 50 mM Tris-HCL buffer solution (pH 7.4). Centrifuge homogenate at 3000 rpm for 30 min, and collect supernatant. Add methanol to collected supernatant (1:1) mix, and centrifuge at 3000 rpm for 30 min. Collect supernatant in fresh vials, and store at −80 °C.
7. *Analysis*: Measure fluorescence intensities in brain and serum samples with a fluorescent reader (Tecam; excitation/emission: Dextran-Cascade Blue, 390/410; sodium fluorescein, 488/510; Dextran-Texas Red, 590/600 nm). Calculate tracer concentrations using standard curves established for each tracer. Determine the K_i for each tracer using the equation developed by Ohno et al. [29]:

$$K_i = (C_{\mathrm{br}} - V_{\mathrm{o}} C_{\mathrm{bl}}) / \int C_{\mathrm{pl}} \tilde{n} dt$$

where C_{br} is the concentration of tracer in brain tissue at the time of decapitation (ng/g), C_{bl} is the concentration of tracer (ng/mL) in the last blood sample, V_o is regional blood volume (mL/g), t is time (min), C_{pl} is the arterial tracer concentration, and $\int C_{pl} \cdot dt$ denotes integration of C_{pl} over time. That integral is determined using the

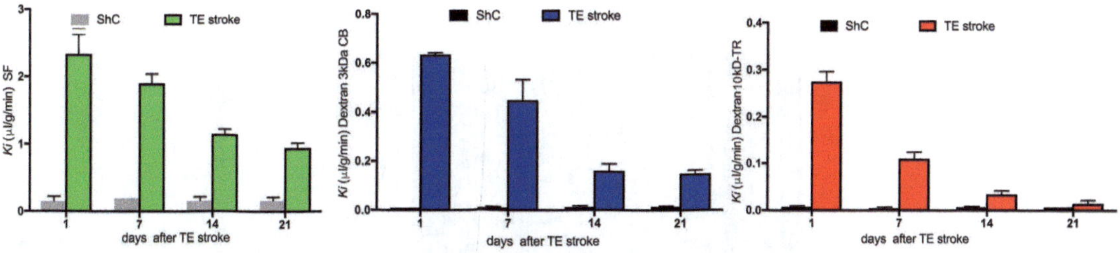

Fig. 4 In vivo permeability assay with selective size tracers: sodium fluorescein (SF), Dextran 3 kDa-Cascade Blue (Dextran-CB), and Dextran 10 kDa-Texas Red (Dextran-TR). Transfer coefficient (K_i) was determined for each tracer at the endpoint of the experiment (different animals at each time point). This assay provides quantitative assessment of BBB permeability, and it also allows determination of "pore size" in chronic stroke, i.e., the apparent size of any pores in the BBB after stroke

trapezoidal method. Figure 4 shows an example of the leakage of different sized tracers after chronic TE stroke (*see* **Note 8**).

3.3 Immunohistology and Qualitative Tracer Study

To visualize changes in TJ expression and BBB leakage after chronic stroke, a combined immunohistochemistry and tracer study is performed (Fig. 5). Our standard markers for examining the TJs are claudin-5 and ZO-1. However, the same methodology can be employed for other TJ proteins (e.g., occludin and claudin-1) or adherens junction proteins. BBB leakage can be simultaneously assessed by using an appropriate fluorescent labelled tracer for tri-color imaging. In our protocol, we combine the tracer Dextran-Cascade Blue and immunofluorescence with monoclonal antibodies claudin-5-Alexa 488 and ZO-1-Alexa 590.

1. Prepare Dextran-Cascade Blue solution in sterile isotonic 0.9% NaCl solution at a final concentration of 10 µg/mL (*see* **Note 9**).
2. Inject ketamine/xylazine to anesthetize mice. After mice become unresponsive to pinch, inject Dextran-Cascade Blue (*ip* bolus 100 µL), and allow tracer to circulate for 30 min.
3. After 30 min, rapidly remove the brain, and fix the brain overnight at 4 °C using paraformaldehyde. Transfer the brain to 30% sucrose for cryoprotection for 2–3 days (until the brain sinks to the bottom of the tube), decant sucrose solution, place the brain in a module cover with OCT, and freeze at −80 °C.
4. Section the brain into 20–25 µm-thick coronal sections on a cryostat. Dry brain sections for 20 min at room temperature, then wash in PBS for 10 min (×3), and permeabilize in 0.05% Triton X-100 in PBS for 10 min. After permeabilization, wash sections once in PBS for 10 min, and incubate in blocking solution (5% goat serum, PBS, and 0.05% Triton X-100) for 1 h at room temperature. Make primary antibodies solution in blocking solution by adding fluorophore labelled primary

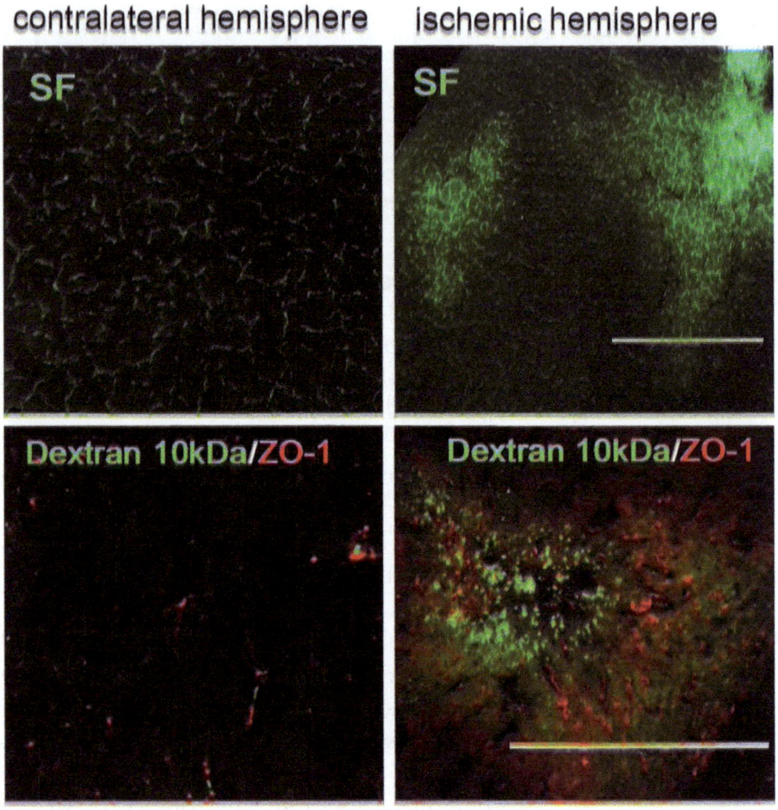

Fig. 5 Visualization of BBB permeability for tracer sodium fluorescein (SF) and Dextran 10 kDa-FITC in contralateral and ischemic hemisphere in chronic stoke condition 14 days after TE stroke induction. To visualize SF, the brain was snap frozen and sectioned on a cryostat. Fixation is not recommended. Tracer Dextran 10 kDa is fixable and can be used as a good marker of permeability in combination with immunofluorescence. The images represent a combination of tracer study (qualitative leakage) and immunohistochemistry (staining with the tight junction protein ZO-1). Scale bar = 400 μm

antibodies (claudin-5-Alexa 488 and ZO-1-Alexa 590 at final 1:100 dilutions). Incubate sections overnight at 4 °C. After that, wash sections with PBS for 10 min (×3), and mount in Vectashield Antifade mounting media. Analyze sections by fluorescence microscopy for both changes in TJ protein expression and BBB leakage.

4 Notes

1. Use sterile nonpyrogenic isotonic 0.9% NaCl for injections in order to avoid any potential infection of mice.
2. Collect blood from the left ventricle of a donor mouse with the same genetic background as the recipient.

3. Obtained thromboemboli with a size range from 1 to 6 μm (approximately 90% are <4 μm).
4. Physiological parameters (P_{O2}, P_{CO2}, glucose, blood pH, body temperature) should be measured before, during, and after surgical procedures. Body temperature should be monitored and maintained at 37 ± 0.5 °C during surgical procedure using a temperature control system.
5. Sham-operated control mice receive 100 μL 0.9% NaCl.
6. MR images are analyzed under the same contrast enhancement to maximize signal intensity of the hyperintense cerebral infarct region. The infarct region is outlined on every section and multiplied by the interslice distance to generate infarct volume [10]. T2-weighted imaging is also accompanied with assessment of BBB permeability (T1-weighted images), and it follows brain infarct resolution.
7. There is a 99% survival rate after a single and after multiple T1 scans.
8. V_o is determined in a second set of animals where mice were sacrificed 1 min after intravenous injection of sodium fluorescein, Dextran-Cascade Blue, and Dextran-Texas Red. Assuming no extravasation during this short circulation time, $V_o = C_{br}/C_{bl}$.
9. Dextran-Cascade Blue is also available in molecular size of 3 and 10 kDa for assessing the leakage of larger molecules across the BBB in chronic stroke. To combine tracer and immunofluorescent staining, the tracer should have an anion charge and be fixable via lysine for use with an aldehyde-based fixative.

Acknowledgments

This work is supported by NIH grants: AG057928 to A.V.A. and NS098211 to S.M.S.

References

1. Dirnagl U, Iadecola C, Moskowitz MA (1999) Pathobiology of ischaemic stroke: an integrated view. Trends Neurosci 22(9): 391–397. https://doi.org/10.1016/s0166-2236(99)01401-0
2. Delavaran H, Jonsson AC, Lovkvist H et al (2017) Cognitive function in stroke survivors: a 10-year follow-up study. Acta Neurol Scand 136(3):187–194. https://doi.org/10.1111/ane.12709
3. Heye AK, Thrippleton MJ, Armitage PA et al (2016) Tracer kinetic modelling for DCE-MRI quantification of subtle blood-brain barrier permeability. NeuroImage 125:446–455. https://doi.org/10.1016/j.neuroimage.2015.10.018
4. Sood RR, Taheri S, Candelario-Jalil E et al (2008) Early beneficial effect of matrix metalloproteinase inhibition on blood-brain barrier permeability as measured by magnetic resonance imaging countered by impaired long-

term recovery after stroke in rat brain. J Cereb Blood Flow Metab 28(2):431–438. https://doi.org/10.1038/sj.jcbfm.9600534

5. Portegies ML, Wolters FJ, Hofman A et al (2016) Prestroke vascular pathology and the risk of recurrent stroke and poststroke dementia. Stroke 47(8):2119–2122. https://doi.org/10.1161/STROKEAHA.116.014094
6. Mok VC, Lam BY, Wong A et al (2017) Early-onset and delayed-onset poststroke dementia - revisiting the mechanisms. Nat Rev Neurol 13(3):148–159. https://doi.org/10.1038/nrneurol.2017.16
7. Tobinick E (2011) Rapid improvement of chronic stroke deficits after perispinal etanercept: three consecutive cases. CNS Drugs 25(2):145–155. https://doi.org/10.2165/11588400-000000000-00000
8. Krafft PR, Bailey EL, Lekic T et al (2012) Etiology of stroke and choice of models. Int J Stroke 7(5):398–406. https://doi.org/10.1111/j.1747-4949.2012.00838.x
9. Gennaro M, Mattiello A, Pizzorusso T (2019) Rodent models of developmental ischemic stroke for translational research: strengths and weaknesses. Neural Plast 2019:5089321. https://doi.org/10.1155/2019/5089321
10. Stamatovic SM, Phillips CM, Keep RF et al (2020) A novel approach to treatment of thromboembolic stroke in mice: redirecting neutrophils toward a peripherally implanted CXCL1-soaked sponge. Exp Neurol 330: 113336. https://doi.org/10.1016/j.expneurol.2020.113336
11. Sladojevic N, Stamatovic SM, Johnson AM et al (2019) Claudin-1-dependent destabilization of the blood-brain barrier in chronic stroke. J Neurosci 39(4):743–757. https://doi.org/10.1523/JNEUROSCI.1432-18.2018
12. Garbuzova-Davis S, Haller E, Williams SN et al (2014) Compromised blood-brain barrier competence in remote brain areas in ischemic stroke rats at the chronic stage. J Comp Neurol 522(13):3120–3137. https://doi.org/10.1002/cne.23582
13. Morimoto T, Yasuhara T, Kameda M et al (2011) Striatal stimulation nurtures endogenous neurogenesis and angiogenesis in chronic-phase ischemic stroke rats. Cell Transplant 20(7):1049–1064. https://doi.org/10.3727/096368910X544915
14. Spychala MS, Venna VR, Jandzinski M et al (2018) Age-related changes in the gut microbiota influence systemic inflammation and stroke outcome. Ann Neurol 84(1):23–36. https://doi.org/10.1002/ana.25250
15. Kim H, Seo JS, Lee SY et al (2020) AIM2 inflammasome contributes to brain injury and chronic post-stroke cognitive impairment in mice. Brain Behav Immun 87:765–776. https://doi.org/10.1016/j.bbi.2020.03.011
16. Choi C, Kim HM, Shon J et al (2018) The combination of mannitol and temozolomide increases the effectiveness of stem cell treatment in a chronic stroke model. Cytotherapy 20(6):820–829. https://doi.org/10.1016/j.jcyt.2018.04.004
17. Tajiri N, Lee JY, Acosta S et al (2016) Breaking the blood-brain barrier with mannitol to aid stem cell therapeutics in the chronic stroke brain. Cell Transplant 25(8):1453–1460. https://doi.org/10.3727/096368916X690971
18. Bihel E, Pro-Sistiaga P, Letourneur A et al (2010) Permanent or transient chronic ischemic stroke in the non-human primate: behavioral, neuroimaging, histological, and immunohistochemical investigations. J Cereb Blood Flow Metab 30(2):273–285. https://doi.org/10.1038/jcbfm.2009.209
19. Stamatovic SM, Johnson AM, Keep RF et al (2016) Junctional proteins of the blood-brain barrier: new insights into function and dysfunction. Tissue Barriers 4(1):e1154641. https://doi.org/10.1080/21688370.2016.1154641
20. Sweeney MD, Zhao Z, Montagne A et al (2019) Blood-brain barrier: from physiology to disease and back. Physiol Rev 99(1):21–78. https://doi.org/10.1152/physrev.00050.2017
21. Mintzopoulos D, Khanicheh A, Konstas AA et al (2008) Functional MRI of rehabilitation in chronic stroke patients using novel MR-compatible hand robots. Open Neuroimaging J 2:94–101. https://doi.org/10.2174/1874440000802010094
22. Taheri S, Gasparovic C, Huisa BN et al (2011) Blood-brain barrier permeability abnormalities in vascular cognitive impairment. Stroke 42(8): 2158–2163. https://doi.org/10.1161/STROKEAHA.110.611731
23. Sladojevic N, Stamatovic SM, Keep RF et al (2014) Inhibition of junctional adhesion molecule-A/LFA interaction attenuates leukocyte trafficking and inflammation in brain ischemia/reperfusion injury. Neurobiol Dis 67:57–70. https://doi.org/10.1016/j.nbd.2014.03.010
24. Stamatovic SM, Shakui P, Keep RF et al (2005) Monocyte chemoattractant protein-1 regulation of blood-brain barrier permeability. J Cereb Blood Flow Metab 25(5):593–606. https://doi.org/10.1038/sj.jcbfm.9600055

25. Beller E, Reuter L, Kluge A et al (2018) Pilot study to assess visualization and therapy of inflammatory mechanisms after vessel reopening in a mouse stroke model. Sci Rep 8(1):745. https://doi.org/10.1038/s41598-017-17533-5
26. Johnson AM, Roach JP, Hu A et al (2018) Connexin 43 gap junctions contribute to brain endothelial barrier hyperpermeability in familial cerebral cavernous malformations type III by modulating tight junction structure. FASEB J 32(5):2615–2629. https://doi.org/10.1096/fj.201700699R
27. Overgaard K, Sereghy T, Boysen G et al (1992) A rat model of reproducible cerebral infarction using thrombotic blood clot emboli. J Cereb Blood Flow Metab 12(3):484–490. https://doi.org/10.1038/jcbfm.1992.66
28. He Z, Wang X, Wu Y et al (2014) Treadmill pre-training ameliorates brain edema in ischemic stroke via down-regulation of aquaporin-4: an MRI study in rats. PLoS One 9(1):e84602. https://doi.org/10.1371/journal.pone.0084602
29. Ohno K, Chiueh CC, Burns EM et al (1980) Cerebrovascular integrity in protein-deprived rats. Brain Res Bull 5(3):251–255. https://doi.org/10.1016/0361-9230(80)90166-5

Chapter 18

Immunohistochemical Analysis of Tight Junction Proteins

Chris Greene and Matthew Campbell

Abstract

Tight junction proteins are integral membrane proteins located apically on epithelial and endothelial cells. They form a selective paracellular barrier restricting the passage of solutes and ions across epithelial and endothelial sheets. In brain endothelial cells, the enrichment of tight junction proteins is one of the unique features of the blood–brain barrier, the physiological boundary that separates the blood from the parenchyma. The predominant tight junction family proteins are the claudins, but several others have been described in recent years including the marvel family, occludin, and lipolysis-stimulated lipoprotein receptor. Together, the tight junctions create a highly electrical-resistant, impermeable paracellular channel that strictly restricts the movement of material from the blood to the parenchyma and vice versa. In this chapter, we will discuss immunohistochemical methods to assess tight junction expression and localization and an ImageJ-based method for quantifying tight junction staining in healthy and diseased states.

Key words Tight junction, Blood–brain barrier, Immunohistochemistry, ImageJ

1 Introduction

Blood vessels in the central nervous system provide oxygen-rich blood and vital nutrients to neurons in the brain to ensure constant homeostasis of the cerebral microenvironment. The proper structure and function of blood vessels are vital for normal brain function. Despite accounting for 2% of body mass, neurons consume ~20% of the body's energy [1]. The complex vascular network in the brain supplies oxygen and nutrients in response to neural activation, a process known as hyperemia, and removes metabolic waste products. Unique to the cerebral vasculature are the properties known as the blood–brain barrier (BBB). The BBB is a semi-permeable barrier that selectively controls the transport of material between the blood and the brain and prevents the entry of pathogenic and neurodegenerative solutes. The BBB is formed by a continuous layer of non-fenestrated endothelial cells which have distinctive properties compared to peripheral endothelial cells. These properties include an enrichment of tight junction proteins,

Nicole Stone (ed.), *The Blood-Brain Barrier: Methods and Protocols*,
Methods in Molecular Biology, vol. 2492, https://doi.org/10.1007/978-1-0716-2289-6_18,

nutrient and efflux transporters such as glucose transporter-1 (GLUT-1) and P-glycoprotein (P-gp) low rates of transcytosis, and reduced expression of cell adhesion molecules to limit immune cell trafficking into the brain [2]. It is the complex interactions of brain endothelial cells, pericytes, astrocytes, and the acellular basement membrane, the so-called neurovascular unit, that confer properties on the cerebral vasculature that creates a dynamic, highly regulatable microenvironment that is vital for central nervous system homeostasis.

Tight junctions are integral membrane proteins that span the intercellular cleft and contact at "kissing points" to create a seal in the paracellular space between adjacent endothelial cells. Located luminally in endothelial cells, they are composed of several families of proteins including the claudins, occludin, and marvel family of proteins [3, 4]. The claudins are the predominant junctional component with up to 27 family members described. Despite sharing minimal sequence homology, tight junctions have similar structures. They consist of four transmembrane domains, a C-terminus, and N-terminus and have two extracellular loops [5]. The carboxy-terminal domain of tight junctions including claudin-5, occludin, and marveld2 binds to intracellular scaffolding proteins called the zonulae occludens which links the tight junction to the actin cytoskeleton [6, 7]. Tight junctions are vital for BBB integrity with size-selective loosening of the BBB observed in claudin-5- and LSR-deficient animals [8, 9]. Furthermore, disruption of the endothelial tight junction is observed in disorders of the central nervous system including stroke [10, 11], traumatic brain injury [12], schizophrenia [13–15], and neuroinflammation [16]. As such, assessment of tight junction structure and expression is routinely used to assess BBB integrity in human tissue and mouse models of CNS disorders. In this chapter, we discuss straightforward strategies to stain junctional proteins and a pipeline to analyze tight junction structure in wild-type and reporter mice and in healthy and disease models.

In this chapter we have described a simple method for the analysis of tight junction and blood vessel structure in healthy and diseased models. This protocol allows a quantitative analysis of important vessel parameters including length, number of branches, number of junctions, and area fraction. A reliable method to quantify vessel parameters is essential to identify pathological changes in various central nervous system disorders. Methods to assess tight junctions include conventional immunohistochemistry as well as with transgenic reporter mice. The protocol outlined in this chapter is simple and inexpensive and does not require commercial software. ImageJ is a freely downloadable software that is useful for many types of analysis including immunohistology.

2 Materials

1. C57BL/6 J mice and Cldn5-eGFP mice.
2. Leica cryostat (Leica), optimal cutting temperature (OCT), section brushes, and poly-lysine-coated slides (all VWR).
3. Methanol.
4. Permeabilization and blocking solution: 5% normal goat serum, 0.1% Triton X-100, 1× phosphate-buffered saline (PBS) (all Sigma).
5. Primary antibody solution—rabbit anti-claudin-5 primary antibody (Invitrogen) diluted in 1% normal goat serum, 0.1% Triton X-100, 1× PBS.
6. Secondary antibody solution—goat anti-rabbit Alexa Fluor 488 conjugated secondary antibody (Abcam) diluted in 1× PBS.
7. Nuclei solution—1 μg/mL Hoechst 33342 diluted in 1× PBS.
8. PAP pen hydrophobic barrier (Mason Technology).
9. Glass coverslips (24 × 50 mm, VWR).
10. Hydromount (Analab).
11. Humidity chamber.
12. 1× PBS.
13. 0.1% Triton X-100 in 1× PBS.

3 Methods

3.1 Tissue Preparation

1. Euthanize animals according to approved regulatory guidelines. Remove the brain, and place into a cryomold filled with optimal cutting temperature (OCT). Brains from Cldn5-eGFP mice are placed in 4% formaldehyde overnight at 4 °C, washed three times in PBS, and cryoprotected in 20% sucrose for 48 h. Brains are subsequently embedded in OCT.
2. Freeze sample in liquid nitrogen-cooled 2-methylbutane. Immediately store samples at −80 °C until ready for processing on cryostat. Store samples in cryostat for a minimum of 30 min prior to sectioning.
3. Section samples to 10–50 μm thickness, and capture on poly-lysine-coated slides. Allow to air-dry for 30 min at room temperature (*see* **Note 1**).
4. Fix samples in alcohol-based fixative (methanol) at −20 °C for 10 min (*see* **Note 2**).
5. Rehydrate samples in PBS for 2 min, and proceed to blocking and permeabilization.

3.2 Blocking and Permeabilization

1. Remove the slides from PBS and dab dry the sides.
2. Draw a circle around the sections with a hydrophobic marker.
3. Place 100 μL of permeabilization and blocking solution onto slides (*see* **Note 3**). Leave slides in humidity chamber for 1 h at room temperature.

3.3 Antibody Incubation

1. Prepare primary antibody in 0.1% Triton X-100, 1× PBS. Remove permeabilization and blocking solution, and place 100 μL of primary antibody solution onto slides. Leave slides in humidity chamber for 18 h at 4 °C or for 1 h at 37 °C.
2. Remove primary antibody solution, and wash slides three times for 5 min each in PBS.
3. Place 100 μL of secondary antibody solution containing vessel marker (*see* **Note 4**), if using, onto slides. Leave slides in humidity chamber for 1 h at room temperature.
4. Remove secondary antibody solution, and wash slides three times for 5 min each in PBS.
5. Place 100 μL of nuclei stain (Hoechst 33342) onto slides for 30 s. Remove nuclei staining solution, and wash slides once for 5 min in PBS.
6. Place one drop of hydromount per section, and cover with a glass coverslip. Leave slides to dry in the dark for a minimum of 1 h prior to visualization.
   (a) Microscopic detection

      Images were acquired on a Zeiss LSM 710 confocal microscope with 10× and 40× objectives with an image size of 1024 × 1024 pixels.

   (b) Image processing
      1. Export images in TIFF format. Open image in ImageJ, convert to 8 bit, and separate channels (Image – Color – Split channels).
      2. Apply median filter (Process – Filters – Median), threshold the image (Image – Adjust – Threshold), and binarize the image (Process – Binary – Make Binary).
      3. Apply skeletonization (Process – Binary – Skeletonize) and analyze skeleton (Analyze – Skeleton). Images with elevated levels of background can be improved by reducing the noise (Process – Noise – Despeckle) prior to converting to binary image.
      4. Copy branch information and results to Excel. Sum branch length to get total vessel length. Divide total vessel length by the sum of #Branches to get average

vessel length. Sum #Junctions for the total number of junctions.

(c) Statistical analysis

Statistical analysis is performed using GraphPad Prism 9. Differences between two groups were tested with a two-tailed unpaired t-test. Statistical significance defined as $*P < 0.05$, $**P < 0.01$, $***P < 0.001$, and $****P < 0.0001$.

3.4 Representative Results

Figure 1a shows immunohistochemistry for claudin-5 in the brain using the Invitrogen rabbit anti-claudin-5 primary antibody (#34–1600), while the image on the right displays GFP expression from a reporter mouse expressing GFP under the control of the claudin-5 promoter (*see* **Note 5**). Figure 1b displays the processing steps performed in ImageJ to analyze vascular parameters including vessel length, number of branches, and percentage of claudin-5 stained area. These steps can be adapted to analyze other junctional proteins such as occludin and tricellulin among others as well as other tissue types. Figure 2 shows the effect of fixative on claudin-5 immunoreactivity. Immersion fixation overnight in 4% formaldehyde results in almost complete loss of vascular claudin-5 specificity with diffuse antibody staining. In contrast, post-fixation of flash frozen samples with methanol results in specific claudin-5 reactivity in blood vessels. While formaldehyde-based fixatives are preferred for retaining morphology of the samples, the ability to stain junctional components is significantly diminished, and as such, alcohol-based fixation is preferred.

Finally, using the methods described, we assessed the effect of traumatic brain injury on claudin-5 immunoreactivity (Fig. 3). Mice subjected to a closed-skull controlled cortical impact had significantly reduced % area of claudin-5 immunoreactivity (−48%, sham: 1.483 ± 0.1699, impact: 0.7703 ± 0.049, $P < 0.01$), reduced number of branches (−48%, sham: 305.2 ± 25.08, impact: 158.7 ± 22.69, $P < 0.01$), and reduced vessel length (−37%, sham: 100 ± 3.863, impact: 63.34 ± 3.142, $P < 0.0001$) in the ipsilateral cortex following a 5 m/s impact.

4 Notes

1. Allowing the slides to dry is important for firm attachment of the sections for subsequent processing steps.
2. Flash frozen samples are preferred for immunostaining of tight junctions as formaldehyde-based fixatives significantly decrease antibody binding resulting in a loss of signal intensity.

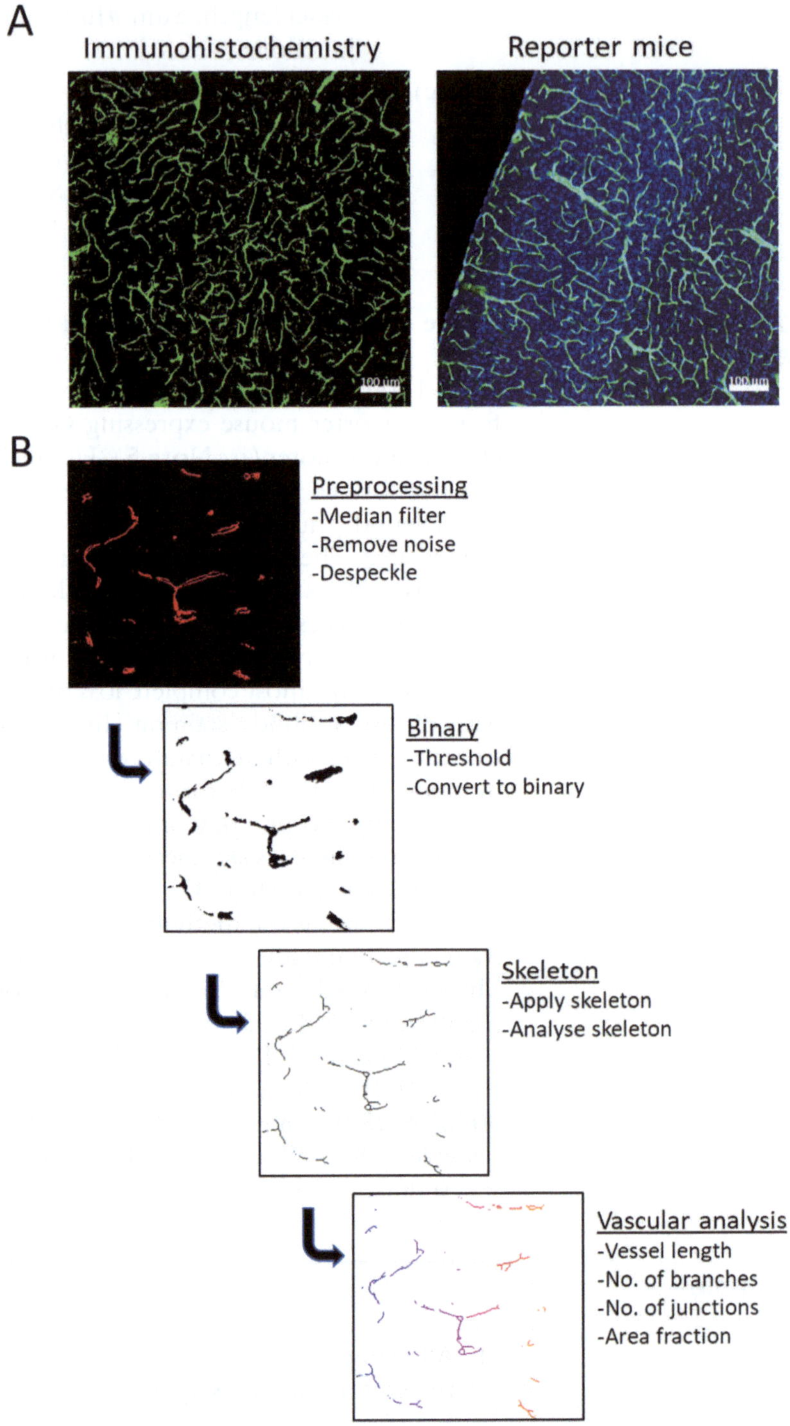

Fig. 1 Immunohistology and analysis steps. (**a**) Assessment of claudin-5 by immunohistochemistry with rabbit anti-claudin-5 polyclonal antibody (Invitrogen: #341600) or in Cldn5-eGFP reporter mice. (**b**) Stepwise processing for vascular analysis of tight junction staining

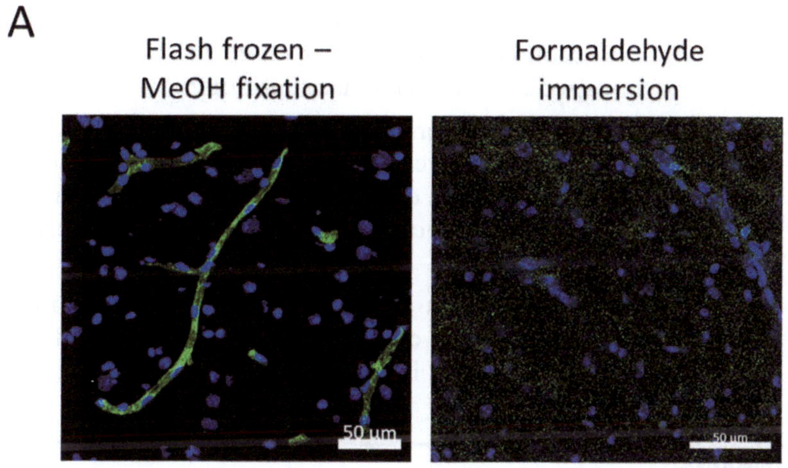

Fig. 2 Effect of fixation on tight junction staining. Immunohistochemistry for claudin-5 in flash frozen brains post-fixed with methanol or in brains fixed by immersion in 4% formaldehyde overnight

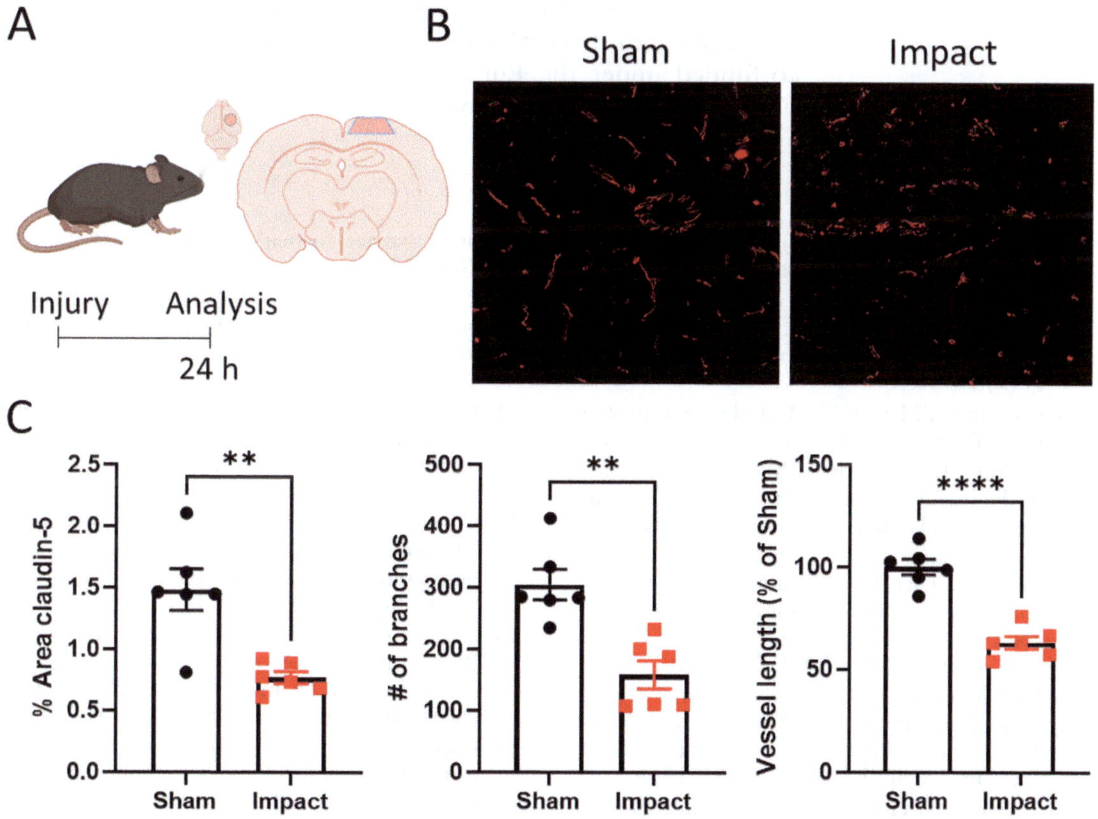

Fig. 3 Vascular analysis in traumatic brain injury. (**a**) Experimental setup—mice were impacted with a closed-skull controlled cortical impact at 5 m/s velocity and 2 mm impact depth. 24 h later animals were sacrificed. (**b**) Immunohistochemistry for claudin-5 in sham and 5 m/s impact groups. (**c**) The percentage area of claudin-5, number of branches, and vessel length were assessed in ImageJ as outlined in the methods section. Data represents mean $\pm$ SEM with $n = 6$ mice per group. $^{**}P < 0.01$, $^{****}P < 0.0001$

3. Serum for blocking buffers depends on the species that the secondary antibody was raised in. Use serum from the same species of the secondary antibody.
4. Tight junction staining can be normalized to other vessel markers; however, it is important to choose the appropriate marker based on experimental hypothesis and model investigated. Commonly used markers include pecam1 (CD31) and isolectin B4.
5. Other claudin-5 antibodies can be used including mouse anti-claudin-5488-conjugate (Invitrogen #352588). These antibodies will work in mouse and human samples.

Acknowledgements

This work was supported by grants from Science Foundation Ireland and the Irish Research Council (IRC). The Campbell lab at TCD is also supported by an SFI Centres grant supported in part by a research grant from SFI under grant number 16/RC/3948 and co-funded under the European Regional Development fund by FutureNeuro industry partners.

References

1. Sweeney MD et al (2019) Blood-brain barrier: from physiology to disease and Back. Physiol Rev 99(1):21–78
2. Abbott NJ et al (2010) Structure and function of the blood-brain barrier. Neurobiol Dis 37(1):13–25
3. Cummins PM (2012) Occludin: one protein, many forms. Mol Cell Biol 32(2):242–250
4. Wolburg H, Lippoldt A (2002) Tight junctions of the blood-brain barrier: development, composition and regulation. Vasc Pharmacol 38(6): 323–337
5. Morita K et al (1999) Claudin multigene family encoding four-transmembrane domain protein components of tight junction strands. Proc Natl Acad Sci U S A 96(2):511–516
6. Itoh M et al (1999) Direct binding of three tight junction-associated MAGUKs, ZO-1, ZO-2, and ZO-3, with the COOH termini of claudins. J Cell Biol 147(6):1351–1363
7. Tornavaca O et al (2015) ZO-1 controls endothelial adherens junctions, cell-cell tension, angiogenesis, and barrier formation. J Cell Biol 208(6):821–838
8. Nitta T et al (2003) Size-selective loosening of the blood-brain barrier in claudin-5-deficient mice. J Cell Biol 161(3):653–660
9. Sohet F et al (2015) LSR/angulin-1 is a tricellular tight junction protein involved in blood-brain barrier formation. J Cell Biol 208(6): 703–711
10. Lv J et al (2018) Focusing on claudin-5: a promising candidate in the regulation of BBB to treat ischemic stroke. Prog Neurobiol 161: 79–96
11. Knowland D et al (2014) Stepwise recruitment of transcellular and paracellular pathways underlies blood-brain barrier breakdown in stroke. Neuron 82(3):603–617
12. Chodobski A, Zink BJ, Szmydynger-Chodobska J (2011) Blood-brain barrier pathophysiology in traumatic brain injury. Transl Stroke Res 2(4):492–516
13. Greene C, Hanley N, Campbell M (2020) Blood-brain barrier associated tight junction disruption is a hallmark feature of major psychiatric disorders. Transl Psychiatry 10(1):373
14. Greene C et al (2018) Dose-dependent expression of claudin-5 is a modifying factor in schizophrenia. Mol Psychiatry 23(11): 2156–2166
15. Nishiura K et al (2017) PKA activation and endothelial claudin-5 breakdown in the schizophrenic prefrontal cortex. Oncotarget 8(55):93382–93391
16. Greene C, Hanley N, Campbell M (2019) Claudin-5: gatekeeper of neurological function. Fluids Barriers CNS 16(1):3

Chapter 19

An In Vitro Blood-Brain Barrier Model to Study Firm Shear Stress-Resistant Leukocyte Adhesion to Human Brain Endothelial Cells

Camilla Cerutti and Ignacio A. Romero

Abstract

Adhesion between leukocytes and brain endothelial cells, which line cerebral blood vessels, is a key event in both physiological and pathological conditions such as neuroinflammatory diseases. Leukocyte recruitment from blood into tissues is described as a multistep process involving leukocyte rolling on endothelial cells, adhesion, crawling, and diapedesis under hemodynamic shear stress. In neuroinflammatory conditions, there is an increase in leukocyte adhesion to the brain endothelial cells, activated by proinflammatory molecules such as cytokines. Here, we describe an in vitro technique to study the interaction between human leukocytes with human brain endothelial cells under shear stress mimicking the blood flow in vivo, coupled to live-cell imaging.

Key words Cytokines, Leukocytes, Brain endothelial cells, Cell adhesion, Shear stress, Live-cell imaging, Neuroinflammation, CNS

1 Introduction

The interaction of circulating immune cells with highly specialized brain endothelial cells forming the blood–brain barrier (BBB) is essential for immunosurveillance and plays an important part in the development and progression of neuroinflammatory diseases and central nervous system (CNS) disorders [1]. The ability of circulating leukocytes to interact, adhere, and infiltrate to and across the brain endothelium [2–4], under blood shear stress [5, 6], is a crucial step in neuroinflammation led by a variety of proinflammatory secreted proteins and factors [7]. Extracellular signals that activate brain endothelium include cytokines, which are well known to regulate cell adhesion molecules (CAM) [8], major regulators of leukocyte–endothelium interactions. Cytokines stimulate brain endothelial cells transcriptionally and/or via their surface receptors to activate downstream signalling networks to regulate

Nicole Stone (ed.), *The Blood-Brain Barrier: Methods and Protocols*,
Methods in Molecular Biology, vol. 2492, https://doi.org/10.1007/978-1-0716-2289-6_19,

cell CAM expression [9]. For example, proinflammatory cytokines such as tumor necrosis factor alpha (TNFα) and interferon gamma (IFNγ) are highly overexpressed in neurodegenerative disorders, leading to increased vascular permeability and expression of endothelial P- and E-selectins and CAMs, VCAM-1 and ICAM-1, on the endothelial membrane surface [4, 7, 10, 11]. Once leukocytes are firmly adhered to the endothelium, they may migrate across the endothelial layer either transcellularly or paracellularly to the perivascular space and from there into the brain parenchyma where they contribute to the overall neuroinflammatory response [2, 12, 13]. Firm adhesion constitutes one of the first steps mediating leukocyte entry into the brain. As a result, leukocyte firm adhesion represents a major therapeutic target to modulate neuroinflammation, a hallmark of many neurodegenerative diseases such as multiple sclerosis [14, 15].

Several studies have investigated the contribution of cytokines in leukocyte adhesion to brain endothelial cells under static conditions, as well as to crossing the endothelium either in endothelial cell-coated transwells or in 3D assays [16–18]. Here, we describe a method to study leukocyte adhesion to brain endothelial cells, by inducing activation of brain endothelial cells with a combination of the proinflammatory cytokines tumor necrosis factor alpha (TNFα) and interferon gamma (IFNγ). The technique described in this chapter was previously used to study the role of microRNAs in cytokine-activated human brain endothelial cells [10, 11]. The shear stress component is similar to in vivo-like mechanical forces of the blood in the CNS microvasculature (hemodynamic forces) which has been shown to be crucial for immune cell adhesion. Using a commercially available Ibidi® flow chamber that accommodates human brain endothelial cell monolayers in channels, we set up a system that mimics circulating leukocyte cells in the microvasculature. In the assay, leukocyte cells are perfused onto the brain endothelial monolayer-coated channels at controlled speeds to study their shear stress-resistant firm adhesion with live-cell imaging. This technique allows the real-time visualization of human leukocytes interacting with and adhering to cytokine-treated and untreated human brain endothelial cells, which is one of the crucial steps in the complex process of leukocyte extravasation into the central nervous system. Furthermore, this technique can be easily adapted to a variety of different studies aimed to quantify immune cell behavior with brain endothelial cells under shear stress conditions produced by controlled flow.

2 Materials

2.1 Cell Culture

1. hCMEC/D3 cell line [19] was used at passages 26–34 (*see* **Note 1**).
2. hCMEC/D3 cell complete medium: endothelial cell basal medium-2 (EGM-2) (Lonza, Walkersville, USA) and supplemented with the following components obtained from the manufacturer: 0.025% (v/v) rhEGF, 0.025% (v/v) VEGF, 0.025% (v/v) IGF, 0.1% (v/v) rhFGF, 0.1% (v/v) gentamycin, 0.1% (v/v) ascorbic acid, 0.04% (v/v) hydrocortisone, and 2.5% (v/v) fetal bovine serum (FBS), hereafter referred to as endothelial complete medium (store at 4 °C).
3. Trypsin–EDTA (store at 4 °C).
4. Phosphate-buffered saline (PBS) without magnesium ($MgCl_2$) and calcium ($CaCl_2$) ($PBS^{-/-}$).
5. Phosphate-buffered saline (PBS) with magnesium ($MgCl_2$) and calcium ($CaCl_2$) ($PBS^{+/+}$).
6. Ibidi® μ-Slide $VI^{0.4}$ (Ibidi® GmbH), tissue culture treated uncoated. For technical features and specifications, see https://ibidi.com/channel-slides/57%2D%2Dslide-vi-04.htmL.
7. Collagen (0.1% solution in 0.1 M acetic acid) type I from calf skin (Sigma, St. Louis, USA) to coat flasks and Ibidi® μ-Slide $VI^{0.4}$ for hCMEC/D3 cell culture (store at 4 °C).
8. T cell line Jurkat from acute T cell leukemia (ATCC) (*see* **Note 2**).
9. Jurkat T cells complete medium: Roswell Park Memorial Institute 1640 medium (RPMI 1640) W/GLUTAMAX I (Gibco® Invitrogen, Paisley, UK) supplemented with 10% (v/v) heat-inactivated fetal calf serum, 100 μg/mL streptomycin, and 100-units/mL penicillin. Store at 4 °C.
10. Tissue culture light microscope.
11. Neubauer chamber to count cells.

2.2 Proinflammatory Cytokine Treatment

1. Recombinant human tumor necrosis factor alpha (TNFα) and interferon gamma (IFNγ) (R&D Systems, Inc., Minneapolis, USA).
2. 0.1% (w/v) bovine serum albumin (BSA) in phosphate-buffered saline (PBS) without magnesium ($MgCl_2$) and calcium ($CaCl_2$) ($PBS^{-/-}$).
3. Resuspend cytokine in PBS with 0.1% (w/v) BSA at a concentration of 50 nM. Store aliquots at −80 °C (*see* **Notes 2** and **3**).

2.3 Jurkat T Cell Fluorescent Labelling

1. 5 mM 5-chloromethylfluorescein diacetate (CMFDA) green (λ_{Ex} = 495 nm, (λ_{Em} = 521 nm), (Life Technologies) in DMSO. Store at −20 °C in 10–20 μl aliquots.
2. Jurkat T cell medium without serum: RPMI-1640 with L-glutamine, 25 mM Hepes, 100 μg/mL streptomycin, and 100 U/mL penicillin (store at 4 °C).

2.4 Flow-Adhesion Assay

1. Inverted time-lapse microscope with environmentally controlled chamber (37 °C, 5% CO_2) and digital camera for fast acquisition. An Olympus IX70 microscope (Tokyo, Japan) with 10×/0.30 Ph1 DL objective and equipped with a Q-IMAGING QICAM FAST 1394 12-bit camera was used to perform the assay described here. Place a clean piece of absorbent paper on top of the environmental chamber or in the space adjacent to the microscope (*see* **Note 4**).
2. Image-Pro Plus software (Media Cybernetics, Inc., Bethesda, USA) for imaging acquisition (*see* **Note 5**).
3. Hamilton glass syringes (5 or 10 mL) coupled to a high-precision pump with withdraw/pull option (*see* **Note 6**) (placed on microscope table/bench on the side/small trolley stand). A PHD ULTRA (Harvard Apparatus) was used here.
4. Two tubing hose clips (Ibidi®).
5. Three timers.
6. Tubing and connectors to assemble the "flow system" as described in Fig. 1:
   (a) Ibidi® μ-Slide channel elbow male Luer connectors for μ-Slide VI 0.4 flow kit (Ibidi®).
   (b) Female Luer Lock Coupler (Ibidi®).
   (c) Y connectors with 400 Series Barbs 3/32″ Natural Polypropylene (Value Plastic, Inc.).
   (d) Female Luer Lug Style to classic series barb 1/16″ Natural Polypropylene (Value Plastic, Inc.).
   (e) Tubing Tygon 3350 Sanitary Silicone Tubing (Saint-Gobain).
   (f) Three-way stopcock with male Luer lock adapter and port covers (Kendall Argyle™ EZ-FLO™).

2.5 Quantification

1. ImageJ (NIH) or Fiji (they are the same program), free to download from https://imagej.nih.gov/ij/, with cell counter plug-in (downloadable from https://imagej.nih.gov/ij/plugins/cell-counter.htmL).

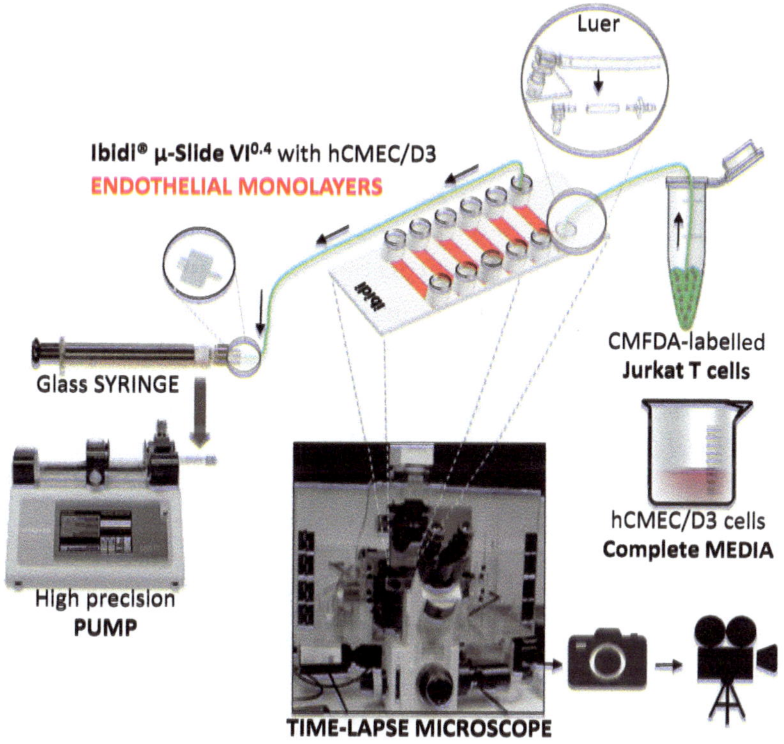

Fig. 1 Flow-based assay of Jurkat T cell adhesion to hCMEC/D3 cells coupled to live-cell imaging. Schematic representation of how the system of connectors/tubing/Ibidi®µ-Slide^{VI} is assembled and connected to the pump/syringe, with an enlargement to illustrate the Ibidi® connectors for the µ-Slide^{VI} channel inlet and glass syringe to create the "flow system" (taken from www.ibidi.com). Once assembled the system is placed inside an inverted time-lapse microscope environmentally controlled chamber for live-cell imaging where 1 frame/s is captured and rendered as a movie

3 Methods

3.1 Leukocyte Cell Culture

1. Thaw Jurkat T cells from frozen aliquots in a water bath at 37 °C.
2. Resuspend the cells in a 25-cm^2 flask containing 5–10 mL of Jurkat T complete medium.
3. Put the 25-cm^2 flask with the cells in suspension at 37 °C in a tissue culture incubator (37 °C, 5% v/v CO_2) for 12 h (*see* **Note 7**).
4. After 12 h gently transfer the cells in a falcon, and centrifuge it for 7 min at 200 × g at room temperature (RT).
5. Resuspend the pellet in 5 mL of Jurkat T complete medium in a new 25-cm^2 flask.

6. For maintenance change medium and count every other day (3 times a week).
7. Every passage, cells must be diluted at 5×10^5 cells/mL with fresh medium (cell density).
8. Once a week count the cell number, and centrifuge for 5 min at $200 \times g$, and pass them to a new flask.
9. Twice a week count the cell number, and supplement with fresh media only to 5×10^5 cells/mL.
10. To improve consistency between experiments, discard Jurkat T cells after 6 weeks of culture, and defrost a fresh cell aliquot.

3.2 hCMEC/D3 Brain Endothelial Cell Culture

1. Prepare a 1/20 solution of collagen from calf skin in ($PBS^{+/+}$) on the day.
2. Coat a 25-cm^2 tissue culture flask with 2 mL of freshly prepared collagen solution for 1 h in a tissue culture incubator (37 °C, 5% CO_2).
3. Aspirate the collagen, and add 5 mL of $PBS^{+/+}$ to the flask.
4. Put in the incubator (37 °C, 5% CO_2) for 2 min and then aspirate.
5. Add 5 mL of endothelial complete medium at 37 °C.
6. Put the flask in the incubator (37 °C, 5% CO_2) to equilibrate the medium.
7. Thaw hCMEC/D3 from frozen aliquots in a water bath at 37 °C.
8. Take the collagen-coated 25-cm^2 flask from the incubator, add the thawed hCMEC/D3 cells, and put the flask back in the incubator.
9. Maintain cells at 37 °C in a tissue culture incubator (37 °C, 5% CO_2), and change medium after 6 h.
10. When cells reach 95% confluence (*see* **Note 8**), passage them by removing the medium from the flask and washing cells with 5 mL of $PBS^{+/+}$.
11. Aspirate the $PBS^{+/+}$ and add 5 mL of $PBS^{-/-}$.
12. Put the flask in the incubator (37 °C, 5% CO_2) for 2–3 min.
13. Aspirate the $PBS^{-/-}$, add 2 mL of trypsin–EDTA, and leave the flask in the incubator (37 °C, 5% CO_2) for 5 min.
14. Add 8 mL of endothelial complete medium (37 °C) to inactivate the trypsin–EDTA.
15. Count the cells with a Neubauer chamber.
16. Centrifuge cells for 5 min at 200 g.
17. Resuspend cells in warm endothelial complete medium, and seed hCMEC/D3 cells onto collagen-coated culture flasks to

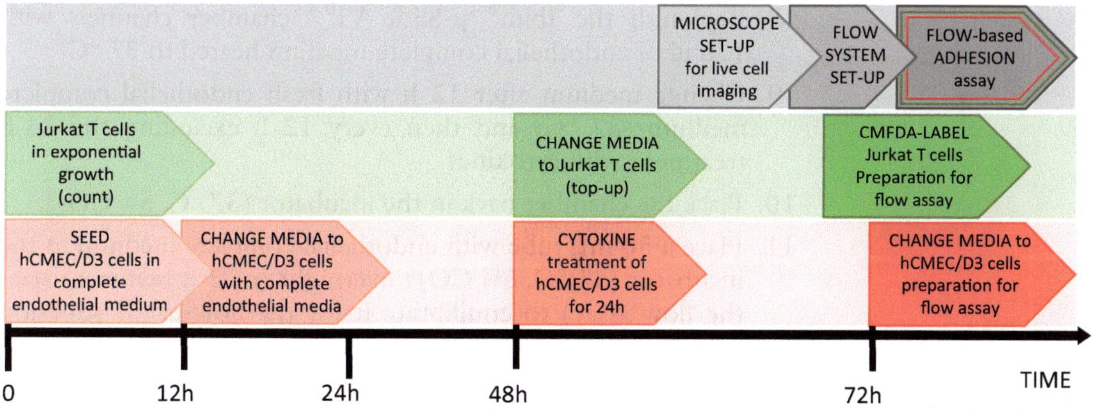

Fig. 2 A schematic of experimental timeline for the flow-based assay. The timeline of the entire experiment setup is 72 h. Experimental timeline of the main steps for hCMEC/D3 cell (red), Jurkat T cell (green), and microscope (gray) preparation for the flow-based adhesion assay using the Ibidi® μ-Slide VI$^{0.4}$

maintain cell culture (at a 5×10^5 cells/mL cell density) or in Ibidi® μ-Slide VI$^{0.4}$ for use in experiments (at a 3×10^4 cells/condition cell density) (*see* Subheading 3.3).

18. Use hCMEC/D3 cells between passage 25–35 (times of cell trypsinization), and change the medium every 2 days.

3.3 hCMEC/D3 Cell Seeding in Ibidi® μ-Slide VI^{04}

72-h pre-assay (*see* scheme in Fig. 2)

1. Coat all six channels of the Ibidi® μ-Slide VI$^{0.4}$ with 30 μL of collagen solution (*see* Subheading 3.2, **step 1**), using a p200 micropipette, and then incubate for 1 h in a tissue culture incubator (37 °C, 5% CO_2).
2. Prepare hCMEC/D3 cells (*see* Subheading 3.2) in endothelial complete medium (warmed to 37 °C) at a cell density of 1×10^6/mL.
3. Wash the Ibidi® μ-Slide VI$^{0.4}$ channels with 120 μL of warm (37 °C) PBS$^{+/+}$.
4. Aspirate the PBS$^{+/+}$ from the channel inlets (to learn how to aspirate, see min 3:21 of the video https://www.youtube.com/watch?v=SPmrvnoD158).
5. Seed 30 μL of hCMEC/D3 cells (1×10^6/mL) per channel making sure to add the cells in one single go, to allow cells to reach the other end of the channel and distribute equally in the channel area (*see* **Note 8**).
6. Check that the cells are distributed equally in the channel area under a microscope, and then repeat **step 5** for the other five channels.
7. Place the Ibidi® μ-Slide VI$^{0.4}$ chamber in the incubator (37 °C, 5% CO_2) for 30 min.

8. Replenish the Ibidi® μ-Slide $VI^{0.4}$ chamber channels with 120 μl of endothelial complete medium heated to 37 °C.
9. Change medium after 12 h with fresh endothelial complete medium (37 °C) and then every 12 h excluding the 24-h treatment with cytokines.
10. Place the chamber back in the incubator (37 °C, 5% CO_2).
11. Place a 50 mL tube with endothelial complete medium in the incubator (37 °C, 5% CO_2) overnight (or 3 h before to start the flow assay) to equilibrate it for the flow-based adhesion assay.

3.4 hCMEC/D3 Cell Treatment with Proinflammatory Cytokines

24-h pre-assay (*see* scheme in Fig. 2)

1. 48 h after seeding the hCMEC/D3 cells in the Ibidi® μ-Slide $VI^{0.4}$ chamber, cells should be almost confluent (80%) and ready for treatment.
2. Aspirate endothelial complete medium from Ibidi® μ-Slide $VI^{0.4}$ chamber channel with endothelial monolayers carefully.
3. Wash the Ibidi® μ-Slide $VI^{0.4}$ channels with 120 μL of warm (37 °C) $PBS^{+/+}$.
4. Aspirate the $PBS^{+/+}$ from the channel inlets.
5. Add 120 μl fresh endothelial complete medium (37 °C), and put it in the incubator (37 °C, 5% CO_2).
6. Thaw a vial each of the cytokines TNFα and IFNγ stock at 50 nM in ice.
7. Prepare a mix TNFα and IFNγ at the desired final concentration in warm endothelial complete medium. We used exponential concentration to perform a dose–response (0, 0.1, 1, 10 ng/mL).
8. Take Ibidi® μ-Slide $VI^{0.4}$ channels from the incubator, and aspirate the medium from the channel inlets carefully.
9. Add 120 μl/Ibidi® μ-Slide $VI^{0.4}$ channel of the cytokine mix at the concentration desired.

3.5 Microscope Setup for Live Imaging

1. Turn on the time-lapse microscope to allow the stage to reach the temperature and the chamber to the indicated environmental conditions (37 °C, 5% CO_2) for at least 2 h.
2. Set the magnification for the time-lapse movie acquisition. Here, we used a 10× objective (Fig. 3).
3. Set the acquisition channels to capture images of Jurkat T cells (labelled with the fluorophore CMFDA green ($\lambda_{Ex} = 495$ nm, $\lambda_{Em} = 521$ nm)) and brain endothelial cells using alternating fluorescence and phase contrast detection. Here we have set a configuration where one single fluorescent filter cube is used to

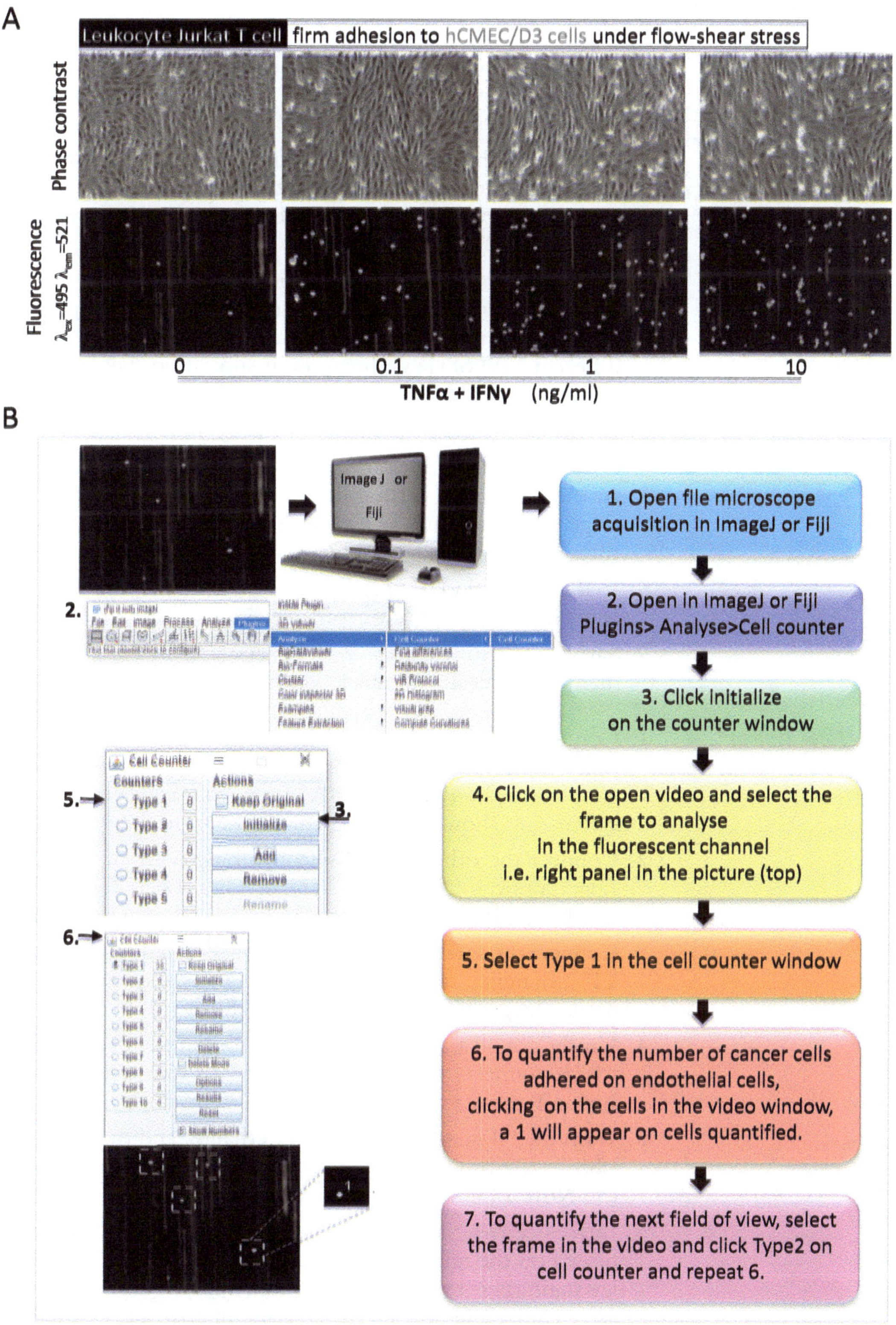

Fig. 3 Representative images of shear-resistant firmly adhered Jurkat T cells to hCMEC/D3 cells and quantification. (**a**) Top panels: phase contrast image of confluent hCMEC/D3 cells. (**b**) Bottom panels: fluorescence images ($\lambda_{ex} = 495$, $\lambda_{em} = 521$) of CMFDA-labelled Jurkat T cells firmly adhered to hCMEC/D3 cells. Field of view (640 × 480 μm). 1–7. Step-by-step instructions to quantify shear-resistant Jurkat T cells firmly adhered to hCMEC/D3 cells, using ImageJ or Fiji

allow rapid acquisition time. The acquisition of fluorescence and phase contrast images could be sequential as long as they are very close in time (1 frame/s).

4. Customize the exposure (ms) time of the bright field and fluorescence channels based on the intensity of your sample. We keep the lamp power between 0 and 18% to minimize fluorophore bleaching.
5. Set the acquisition time at 1 frame/s for 10 min (the adhesion assay is 6 min in total; then after the end of the experiment, some extra acquisition time is required to take pictures of different field of views; however, it should not take more than 1 min) (*see* **Note 9**).
6. In the acquisition setting of memory function, set "record while acquiring" (*see* **Note 10**).
7. In the acquisition setting of live stream recording, set "allow stage movement while acquiring"—to take ten different pictures along the Ibidi® μ-Slide channel.
8. In the acquisition setting, set "save as" always in the acquisition software file format (e.g., .ome or .lif).
9. Save these settings (**steps 2–8**), to apply them to all the experimental technical replicate acquisitions.

3.6 "Flow System" Setup

1. Connect the silicon tubing with female Luer Lug Style classic series barb 1/16 connectors to the Hamilton glass syringe, coupled to the PHD ULTRA pump, to start create the "flow system," as shown in Fig. 1, enlargement on the syringe.
2. Connect the other end of silicon tubing with an elbow male Luer connector, then the female Luer Lock Coupler, and another male Luer connector (Ibidi®), as shown in Fig. 1.
3. Connect the Luers (enlargement in Fig. 1) to the Ibidi® μ-Slide $VI^{0.4}$ chamber channel inlet and outlet as shown in the Ibidi movie: https://ibidi.com/content/150-mv-21.
4. Finally, connect another piece of tubing of the desired length, which connects the "flow system" with the tube containing Jurkat T cells or medium (Fig. 1).
5. Wash the "flow system" with 70% ethanol, and then wash with $PBS^{+/+}$ (37 °C) withdrawing with the pump at 1 or 2 dyn/cm^2 (*see* **Note 11**).
6. Introduce the "flow system" (except the pump, which can be placed on the microscope table/bench on the side/small trolley stand; Fig. 1) into the environmentally controlled chamber (37 °C, 5% CO_2) of the time-lapse microscope, and perform another wash with warm $PBS^{+/+}$ at 0.5 dyn/cm^2 (*see* **Note 12**). After the wash, leave the "flow system" inside the

environmentally controlled chamber, and switch off the pump (*see* **Note 13**).

7. Take the 50 mL tube with endothelial complete medium equilibrated overnight in the incubator (*see* Subheading 3.4, **step 3**), and use it to perform another two washes of the "flow system" in the time-lapse microscope environmentally controlled chamber.
8. Set the flow rate (θ) applied to produce the required shear stress τ (dyn/cm^2).
   (a) For the Ibidi® μ-Slide^{VI}, the flow rate is calculated according to the equation:

$$\tau\left[\mathrm{dyn/cm^2}\right] = \eta\left[\mathrm{dyn} * s/\mathrm{cm^2}\right]\tilde{n}176.1\ \Phi\ \mathrm{mL/\min}]$$

where the relationship between shear stress (τ) and flow rate (Φ) is based on the dynamic viscosity (η) of water at 22 °C, $\eta = 0.01$ dyn·s/cm^2 and other parameters specific to the geometry of the system based on the Ibidi® μ-Slide geometry.

   (b) Shear stress (τ) is calculated according to $\tau = 176.1\ \eta\Phi$ ($\eta = 0.01$, $\Phi =$ volumetric flow rate). For more detailed information on the parameters to determine shear stress for vessel blood flow see Table S1 in Additional file 2 from [11].
   (c) Based on the Ibidi® μ-Slide VI$^{0.4}$ geometries, Jurkat T cells are pulled through the system for 5 min (we define it as ACCUMULATION TIME) at 0.5 dyn/cm^2 (0.28 mL/min) to allow Jurkat T cells to enter the channel and adhere. Then the speed of the pump is increased for 1 min (we define it as CELL CHALLENGE TIME) at 1.5 dyn/cm^2 (0.85 mL/min) to challenge the adhered cells, in order to study only the firmly adhered Jurkat T cells that resist the increased flow shear stress.

3.7 Jurkat T Cell Preparation for the Flow Assay

1-h pre-assay (*see* scheme in Fig. 2)

1. Dye-label Jurkat T cells: count the cells (*see* **Note 14**), and centrifuge the number of cells for labelling for 7 min at 200 g.
2. Resuspend the cells gently in pre-warmed Jurkat T cells medium without serum, at a 2×10^6 cells/mL density.
3. Add 5 mM of CMFDA CellTracker™ to the cells.
4. Incubate 30 min in the incubator (37 °C, 5% CO_2).
5. Centrifuge the cells 5 min at 200 g, and remove the CellTracker™.
6. Resuspend the cells in complete Jurkat T cells medium, and put in the incubator (37 °C, 5% CO_2) for 30 min.

7. Confirm that the PC3 cells treated with CMFDA fluoresce green when excited at 488 nm under the time-lapse microscope.
8. Centrifuge the cells 5 min at 200 g, and resuspend the cells in in endothelial complete medium (warmed to 37 °C) at a concentration of 2×10^6 cells/mL for the assay (*see* **Note 15**).
9. Place the fluorescently labelled Jurkat T cells in the time-lapse microscope environmentally controlled chamber for the flow assay.

3.8 Endothelial Cell Preparation for the Flow Assay

24-h post-cytokine treatment (*see* scheme in Fig. 2)

1. Take the Ibidi® μ-Slide $VI^{0.4}$ chamber with hCMEC/D3 cell from the tissue culture incubator, and change medium with 120 μl fresh warm (37 °C) endothelial complete medium.
2. Wash twice with warm $PBS^{+/+}$, and add 100 μl of fresh warm (37 °C) endothelial complete medium.
3. Connect the Ibidi® μ-Slide $VI^{0.4}$ chamber to the "flow system" set in Subheading 3.6 (*see* **Note 16**) as follows:
   (a) Take the "flow system" out of the time-lapse microscope environmentally controlled chamber, and place it on top of a clean piece of absorbent paper (*see* **Note 17**).
   (b) Clip the tubing near the Ibidi® μ-Slide $VI^{0.4}$ chamber inlet and outlet with the two tubing hose clips (Ibidi®) as per the manufacturer's instructions (see movie at min 7:46 https://ibidi.com/content/150-mv-21).
   (c) Open the Ibidi® μ-Slide $VI^{0.4}$ chamber containing hCMEC/D3 cells, and add 30 μl of fresh endothelial complete medium to the first channel, closing the lids on the five remaining channels (in this step you lose the sterility working in the microscope area) (*see* **Note 18**).
   (d) Carefully detach the inlet and outlet Luer connectors (as shown in the manufacturer's instructions movie at min 8), and plug it into the Ibidi® μ-Slide $VI^{0.4}$ chamber with hCMEC/D3 cells first channel inlets (*see* **Note 19**).
   (e) Make sure the system is without air bubbles. If there are bubbles (1) in the channel inlet or outlet before to plug the Luer connectors, aspirate the bubbles with p1000, and then refill the inlet and outlet with 30 μl of fresh endothelial complete medium; (2) in the tubing repeat the washes as in **steps 6** and 7 in Subheading 3.6, and then proceed with **step 3** (b) in Subheading 3.8 again.
   (f) Unclip the tubing hose clips.
   (g) Make sure, again, that the system is without air bubbles. If there are bubbles in the chamber with the endothelial

monolayer, that channel cannot be used, because the bubble may have affected the cells and in particular the CAMs on the surface. In case the bubbles are very small (they tend to stay on the top part of the channel), then flow endothelial complete medium at 0.5 dyn/cm^2 to expel the bubbles.

(h) Put the system back with the Ibidi® μ-Slide VI$^{0.4}$ chamber containing hCMEC/D3 cells inside the time-lapse microscope environmentally controlled chamber making sure that the chamber is not wet externally, to avoid issues of condensation in the microscope that would affect the imaging.

(i) Secure the Ibidi® μ-Slide VI$^{0.4}$ chamber containing hCMEC/D3 cells on the microscope stage, and position the stage in order to visualize the middle of the channel using a 5× objective first, then with 10× objective. This will be the field of view of the experiment for Jurkat T cell adhesion to hCMEC/D3 cells under flow.

3.9 Flow-Based Adhesion Assay Coupled to Live-Cell Imaging

1. In the time-lapse microscope environmentally controlled chamber, connect the tube containing CMFDA-labelled Jurkat T cells (2 × 10^6 cells/mL) to the "flow system" (Fig. 1) and the endothelial complete medium tube left overnight in the incubator to equilibrate (*see* Subheading 3.3, **step 3**), beside the Jurkat T cell tube.
2. Set the inverted fluorescence time-lapse microscope for acquisition (setting in Subheading 3.5).
3. Set the pump in withdraw/pull direction at 0.5 dyn/cm^2 (*see* **Note 20**).
4. Set three timers: one for the assay total time of microscope acquisition (6 min, timer 1) (*see* **Note 21**), two for the pump (5 min, timer 2; and 1 min; timer 3).
5. Start the pump, the microscope acquisition and timer 1 (6 min) and timer 2 (5 min).
6. While the pump is pulling cells at 0.5 dyn/cm^2, set the pump to 1.5 dyn/cm^2 (Harvard pumps allow this option). Jurkat T cells should reach the channel of the Ibidi® μ-Slide and be observed to flow into the channel and interact with the endothelial cell monolayer in the selected field of view.
7. Just before 5 min (timer 2), switch the connection from Jurkat T cells tube to the endothelial complete medium tube.
8. After 5 min (timer 2), switch to 1.5 dyn/cm^2 (start timer 3), to increase the speed of the pump to wash off loosely adhered Jurkat T cells (*see* **Notes 22** and **23**) for 1 min, until timer 3 goes off.

9. The frames captured for the movie ends at 6 min (timer 3), resulting in 360 frames (1/s) until this point.
10. At 6 min (timer 3), move the stage/chamber vertically along the middle of the Ibidi® μ-Slide channel to acquire images of ten different fields of view, while the microscope acquisition is still ongoing (we set the acquisition for 10 min; *see* **step 5** in microscope setup Subheading 3.5 for live imaging).
11. When ten different fields of view are acquired, stop the acquisition in the microscope software and the pump. Repeat step **steps 1–10** washing the flow system as previously described for each Ibidi® μ-Slide $VI^{0.4}$ chamber channel/condition and/or technical replicate (we normally perform two to three technical replicates for each treatment).
12. Perform strong washes with warm (37 °C) $PBS^{-/-}$ between each Ibidi® μ-Slide $VI^{0.4}$ chamber channel/experimental replicate.

3.10 Quantification of Images

1. Keep the file in the acquisition software file format (e.g., .seq or .ips), and then save a copy in TIFF if you have limited data storage space (*see* **Note 24**).
2. For quantification, firm adhesion of Jurkat T cells to brain endothelial cells is defined by T cells that remain adhered to hCMEC/D3 cells in the fields of view throughout the accumulation time (5 min) and after increasing the flow to 1.5 dyn/cm^2 (1 min), the cell challenge time.
3. Firmly adhered cells (as shown in Fig. 3) are manually counted on fluorescence images with ImageJ software in ten different fields of view per condition in a single Ibidi® μ-Slide $VI^{0.4}$ chamber channel (for each of the two technical replicates per treatment).
4. Open the movie to analyze with Image J (Fig. 3-1).
5. Open > Plugins > Analyse > Cell counter (Fig. 3-2).
6. Click initialize on the counter window (Fig. 3-3).
7. In the movie window, go to the frame number 360 (last frame at min 6) of the acquisition in the FITC channel (Fig. 3-4).
8. Select cell type 1 in the cell counter (Fig. 3-5).
9. Click on each of the cells in the frame to count them (Fig. 3-6).
10. After finishing the first frame, go to the next frame (field of view), click on cell count type 2, and repeat **step 9**. Repeat to quantify cells in all ten fields of view (Fig. 3, 1–7).

To quantify the number of Jurkat T cells firmly adhered to hCMEC/D3 cells, calculate the mean number of Jurkat T cells firmly adhered per field of view from the number in each of the ten different fields of view of one Ibidi® μ-Slide channel (one condition/one technical replicate). You can perform two or three technical replicates per treated condition with cytokines. For quantification, average the means of each technical replicate.

4 Notes

1. We use a human brain endothelial cell line, hCMEC/D3, but you can use any brain endothelial cells of your interest, provided that you may need to titrate the number of cell to seed.
2. All the procedures used for the Jurkat T cells are valid for the THP-1 monocytic cells and peripheral blood mononuclear cells. We have not tested other leukocyte cell lines.
3. Prepare small-volume aliquots to avoid cycles of freeze and thaw that may interfere with the integrity of the cytokines.
4. Keep the aliquots of cytokines in ice before storage at −80 °C and to thaw to treat the cells.
5. For the acquisition it is recommended to select long working distance objectives to image the endothelial monolayer and the adhered Jurkat T cells within one single focal plane. Images can be captured with either a 10× or 20× objective, depending on the microscope and camera you use for acquisition. The field of view size can change drastically depending on the camera you use, for example.
6. Image-Pro Plus software can be used with any microscope; however, many time-lapse microscopes have their own software. Alternatively you can use the micromanager software free to download from https://micro-manager.org/.
7. It is possible to use any precision pump of your choice and to use two separate pumps to change the flow rate if the pump you have does not have this feature.
8. Always keep the flask vertical in the incubator.
9. As described in the manufacturer's instructions at min 4:56 in this movie: https://ibidi.com/content/150-mv-21.
10. If the microscope available to you has high performance specifications, it is possible to acquire multiple positions within a channel or z-stacks in a single position. The assay can be modified based on the scientific question and/or the available instrumentation.
11. In the experimental setup, enough computer memory and space to record the movies should be made available as files are typically three to four times larger than those required to record one single condition. Please take time to set this part before you prepare your biological sample.

12. In this process no air bubbles should enter in the system. If air bubbles form along the system, in the tubing, for example, they must be removed from the system with continuous washouts, until they reach the glass syringe. A three-way stopcock can be used to connect the tubing to the syringe to empty the syringe, without removing it from the pump. See application note 31 of Ibidi® to learn how to avoid forming bubbles during the slide preparation and how to recognize air bubbles trapped in the system reservoirs: https://ibidi.com/img/cms/support/AN/AN31_Serial_Connection_muSlideVI04.pdf.
13. Based on the tubing length, you can calculate how long one a wash cycle takes to pass through all the connectors/tubing/chamber system.
14. Keep an eye on the syringe capacity: empty it via the Luer connector into a "waste" tube with 70% (v/v) ethanol or 1% (w/v) Virkon (Dupont).
15. Depending on the assay, you may need to prepare and label different numbers of Jurkat T cells. To perform the flow-based adhesion assay in all Ibidi® μ-Slide VI$^{0.4}$ channels (six conditions), it is recommended to prepare an excess of cells (include 20–25% more cells if you are unfamiliar with the technique to allow for technical errors). We initially used 20 million of Jurkat T cells per Ibidi chamber with 6 channels.
16. For the flow-based adhesion assay, only endothelial complete medium is used as endothelial cells require more supplements than leukocytes and are more susceptible to changes in medium composition.
17. It is recommended to take the system out of the environmentally controlled chamber of the time-lapse microscope as it may be difficult to operate inside with precision. We recommend setting a secure small area near the microscope chamber doors or on top of the environmentally controlled chamber.
18. Make sure the pump is off (!). If you clip the tubing while the pump is pulling medium, once the clips are released, the endothelial monolayer in that channel will be compromised/detached. Always stop the pump before clipping the tubing.
19. As described in the manufacturer's instructions at min 7:50 in the movie: https://ibidi.com/content/150-mv-21.
20. To avoid excess medium entering the chamber, it is possible to use sterile cotton buds or absorbent paper, without touching the inlet borders. This is shown in the manufacturer's instructions at min 8 in the movie: https://ibidi.com/content/150-mv-21.
21. Make sure that the syringe in the pump is empty.

22. These times are optimized for the tubing size/length and cell types. Here you can resuspend the Jurkat T cells again to make sure they are not in groups and to avoid those forming clumps that may spoil the adhesion experiment.
23. It is possible to use a second pump to perform the washes with medium. It is also possible to use a Y connector and/or a three-way stopcock Luer for the Jurkat T cells and medium connection to the system.
24. Movies may be larger than 3GB each depending on a range of parameters like pixel binning.

References

1. Marchetti L, Engelhardt B (2020) Immune cell trafficking across the blood-brain barrier in the absence and presence of neuroinflammation. Vasc Biol 2(1):H1–H18
2. Vestweber D (2015) How leukocytes cross the vascular endothelium. Nat Rev Immunol 15(11):692–704
3. Lopes Pinheiro MA et al (2016) Immune cell trafficking across the barriers of the central nervous system in multiple sclerosis and stroke. Biochim Biophys Acta 1862(3):461–471
4. Engelhardt B, Ransohoff RM (2012) Capture, crawl, cross: the T cell code to breach the blood-brain barriers. Trends Immunol 33(12):579–589
5. Alon R, Ley K (2008) Cells on the run: shear-regulated integrin activation in leukocyte rolling and arrest on endothelial cells. Curr Opin Cell Biol 20(5):525–532
6. Molteni R et al (2006) Pathophysiology of leukocyte-tissue interactions. Curr Opin Cell Biol 18(5):491–498
7. Becher B, Spath S, Goverman J (2017) Cytokine networks in neuroinflammation. Nat Rev Immunol 17(1):49–59
8. Allan SM, Rothwell NJ (2001) Cytokines and acute neurodegeneration. Nat Rev Neurosci 2(10):734–744
9. Pober JS, Sessa WC (2014) Inflammation and the blood microvascular system. Cold Spring Harb Perspect Biol 7(1):a016345
10. Cerutti C et al (2017) MiR-126 and miR-126* regulate shear-resistant firm leukocyte adhesion to human brain endothelium. Sci Rep 7: 45284
11. Cerutti C et al (2016) MicroRNA-155 contributes to shear-resistant leukocyte adhesion to human brain endothelium in vitro. Fluids Barriers CNS 13(1):8
12. Dejana E (2006) The transcellular railway: insights into leukocyte diapedesis. Nat Cell Biol 8(2):105–107
13. Greenwood J et al (2011) Review: leucocyte-endothelial cell crosstalk at the blood-brain barrier: a prerequisite for successful immune cell entry to the brain. Neuropathol Appl Neurobiol 37(1):24–39
14. Steinman L (2008) Nuanced roles of cytokines in three major human brain disorders. J Clin Invest 118(11):3557–3563
15. Sospedra M, Martin R (2005) Immunology of multiple sclerosis. Annu Rev Immunol 23: 683–747
16. Weksler B, Romero IA, Couraud PO (2013) The hCMEC/D3 cell line as a model of the human blood brain barrier. Fluids Barriers CNS 10(1):16
17. Sreekanthreddy P et al (2015) A three-dimensional model of the human blood-brain barrier to analyse the transport of nanoparticles and astrocyte/endothelial interactions. F1000Res 4:1279
18. Duperray A et al (2015) Inflammatory response of endothelial cells to a human endogenous retrovirus associated with multiple sclerosis is mediated by TLR4. Int Immunol 27(11):545–553
19. Weksler BB et al (2005) Blood-brain barrier-specific properties of a human adult brain endothelial cell line. FASEB J 19(13): 1872–1874

Chapter 20

An In Vitro Model of the Blood–Brain Barrier to Study Alzheimer's Disease: The Role of β-Amyloid and Its Influence on PBMC Infiltration

Simona Federica Spampinato, Yukio Takeshita, and Birgit Obermeier

Abstract

The blood–brain barrier (BBB) is a highly specialized structure, constituted by endothelial cells that together with astrocytes and pericytes provide a functional interface between the central nervous system and the periphery. Several pathological conditions may affect its functions, and lately BBB involvement in the pathogenesis of Alzheimer's disease has been demonstrated. Both endothelial cells and astrocytes can be differentially affected during the course of the disease. In vitro BBB models present a powerful tool in evaluating the effects that β-amyloid (Aβ), or other pathogenic stimuli, play on the BBB at cellular level. In vitro BBB models derived from human cell sources are rare and not easily implemented. We generated two conditionally immortalized human cell lines, brain microvascular endothelial cells (TY10), and astrocytes (hAST), that, when co-cultured under appropriate conditions, exhibit BBB-like characteristics. This model allowed us to evaluate the transmigration of peripheral blood mononuclear cells (PBMCs) through the in vitro barrier exposed to Aβ and the role played by astrocytes in the modulation of this phenomenon. We describe here the methodology used in our lab to set up our in vitro model of the BBB and to carry out a PBMC transmigration assay.

Key words Blood–brain barrier, Transmigration assay, In vitro model, Endothelium/astrocyte interaction, β-Amyloid

1 Introduction

1.1 The Blood–Brain Barrier

The human brain's complex neuronal network demands a stable microenvironment that is shielded from fluctuations and xenobiotics in the blood circulation. A specialized vascular structure, the blood–brain barrier (BBB), provides the appropriate anatomic and functional interface between the periphery and the central nervous system (CNS), ensuring tightly controlled passage between these two compartments [1]. Endothelial cells of the cerebrovasculature are key in fulfilling this requirement: they feature extraordinarily tight continuous interendothelial junctions as well as low transcytotic activity which greatly restricts passive para- and transcellular

Nicole Stone (ed.), *The Blood-Brain Barrier: Methods and Protocols*,
Methods in Molecular Biology, vol. 2492, https://doi.org/10.1007/978-1-0716-2289-6_20,

flux [2]. At the same time, brain endothelial cells express a unique array of transporters and influx/efflux pumps that actively shuttle across only those substrates that either are required by or need to get cleared from the neural tissue [3]. This paradigm also applies to immune cells. In contrast to peripheral endothelial cells, neurovascular endothelial cells are unique in that they express very low levels of leukocyte adhesion molecules, ensuring that there is no uncontrolled extravasation of cells across the BBB under physiological conditions [4, 5]. Only when leukocyte adhesion molecules are upregulated under pathological conditions the multistep process of leukocyte extravasation is initiated and a neuroinflammatory reaction at and behind the BBB is triggered [6, 7].

During development and maturation of the BBB, the establishment of those barrier and gateway functions of the brain endothelium is implemented by a multitude of pathways which are mediated by physical cell–cell interactions between the core elements of the BBB, as well as by soluble factors that are secreted by astrocytes and pericytes acting on brain endothelial cells [8]. Pericytes are situated adjacent to the abluminal vessel surface, separated from the endothelium by only a thin layer of basement membrane. Astrocytes extend processes and cover blood vessels with their specialized perivascular end feet. While endothelial cells, pericytes, and astrocytes represent the key elements of the BBB, additional cell types, such as microglia, neurons, and circulating immune cells, can further influence the barrier's permeability upon demand or under pathophysiological circumstances. This sophisticated cellular interplay is commonly referred to as neurovascular unit (NVU) [9].

It is well established that a compromised BBB can initiate disease or exacerbate neurological outcomes. In multiple sclerosis, for example, BBB breakdown clearly coincides with the infiltration of autoreactive immune cells [10]. In stroke, oxidative stress during ischemia–reperfusion damages the barrier's integrity leading to uncontrolled influx of neurotoxic blood-derived substances into the nearby neural tissue [11]. Moreover, in recent years the field has provided a growing body of evidence that also subtle changes in barrier function can contribute to pathology, notably at very early stages, in certain cases even before the onset of disease. Impaired clearance of β-amyloid across the BBB has been identified as an early event in Alzheimer's disease (AD), for instance, and luminal leukocyte-endothelial interactions can lower the threshold for seizures in epilepsy [12, 13]. To date, neurovascular deficits have been linked to a majority of neurological diseases [14]. Therefore, more research is needed to better understand how genetic predisposition, pathological insults, or general aging, among others, affect BBB function and to leverage this growing knowledge base to explore novel therapeutic approaches to protect BBB integrity or accelerate recovery.

1.2 BBB in Alzheimer's Disease

There is a sufficient body of evidence for the involvement of the BBB in neurological disorders in general and in AD in particular. AD is the main cause of dementia affecting a large portion of the aged population. It is a neurodegenerative disorder characterized by the presence of β-amyloid (Aβ) aggregates and hyperphosphorylated tau protein that lead to progressive neuronal degeneration, synaptic loss, and plaque formation. Although mainly considered a neuronal disease, all CNS cellular components are strongly involved. Among others, gliosis is a typical hallmark of the disease, as both microglia and astrocytes lose their physiological functions and initiate a neuroinflammatory response that, in the long term, further exacerbates neuronal damage. The BBB is an important player in mediating neuroinflammation. As already pointed out, the BBB is involved in tightly regulating the composition of the neuronal milieu required for proper neuronal and synaptic functioning [15]. The influx of chemicals, blood components, and cells into the CNS is prevented under physiological conditions, but when the BBB loses its barrier properties, a condition often observed in AD, further neuroinflammation and degeneration can be triggered.

Whether BBB loss of function is a consequence or cause of AD is still debated. It has been shown that the BBB can be affected during the early phases of the disease, and its dysfunction may contribute to increased Aβ load and neuroinflammation [16]. BBB involvement in AD could occur before neurodegenerative events, leading to the two-hit vascular hypothesis [17, 18]; first, a cerebrovascular damage causes endothelial dysfunction at the BBB, leading to Aβ overload in the CNS (second hit) which ultimately results in neurodegeneration [19].

Dysfunction of the microcirculation and endothelial cell impairment, occurring before the accumulation of Aβ, is also supported by data obtained in mouse models of AD, neuroimaging studies [20–22], and studies on postmortem human brain tissue [16]. Further, an early BBB breakdown and vascular dysregulation have been described in AD prior to cognitive decline [16, 23, 24].

Alterations in the expression of BBB transporters and/or receptors in cerebral microvessels may account for the observed dysregulation. In particular, low levels of low-density lipoprotein receptor-related protein 1 (LRP-1) [25], as well as increased expression of receptor for advanced glycation end products 1 (RAGE-1) [26], have been reported in animal models of AD [27]. Accordingly, neuroimaging studies in humans have shown an increased uptake of verapamil in selected brain regions, suggesting a reduced active efflux of xenobiotics and drugs at the BBB [28, 29]. A dysregulated transport system across the BBB leads to a loss of the equilibrium between Aβ efflux and influx, which ultimately increases Aβ load in the CNS [16].

In addition, the transmigration of peripheral immune cells may also be facilitated by a compromised BBB in AD. The presence of peripheral macrophages [30] and neutrophils [31] in the brain of AD patients suggests the failure of the BBB in regulating cellular infiltration, with neuroinflammation as a consequence.

The mechanisms involved in BBB damage in AD are still not completely understood. Imbalanced Aβ production and clearance facilitate its deposition at the vascular wall of small brain arteries and capillaries, resulting in cerebral amyloid angiopathy (CAA) [23, 32, 33], a condition observed in over 90% of AD brains [34, 35]. This can lead to BBB breakdown at the arterial and/or arteriolar level [36, 37]. Several mutations occurring in familial AD may affect barrier integrity (reviewed in [14]). In animal models, mutated variants of Aβ peptides are poorly cleared across the BBB [25] so that they rapidly accumulate along the vessel walls. Severe BBB damage associated with microbleeds and endothelial degeneration have been observed in both human PSEN1 mutation carriers and AD mouse models [38]. Further, in carriers of the ApoE4 genotype, the BBB is dysfunctional early on [39–41], characterized by impaired cerebrovascular integrity [42, 43], which may be a consequence of pericyte dysfunction in ApoE4 carriers [44].

Aβ oligomers may directly affect BBB function as well. Using an in vitro approach, it is possible to discriminate the effects of Aβ oligomers on the different components of the BBB. Endothelial cell viability is affected by high Aβ concentrations (10–20 μM oligomers) [45], which also induce modifications in endothelial permeability and/or on the expression of tight proteins, as well as increased expression of adhesion molecules like intracellular adhesion molecule-1 (ICAM-1) and vascular cell adhesion molecule-1 (VCAM-1) [46–48].

Astrocytes are known to provide essential support for the maintenance of endothelial properties at the BBB [49] and play a critical role in the transport of Aβ across the BBB due to their regulation of RAGE and LRP-1 activity [50, 51]. Further, while astrocyte end feet are tightly associated with the abluminal aspect of cerebral vessels under physiological conditions, a recent study confirmed that end feet degeneration occurs in AD [52], and their retraction from vessels contributes to an increase in BBB permeability [53].

Through the release of trophic factors, astrocytes may differently modulate the endothelial barrier response when challenged with Aβ. While facilitating the degradation of tight junction protein claudin-5 and thus barrier leakiness through a VEGF-mediated mechanism [54], astrocytes may also reduce Aβ-induced BBB damage, at least as an early response. In the presence of astrocytes, Aβ-induced endothelial expression of ICAM-1 was prevented, an effect no longer visible when, in the same conditions, endothelial monocultures were challenged with Aβ nor when endothelial/astrocytes co-cultures were exposed to Aβ for a longer period (18 h) [55]. Importantly, the expression of the adhesion molecule ICAM-1 is critical for PBMCs to interact with the endothelium, ultimately leading to transmigration following the multistep paradigm [56].

1.3 Cell Models

1.3.1 Cell Lines

Primary cultures of brain microvascular endothelial cells (BMECs) represent the closest approximation to the in vivo BBB [57]. The most widely used primary BMECs originate from rat, mouse, pig, and cow [58]. The use of human BMECs is rare and limited [59] due to the restricted availability of human brain tissue as well as the high cost and special skills necessary for isolation and culture of primary human BMECs. In addition, most primary BMECs lose their specific characteristics in culture within limited passages and rapidly cease being deemed useful as in vitro models of the human BBB [60, 61].

In order to address these issues, immortalized human BMECs were generated by expressing simian virus 40 large T antigen (SV40 LT) [62]. Other well-characterized human BMEC lines, such as human cerebral microvascular endothelial cells (hCMEC/D3) and transfected human brain microvascular endothelial cells (THBMECs), were produced [62, 63]. hCMEC/D3 were established by transducing primary human brain endothelial cells with lentiviral vectors incorporating human telomerase and SV40 LT. They have high expression of junctional proteins and have been widely used for cell signaling and drug transport studies [64]. THBMECs were isolated from human brain microvessels and immortalized by transfection with SV40 LT [65]. However, these human BMEC lines lack contact inhibition and can lose the morphological and physiological properties of their in vivo counterparts particularly at high passage number or super-confluence. Under those conditions, they can present transudative intercellular junctions and lack paracellular barrier properties, which limit their effective use as an in vitro BBB model [65]. Moreover, complex karyotype changes were reported in immortalized BMECs, highlighting the importance of genetic testing of cell lines before their application to in vitro studies, and on a regular basis thereafter during continued experimentation [66]. Many researchers also have reported the differentiation of human pluripotent stem cells (PSCs) into ECs using various strategies [67–69], but the use of PSCs is not suitable for every lab. As a general statement, there are few cell sources of human origin that are sufficiently robust and therefore appropriate for in vitro BBB experiments.

Here, we took advantage of a conditionally immortalized human BMEC cell line (TY10) and co-cultured them with a conditionally immortalized human astrocyte cell line (hAST) to increase the physiological relevance of the in vitro model. Both cell lines were established using a temperature-sensitive SV40 LT (Ts- SV40 LT) in order to improve BBB-like characteristics of these immortalized cell lines [70, 71]. At 33 °C, Ts-SV40 LT antigen binds and inhibits p53 and retinoblastoma (Rb), which are strong tumor suppressors, leading to continuous cell proliferation (i.e., immortalized phenotype). At 37 °C, Ts-SV40 LT is inactivated, and the cells exhibit growth arrest and are subsequently allowed to mature

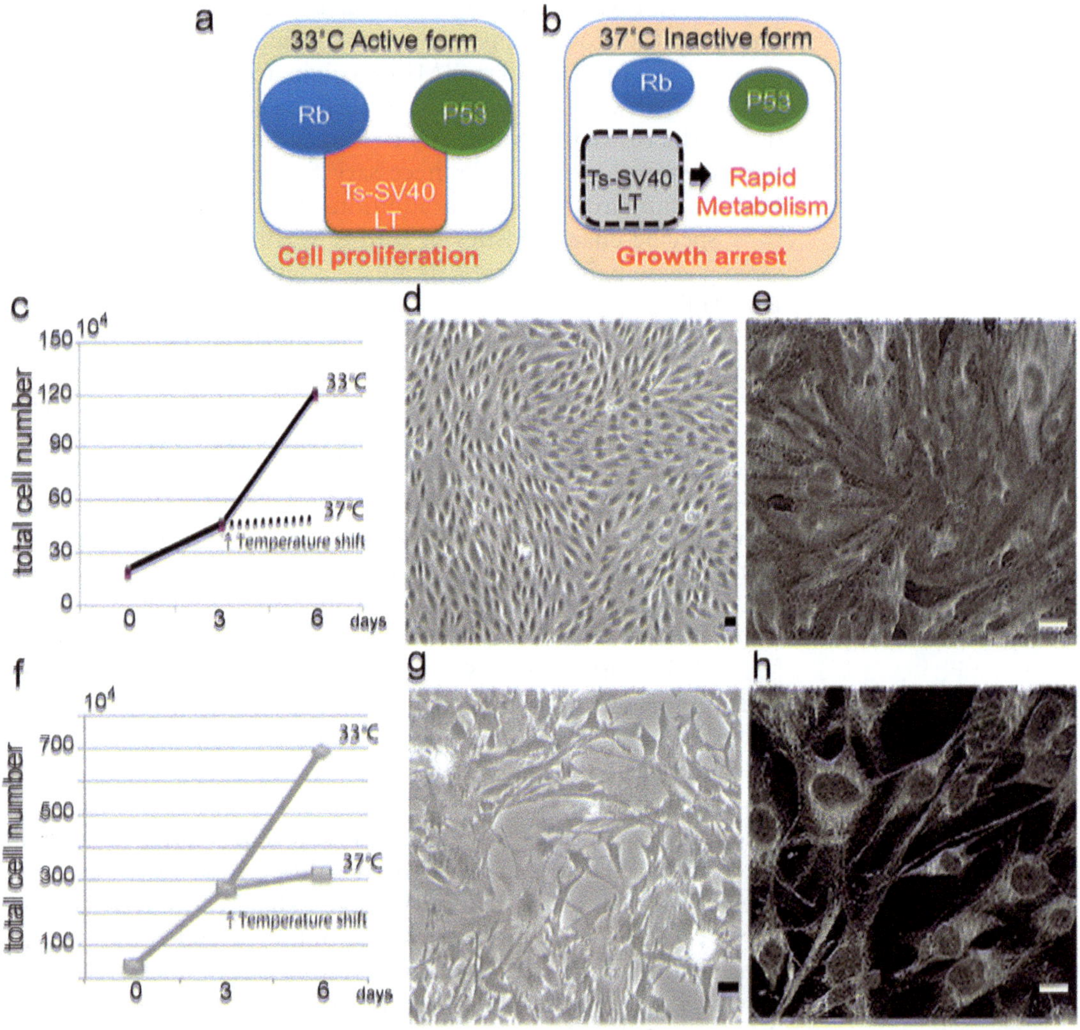

Fig. 1 Establishment of conditionally immortalized human BBB cell lines, transfected with temperature-sensitive SV40 large T antigen (Ts-SV40 LT). (**a**) At 33 °C, Ts-SV40 LT binds and inhibits p53 and Rb, which are strong tumor suppressors, leading to continuous cell proliferation. (**b**) At 37 °C, Ts-SV40 LT is inactivated, and cell growth is halted, initiating differentiation into mature cells. (**c**) Comparison of proliferation rates of TY10 at 33 °C and 37 °C. (**d**) Morphology of TY10 is spindle-shaped. (**e**) TY10 express vWF as lineage marker for endothelium. (**f**) Comparison of proliferation rates of hAST at 33 °C and 37 °C. (**g**) Morphology of hAST is star-shaped. (**h**) hAST express GFAP as lineage marker for astrocytes. Results were previously reported

into endothelial cells (i.e., non-immortalized, differentiated phenotype) (Fig. 1a, b). These conditionally immortalized cells express occludin and claudin-5 at intercellular boundaries as well as BBB characteristic influx and efflux transporters. At 37 °C, conditionally immortalized TY10 retain their physiological and morphological properties (Fig. 1c–h) and represent a useful cellular model for in vitro experiments [72–78].

1.3.2 In Vitro BBB Model

Many currently used in vitro BBB models apply a co-culture system by incorporating astrocyte, or via the addition of astrocyte-conditioned medium, with the goal to further enhance BBB characteristics of BMECs [79, 80]. Complex cell culture models consisting of different combinations of astrocytes, pericytes, neurons, and neural stem cells enable detailed studies of cellular interactions at the BBB, with an emphasis on the modulation of brain endothelial cells by these other members of the NVU [81, 82]. Reconstructing the unique cellular interaction between endothelial cells and astrocyte end feet in particular has been proven difficult, and there were no ideal co-cultured in vitro BBB models in which the end feet of astrocytes can directly contact endothelial cells. To this end, we established an in vitro BBB model that incorporates TY10 and hAST, which maintains BBB properties and allows leukocyte transmigration assays [55, 74, 75, 83]. Confocal 3D analysis with differential live staining of each cell line showed that co-cultured Transwell inserts constitute multilayer structures which consisted of an endothelial monolayer (Fig. 2a), astrocytic end feet (Fig. 2a,

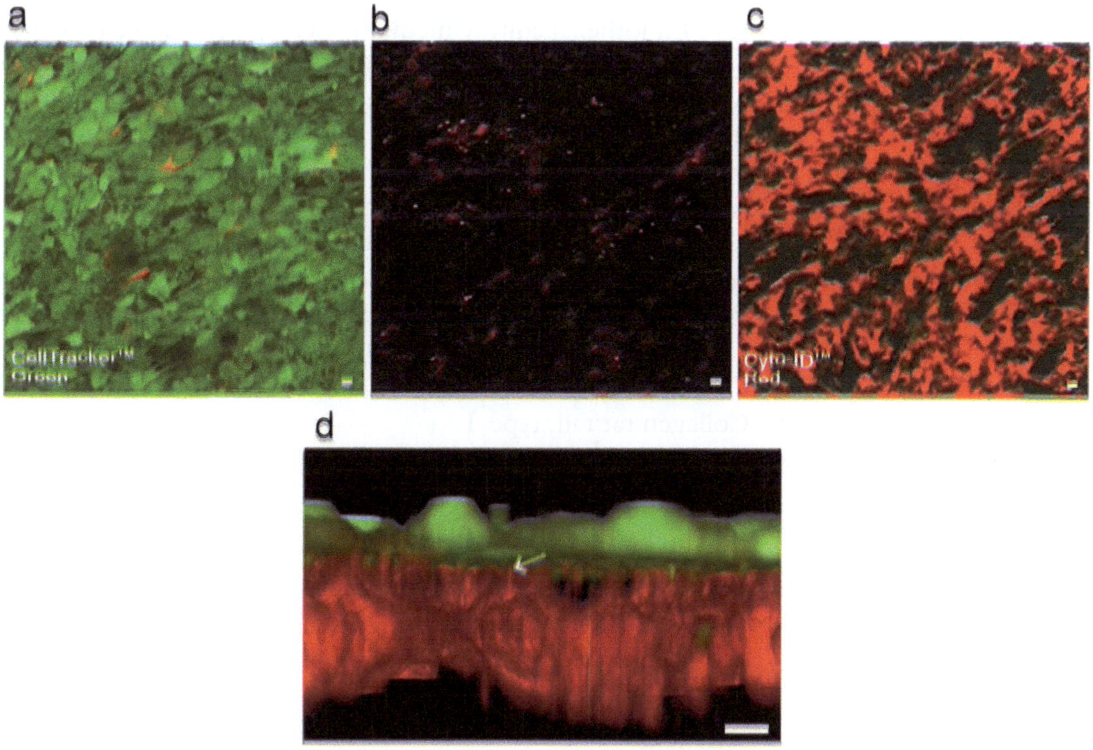

Fig. 2 3D structure of TY10/hAST co-cultures on Transwell membranes. (**a**) TY10 were stained with CellTracker™ Green. (**b**) hAST were stained with Cyto-ID™ Red. Stack file was generated spanning from the TY10 layer (luminal side) through the hAST layer (abluminal side). (**a**) TY10; luminal side. (**b**) hAST protrusions inside membrane pores. (**c**) hAST; abluminal side. (**d**) 3D reconstruction generated from stack files. Scale bar represents 10 μm. Arrow: hAST protrusions through the membrane pores. Results were previously reported [1]

b), Transwell membrane (Fig. 2b), and hAST (Fig. 2c). Some hAST end feet protruded through the membrane pores and were in close proximity to the TY10 cell layer (Fig. 2d). In the following, we describe how our in vitro BBB model allowed us to evaluate the effects of astrocytes in the modulation of PBMC transmigration through the endothelial layer exposed to Aβ, recapitulating one aspect of AD pathogenesis.

2 Materials

1. Human endothelial cell line (TY10).
2. Human astrocyte cell line (hAST).
3. Human peripheral blood mononuclear cells (PBMCs).
4. Phosphate buffer saline (PBS).
5. PBS/EDTA 0.5 M.
6. Fetal bovine serum (FBS), heat inactivated.
7. Culture media:
   (a) Endothelial culture medium (ECM): MCDB-131 supplemented with EGM-2 SingleQuots™ (Lonza) and 20% heat-inactivated FBS.
   (b) Astrocyte culture medium (ACM): astrocyte media containing 2% heat-inactivated FBS, astrocyte growth supplement, and penicillin/streptomycin solution as provided with the astrocyte media kit (CliniSciences).
   (c) Transendothelial migration medium (TEM): RPMI-1640 without phenol red, 1% bovine serum albumin (BSA), Hepes, L-glutamine, Na-pyruvate, MEM non-essential amino acids.
8. Collagen rat tail, type-1.
9. Glacial acetic acid.
10. 6.5 mm polycarbonate membrane cell culture inserts with 8.0 μm pore (Corning Transwell®).
11. 6-well multiwell (MW) plates.
12. 50 and 15 mL conical polystyrene centrifuge tubes.
13. Hematocytometer.
14. 0.4% trypan blue solution.
15. CXCL12.
16. Aβ oligomers: to enrich Aβ solution of oligomers, lyophilized Aβ was solubilized in dimethyl-sulfoxide (DMSO) as a 5 mM stock solution. Subsequent dilutions were made in DMEM medium without phenol red. A concentrated solution of Aβ (1–42), 100 μM, was aggregated by overnight incubation

at RT, followed by three freeze–thaw cycles for enrichment in oligomers. The solution was aliquoted and stored at −20 °C. For experiments, Aβ (1–42) was diluted in ACM to a concentration in the range of 0.5–2.5 μM.

3 Methods

3.1 Setting Up the In Vitro BBB Model

3.1.1 Coating

1. Collagen rat tail type I in a final concentration of 50 μg/mL is added to 1% glacial acetic acid and diluted in sterile deionized water. In the 6-well MW plates, 2 ml solution of collagen is added/well.
2. 6.5 mm polycarbonate membrane cell culture inserts are placed into each well, allowing contact between the collagen solution and the bottom of the insert (abluminal side). 1.5 mL collagen solution is added in the upper side of insert (luminal side).
3. Coated plates are incubated for 1 h at 37 °C in a sterile incubator.
4. Collagen solution is removed, and surfaces are washed three times with sterile PBS.
5. After the last wash, surfaces are kept in PBS until cells are plated.

3.1.2 Astrocyte Plating

1. hAST cells are kept in culture in a T75 cell culture flask in a 5% CO_2 incubator at a fixed temperature of 33 °C until 85% confluence is reached.
2. Detach hAST from their support vessel using a 0.04% trypsin/EDTA solution in PBS.
3. Centrifuge suspended cells at 200 x g for 5 minutes, and resuspend them in ACM before counting with a hemocytometer.
4. Prepare a suspension containing 3×10^5 cells/350 μl ACM.
5. Using a sterile tweezer, flip the 6.5 mm polycarbonate membrane cell culture inserts upside down, and put in a sterile 150 mm petri dish plate (*see* **Note 1**).
6. Plate the suspension of hAST on the bottom of the insert, and keep for 20 minutes in the laminar flow hood, adding a little volume of ACM to avoid the surface to dry (Fig. 3).
7. After 20 minutes, transfer inserts to a 5% CO_2 incubator at 33 °C, and let hAST settle on the bottom surface of the insert for at least 1 h.
8. Add 2 mL of warm ACM in each well of a sterile 6-well MW plate, and flip inserts into the well, so that their bottom side, where hAST are plated, is completely covered by medium (Fig. 3).
9. Keep the plates in the sterile incubator at 33 °C.

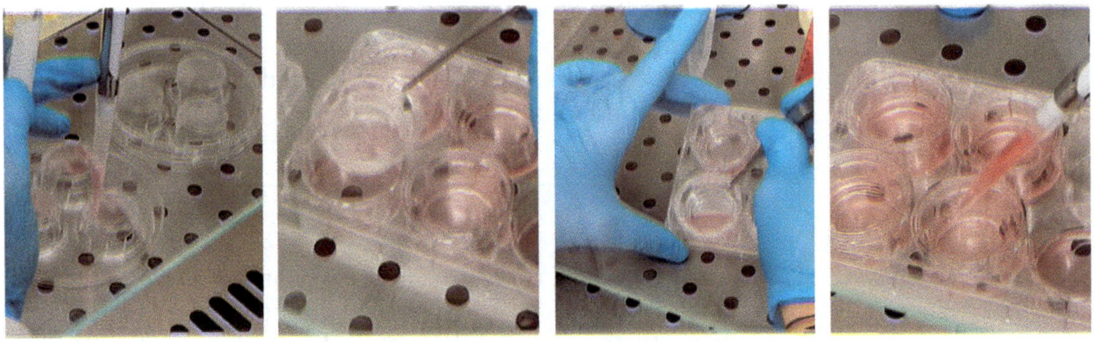

Fig. 3 Illustration of experimental steps of setting up endothelial/astrocyte co-cultures. Astrocytes are plated and allowed to settle on the bottom surface of the Transwell insert (abluminal side) in a laminar hood. After 1 h, inserts are flipped into the well, and astrocytes are completely covered by medium. Endothelial cells are plated on the upper side (luminal side) of the insert

3.1.3 Endothelial Cell Plating

1. TY10 are kept in culture in a T75 cell culture flask in a 5% CO_2 incubator at a fixed temperature of 33 °C until 95% confluence is reached.
2. Detach TY10 from their support using a 0.04% trypsin/EDTA solution in PBS.
3. Centrifuge suspended cells at 200 × g for 5 min, and resuspend them in ECM before counting with a hemocytometer.
4. Prepare a suspension containing 5×10^5 cells/1500 μL ECM.
5. Plate cell suspension on the upper side (luminal) of the insert (Fig. 3).

3.1.4 Maintaining Endothelial/Astrocyte Co-culture

1. Plates are kept in a 5% CO_2 incubator at a temperature of 33 °C.
2. After 24 h, carefully, avoiding damage to the membrane, inserts are washed in PBS to eliminate non-adherent dead cells, followed by addition of fresh ACM to the co-cultures (*see* **Note 2**).
3. Keep co-cultures in the incubator at 33 °C until confluence is reached, about 2–3 days (*see* **Note 3**).
4. After 3 days, fresh ACM is added to both endothelial and astrocyte cultures, and plates are kept at 37 °C for 2 days to allow cell maturation.

3.2 Static PBMC Transendothelial Migration

3.2.1 Treatments of the In Vitro BBB (β-Amyloid ± TNFα & IFNγ)

1. When BBB properties are developed, co-cultures are exposed to treatments.
2. Prepare a solution containing Aβ at the final concentration of 0.5–2.5 μM in ACM. When required by experimental settings, add TNFα at the final concentration of 10 U.I. and IFNγ, 5 U. I. in ACM (T & I).

3. Rinse co-cultures in warm PBS.
4. Add 1.5 mL/well of ACM with or without treatments in a new 6-well MW plate.
5. Carefully, avoiding damage to the membrane, aspirate PBS from the luminal side of the insert, and slowly transfer the insert into the well containing either plain ACM or the treatments.
6. Add 750 μL of either plain ACM or the ACM supplemented with treatment compounds to the insert.
7. All conditions are maintained at 37 °C for either 5 or 18 h, according to experimental settings.

3.2.2 Activation of the In Vitro BBB Model

1. Rinse co-cultures twice in warm PBS to remove any trace of treatment.
2. In a new MW plate, add 1.5 mL/well of warm TEM, and carefully transfer the insert into the well. Add 750 μL of TEM into the insert.
3. Allow co-cultures to equilibrate in the new culture medium for 10 minutes at room temperature (RT) inside the hood.
4. Prepare a solution containing CXCL12 at the final concentration of 50 ng/mL in warm TEM.
5. Carefully remove all the TEM from both the bottom and the upper side of the membrane.
6. In a new 6-well MW plate, add 1.5 mL of warm TEM containing 1% FBS, which will act as a chemoattractant factor for PBMCs through the endothelial layer (*see* **Note 4**).
7. Transfer inserts into an empty MW plate, and add, only to the apical endothelial layer, 700 μL of the solution containing CXCL12 (50 ng/ml), and incubate at 37 °C for 15 minutes (*see* **Note 5**).
8. Remove all the solution containing CXCL12, and add PBMCs to perform the transendothelial migration assay.

3.2.3 PBMC Isolation

1. Collect the blood into BD Vacutainer CPT Tube with Sodium Heparine. Store the tubes at RT until centrifugation (within *2 hours* for best results).
2. Mix the blood sample immediately prior to centrifugation by gently inverting the tube 8–10 times.
3. Centrifuge blood sample at room temperature in a horizontal rotor for 25' at 400 × *g*.
4. Slowly aspirate the supernatant, and to avoid without perturbing the pellet, leave at least 5 mL of plasma.
5. Collect the blood from 2 tubes and put together in a 50 mL, to reach a volume of 15 mL.

6. Fill the tube with sterile PBS until it reaches 30 mL and then mix the solution carefully.
7. Fill another 50 mL with 15 mL of Lymphocyte Separation Medium at room temperature.
8. Gently overlay the diluted blood onto the Lymphocyte Separation Medium using a sterile plastic pipette to allow phase separation.
9. Centrifuge the samples for 20 min at 600 × *g* at room temperature with brake function of the centrifuge system turned OFF. The centrifugation will allow the separation of different layers. The plasma layer, a thin fluffy layer containing mononuclear cells, a layer containing the Lymphocyte Separation Medium and a layer containing all the other blood components (see Fig. 4).
10. Carefully, get rid of the plasma layer, always leaving at least 5 mL of plasma, so that you are not going to lose a lot of monocytes.
11. With a sterile transfer pipette carefully collect all the mononuclear cells and transfer into a clean centrifuge tube.
12. Dilute with at least an equal amount of PBS (generally 15 mL).
13. Centrifuge for 4 min at 600 × *g*.

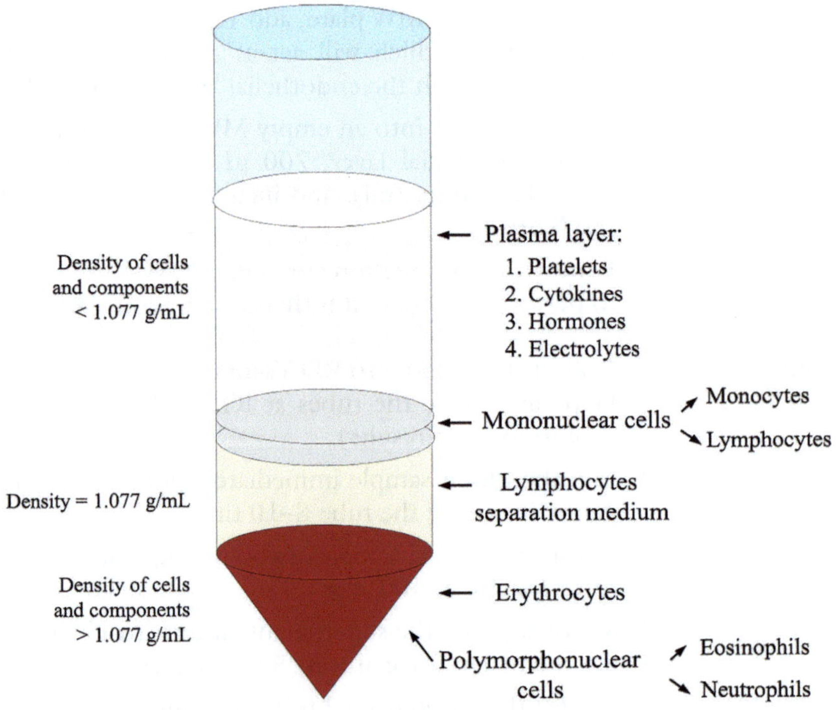

Fig. 4 Schematic illustration of PBMCs isolation

14. Aspirate the supernatant and combine all the pellet in a new 50 mL tube centrifuge for 4 min at RT
15. Remove supernatant and proceed with the lysis of residual RBC: resuspend washed PBMC pellet in 9 ml sterile H_2O for 15 s while pipetting (3 times). Then, add 1 mL of 10x PBS and mix carefully.
16. Centrifuge for 5 min at 400 × *g*.
17. Wash cells again and resuspend in 10 mL TEM.

3.2.4 PBMC Transendothelial Migration

1. Isolate PBMCs from whole blood (previously described), and resuspend them in TEM.
2. Centrifuge for 5 min at 400 × g.
3. Count live cells in a hematocytometer by mixing 10 μL of the cell suspension with an appropriate volume of 0.4% trypan blue/PBS solution (generally 1:500).
4. Prepare a suspension containing 2.8 × 10^6 PBMCs/750 μL TEM on the top of the endothelial layer, on the upper side of the membrane.
5. Incubate the co-cultures for 18 h at 37 °C (*see* **Note 6**).
6. Carefully remove the insert from the well, and transfer it into a new MW plate if you need to process it further.
7. Collect all the TEM contained in the well where the assay was performed, and rinse the bottom of the well with PBS supplemented with 0.5 mM EDTA to allow the recovery of any PBMCs that may be attached to the bottom of the well (*see* **Note 7**). Observe the plate at the microscope to ensure all the cells have been removed from the well.
8. Centrifuge for 5 minutes at 400 × g.
9. Count live cells in a hematocytometer by mixing 10 μL of the cell suspension with an appropriate volume of 0.4% trypan blue/PBS solution (generally 90 μl).
10. The absolute number of migrated PBMCs is related to the input of PBMCs seeded on the endothelial layer at the beginning of the assay (2.8 × 10^6 PBMCs/well).
11. Recovered PBMCs may be further processed (flow cytometry, protein and nucleic acid extraction, etc.).

3.2.5 Results

The results of the transendothelial migration in these experimental settings were previously reported and are summarized in Fig. 5.

1. Differences were observed when endothelial/astrocyte co-cultures were exposed to Aβ for 5 h or 18 h. Inflammatory cytokines induced increased transendothelial migration of PBMCs. Aβ prevented this effect at 5 h, but not at the longer time point (18 h).

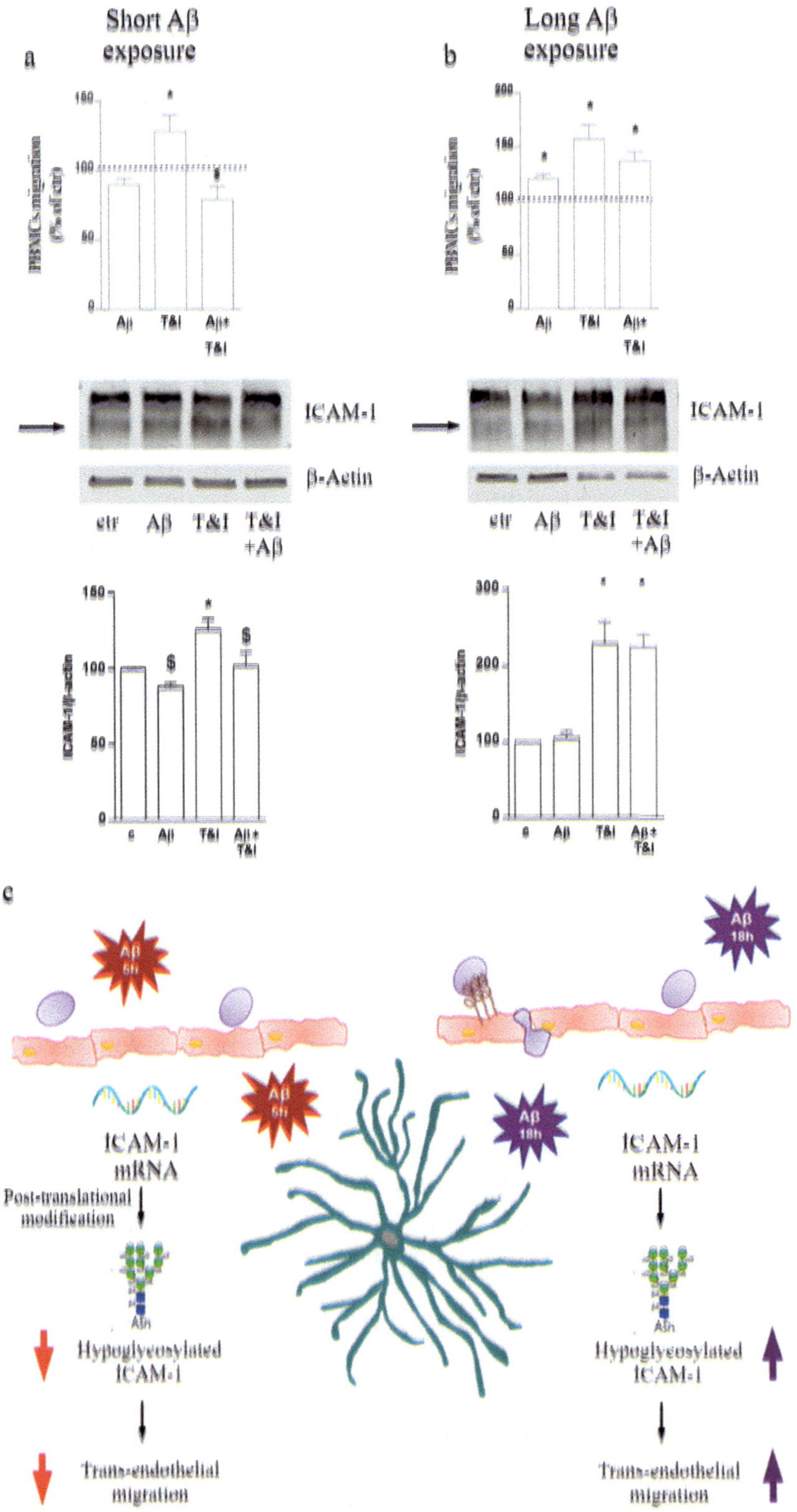

Fig. 5 Endothelial/astrocyte co-cultures were exposed for 5 h (short) or 18 h (long) to TNFα and INFγ (T&I, 10 UI, and 5 UI respectively), Aβ (2.5 μM), or in combination. Barrier properties were investigated by evaluating migration of PBMCs through the in vitro barrier. T&I induced transmigration, while both Aβ and Aμ + T&I

2. In endothelial monocultures, Aβ treatment for 5 h was not able to prevent cytokine-induced transendothelial migration, simulating the effect observed in co-cultures after a longer time exposure to Aβ.
3. After exposure to Aβ, in a time-dependent manner, astrocytes were able to modulate the glycosylation of the ICAM-1, strongly associated with leukocyte adhesion, thus affecting PBMC transmigration.

These results strengthen the important role played by astrocytes in modulating the endothelial response at the BBB under pathological conditions.

4 Notes

1. Keep a sterile tweezer under the hood. Sterilize it in 70% ethanol, and let it dry completely before taking up the insert.
2. When taking up the insert, be very careful to avoid damage to the membrane. If damaged, the amount of PBMCs crossing may be the consequence of the physical damage of the insert and not of changes in endothelial barrier properties.
3. When plated on an insert, it may be very difficult to observe both endothelial cells and astrocytes under the microscope, due to the physical interference of membrane pores. To visualize the state of confluency of the seeded cells, you may daily measure the transendothelial electrical resistance (TEER) for each insert. The value should increase till a plateau is reached (in our settings $34.05 \pm 0.5\ \Omega\ cm^2$). Alternatively, seed the same amount of both hAST (3×10^5 cells/well) and TY10 (5×10^5 cells/well) in a 6-well MW plate, so you can easily visualize cell confluency. The state of the endothelial and astrocytes layer may be still examined at the end of the transmigration assay. This can be done by labelling the membrane with either non-specific live cell staining dyes (CellMask™) or using specific cell markers (for instance, CD31 for endothelial cells

Fig. 5 (continued) prevented transmigration (a, upper panel). This is a consequence of the reduced expression of a specific glycoform of the adhesion molecule ICAM-1 (**a**, lower panel). Longer Aβ exposure caused the loss of its protective functions. Both PBMC transmigration (**b**, upper panel) and ICAM-1 expression (b, lower panel) were induced in the presence of inflammatory stimuli. In conclusion, illustrated in (**c**), in the presence of an inflammatory stimulus, a short Aβ exposure reduces ICAM-1 posttranslational modification (hypoglycosylated species), thus preventing PBMC-endothelial interaction, and consequently, transendothelial migration of PBMCs. Conversely, after longer Aβ exposure, the hypoglycosylated form of ICAM-1 is upregulated, and increased PBMC transmigration is observed. Modulation of ICAM-1 glycosylation was mediated by an astrocytic-derived microRNA (miR-200b). Results were previously reported [55]

and S100β for astrocytes) after cell fixation and permeabilization.

Live cell staining dyes may also be used to stain both TY10 and hAST before plating.

4. FBS may be replaced by a solution of CCL2 (25 ng/mL) in TEM. CCL2 is generally produced by reactive astrocytes and microglia in the CNS recruiting monocytes and other immune cells. As for FBS, its role in the assay is to facilitate PBMC migration.
5. The role of CXCL12 is to induce an inflammatory phenotype in endothelial cells and to facilitate their interaction with PBMCs. In physiological conditions, the cytokine is exposed on the abluminal side of the endothelial layer, while during inflammation the cytokine translocates to the luminal side where its role is to mediate PBMC adhesion.
6. A first readout of PBMC transmigration can be obtained after 3 h. Results at this time point closely reproduce those observed at 18 h of the assay, but the number of migrated cells is inferior, thus increasing the risk of miscalculation of the total amount of migrated cells. For this reason, it is not recommended to stop the assay after 3 h.
7. Monocytes are sticky and tend to attach to the plastic. Washing the bottom of the well with 0.5 mM PBS/EDTA facilitates the recovery of any PBMCs that may be still attached.

Acknowledgments

The authors are sincerely thankful to Dr. Richard Ransohoff. Under Dr. Ransohoff's mentorship, the authors together completed their postdoctoral trainings at Cleveland Clinic Lerner Research Institute, Cleveland, Ohio, and the exceptional work environment during that time fostered this collaboration.

Conflicts of Interest Birgit Obermeier is employee and shareholder of Biogen, but Biogen was not a sponsor of this research. Yukio Takeshita and Simona Federica Spampinato have no conflicts of interest.

References

1. Profaci CP, Munji RN, Pulido RS et al (2020) The blood-brain barrier in health and disease: important unanswered questions. J Exp Med 217(4):e20190062
2. Abbott NJ, Patabendige AA, Dolman DE et al (2010) Structure and function of the blood-brain barrier. Neurobiol Dis 37(1):13–25
3. Daneman R (2012) The blood-brain barrier in health and disease. Ann Neurol 72(5):648–672

4. Daneman R, Zhou L, Agalliu D et al (2010) The mouse blood-brain barrier transcriptome: a new resource for understanding the development and function of brain endothelial cells. PLoS One 5(10):e13741
5. Ransohoff RM, Engelhardt B (2012) The anatomical and cellular basis of immune surveillance in the central nervous system. Nat Rev Immunol 12(9):623–635
6. Carrithers MD, Visintin I, Kang SJ et al (2000) Differential adhesion molecule requirements for immune surveillance and inflammatory recruitment. Brain 123(Pt 6):1092–1101
7. Marchetti L, Engelhardt B (2020) Immune cell trafficking across the blood-brain barrier in the absence and presence of neuroinflammation. Vasc Biol 2(1):H1–H18
8. Obermeier B, Daneman R, Ransohoff RM (2013) Development, maintenance and disruption of the blood-brain barrier. Nat Med 19(12):1584–1596
9. Neuwelt EA, Bauer B, Fahlke C et al (2011) Engaging neuroscience to advance translational research in brain barrier biology. Nat Rev Neurosci 12(3):169–182
10. Lindner M, Klotz L, Wiendl H (2018) Mechanisms underlying lesion development and lesion distribution in CNS autoimmunity. J Neurochem 146(2):122–132
11. Ludewig P, Winneberger J, Magnus T (2019) The cerebral endothelial cell as a key regulator of inflammatory processes in sterile inflammation. J Neuroimmunol 326:38–44
12. Chakraborty A, de Wit NM, van der Flier WM et al (2017) The blood brain barrier in Alzheimer's disease. Vasc Pharmacol 89:12–18
13. Fabene PF, Laudanna C, Constantin G (2013) Leukocyte trafficking mechanisms in epilepsy. Mol Immunol 55(1):100–104
14. Sweeney MD, Zhao Z, Montagne A et al (2019) Blood-brain barrier: from physiology to disease and back. Physiol Rev 99(1):21–78
15. Zhao Z, Nelson AR, Betsholtz C et al (2015) Establishment and dysfunction of the blood-brain barrier. Cell 163(5):1064–1078
16. Montagne A, Zhao Z, Zlokovic BV (2017) Alzheimer's disease: a matter of blood-brain barrier dysfunction? J Exp Med 214(11): 3151–3169
17. Wang JZ, Zhu WD, Xu ZX et al (2014) Pin1, endothelial nitric oxide synthase, and amyloid-beta form a feedback signaling loop involved in the pathogenesis of Alzheimer's disease, hypertension, and cerebral amyloid angiopathy. Med Hypotheses 82(2):145–150
18. Lamoke F, Mazzone V, Persichini T et al (2015) Amyloid beta peptide-induced inhibition of endothelial nitric oxide production involves oxidative stress-mediated constitutive eNOS/HSP90 interaction and disruption of agonist-mediated Akt activation. J Neuroinflamm 12:84
19. Sagare AP, Bell RD, Zlokovic BV (2012) Neurovascular dysfunction and faulty amyloid beta-peptide clearance in Alzheimer disease. Cold Spring Harb Perspect Med 2(10):a011452
20. Jian WX, Zhang Z, Chu SF et al (2018) Potential roles of brain barrier dysfunctions in the early stage of Alzheimer's disease. Brain Res Bull 142:360–367
21. van de Haar HJ, Jansen JFA, van Osch MJP et al (2016) Neurovascular unit impairment in early Alzheimer's disease measured with magnetic resonance imaging. Neurobiol Aging 45: 190–196
22. van de Haar HJ, Jansen JFA, Jeukens C et al (2017) Subtle blood-brain barrier leakage rate and spatial extent: considerations for dynamic contrast-enhanced MRI. Med Phys 44(8): 4112–4125
23. Kisler K, Nelson AR, Montagne A et al (2017) Cerebral blood flow regulation and neurovascular dysfunction in Alzheimer disease. Nat Rev Neurosci 18(7):419–434
24. Snyder HM, Corriveau RA, Craft S et al (2015) Vascular contributions to cognitive impairment and dementia including Alzheimer's disease. Alzheimers Dement 11(6):710–717
25. Deane R, Wu Z, Sagare A et al (2004) LRP/amyloid beta-peptide interaction mediates differential brain efflux of Abeta isoforms. Neuron 43(3):333–344
26. Deane R, Du Yan S, Submamaryan RK et al (2003) RAGE mediates amyloid-beta peptide transport across the blood-brain barrier and accumulation in brain. Nat Med 9(7):907–913
27. Cirrito JR, Deane R, Fagan AM et al (2005) P-glycoprotein deficiency at the blood-brain barrier increases amyloid-beta deposition in an Alzheimer disease mouse model. J Clin Invest 115(11):3285–3290
28. Deo AK, Borson S, Link JM et al (2014) Activity of P-glycoprotein, a beta-amyloid transporter at the blood-brain barrier, is compromised in patients with mild Alzheimer disease. J Nucl Med 55(7):1106–1111
29. van Assema DM, Lubberink M, Bauer M et al (2012) Blood-brain barrier P-glycoprotein function in Alzheimer's disease. Brain 135 (Pt 1):181–189
30. Hultman K, Strickland S, Norris EH (2013) The APOE varepsilon4/varepsilon4 genotype potentiates vascular fibrin(ogen) deposition in amyloid-laden vessels in the brains of

Alzheimer's disease patients. J Cereb Blood Flow Metab 33(8):1251–1258

31. Zenaro E, Pietronigro E, Della Bianca V et al (2015) Neutrophils promote Alzheimer's disease-like pathology and cognitive decline via LFA-1 integrin. Nat Med 21(8):880–886
32. Montagne A, Barnes SR, Sweeney MD et al (2015) Blood-brain barrier breakdown in the aging human hippocampus. Neuron 85(2): 296–302
33. Tarasoff-Conway JM, Carare RO, Osorio RS et al (2016) Clearance systems in the brain--implications for Alzheimer diseaser. Nat Rev Neurol 12(4):248
34. Hartz AM, Bauer B, Soldner EL et al (2012) Amyloid-beta contributes to blood-brain barrier leakage in transgenic human amyloid precursor protein mice and in humans with cerebral amyloid angiopathy. Stroke 43(2): 514–523
35. del Valle J, Duran-Vilaregut J, Manich G et al (2011) Cerebral amyloid angiopathy, blood-brain barrier disruption and amyloid accumulation in SAMP8 mice. Neurodegener Dis 8(6):421–429
36. Saito S, Ihara M (2016) Interaction between cerebrovascular disease and Alzheimer pathology. Curr Opin Psychiatry 29(2):168–173
37. Weller RO, Boche D, Nicoll JA (2009) Microvasculature changes and cerebral amyloid angiopathy in Alzheimer's disease and their potential impact on therapy. Acta Neuropathol 118(1):87–102
38. Wen PH, De Gasperi R, Sosa MA et al (2005) Selective expression of presenilin 1 in neural progenitor cells rescues the cerebral hemorrhages and cortical lamination defects in presenilin 1-null mutant mice. Development 132(17):3873–3883
39. Reiman EM, Chen K, Alexander GE et al (2004) Functional brain abnormalities in young adults at genetic risk for late-onset Alzheimer's dementia. Proc Natl Acad Sci U S A 101(1):284–289
40. Sheline YI, Morris JC, Snyder AZ et al (2010) APOE4 allele disrupts resting state fMRI connectivity in the absence of amyloid plaques or decreased CSF Abeta42. J Neurosci 30(50): 17035–17040
41. Thambisetty M, Beason-Held L, An Y et al (2010) APOE epsilon4 genotype and longitudinal changes in cerebral blood flow in normal aging. Arch Neurol 67(1):93–98
42. Hajjar I, Sorond F, Lipsitz LA (2015) Apolipoprotein E, carbon dioxide vasoreactivity, and cognition in older adults: effect of hypertension. J Am Geriatr Soc 63(2): 276–281
43. Suri S, Mackay CE, Kelly ME et al (2015) Reduced cerebrovascular reactivity in young adults carrying the APOE epsilon4 allele. Alzheimers Dement 11(6):648–57 e1
44. Bell RD, Winkler EA, Singh I et al (2012) Apolipoprotein E controls cerebrovascular integrity via cyclophilin A. Nature 485(7399): 512–516
45. Wan W, Cao L, Liu L et al (2015) Abeta(1-42) oligomer-induced leakage in an in vitro blood-brain barrier model is associated with up-regulation of RAGE and metalloproteinases, and down-regulation of tight junction scaffold proteins. J Neurochem 134(2): 382–393
46. Gheorghiu M, Enciu AM, Popescu BO et al (2014) Functional and molecular characterization of the effect of amyloid-beta42 on an in vitro epithelial barrier model. J Alzheimers Dis 38(4):787–798
47. Marco S, Skaper SD (2006) Amyloid beta-peptide1-42 alters tight junction protein distribution and expression in brain microvessel endothelial cells. Neurosci Lett 401(3): 219–224
48. Liu R, Zhang TT, Wu CX et al (2011) Targeting the neurovascular unit: development of a new model and consideration for novel strategy for Alzheimer's disease. Brain Res Bull 86(1–2):13–21
49. Spampinato SF, Bortolotto V, Canonico PL et al (2019) Astrocyte-derived paracrine signals: relevance for neurogenic niche regulation and blood-brain barrier integrity. Front Pharmacol 10:1346
50. Cai Z, Wan CQ, Liu Z (2017) Astrocyte and Alzheimer's disease. J Neurol 264(10): 2068–2074
51. Askarova S, Yang X, Sheng W et al (2011) Role of Abeta-receptor for advanced glycation endproducts interaction in oxidative stress and cytosolic phospholipase A(2) activation in astrocytes and cerebral endothelial cells. Neuroscience 199:375–385
52. Gondo A, Shinotsuka T, Morita A et al (2014) Sustained down-regulation of beta-dystroglycan and associated dysfunctions of astrocytic endfeet in epileptic cerebral cortex. J Biol Chem 289(44):30279–30288
53. Savettieri G, Di Liegro I, Catania C et al (2000) Neurons and ECM regulate occludin localization in brain endothelial cells. Neuroreport 11(5):1081–1084
54. Spampinato SF, Merlo S, Sano Y et al (2017) Astrocytes contribute to Abeta-induced blood-

brain barrier damage through activation of endothelial MMP9. J Neurochem 142(3): 464–477

55. Spampinato SF, Merlo S, Fagone E et al (2019) Astrocytes modify migration of PBMCs induced by beta-amyloid in a blood-brain barrier in vitro model. Front Cell Neurosci 13:337
56. Scott DW, Patel RP (2013) Endothelial heterogeneity and adhesion molecules N-glycosylation: implications in leukocyte trafficking in inflammation. Glycobiology 23(6): 622–633
57. Smith M, Omidi Y, Gumbleton M (2007) Primary porcine brain microvascular endothelial cells: biochemical and functional characterisation as a model for drug transport and targeting. J Drug Target 15(4):253–268
58. Deli MA, Abraham CS, Kataoka Y et al (2005) Permeability studies on in vitro blood-brain barrier models: physiology, pathology, and pharmacology. Cell Mol Neurobiol 25(1): 59–127
59. Bernas MJ, Cardoso FL, Daley SK et al (2010) Establishment of primary cultures of human brain microvascular endothelial cells to provide an in vitro cellular model of the blood-brain barrier. Nat Protoc 5(7):1265–1272
60. Gribkoff VK, Kaczmarek LK (2017) The need for new approaches in CNS drug discovery: why drugs have failed, and what can be done to improve outcomes. Neuropharmacology 120:11–19
61. Osaki T, Shin Y, Sivathanu V et al (2018) In vitro microfluidic models for neurodegenerative disorders. Adv Healthc Mater 7(2): 1700489
62. Stins MF, Gilles F, Kim KS (1997) Selective expression of adhesion molecules on human brain microvascular endothelial cells. J Neuroimmunol 76(1–2):81–90
63. Weksler BB, Subileau EA, Perriere N et al (2005) Blood-brain barrier-specific properties of a human adult brain endothelial cell line. FASEB J 19(13):1872–1874
64. Carl SM, Lindley DJ, Das D et al (2010) ABC and SLC transporter expression and proton oligopeptide transporter (POT) mediated permeation across the human blood--brain barrier cell line, hCMEC/D3 [corrected]. Mol Pharm 7(4):1057–1068
65. Takeshita Y, Ransohoff RM (2012) Inflammatory cell trafficking across the blood-brain barrier: chemokine regulation and in vitro models. Immunol Rev 248(1):228–239
66. Mkrtchyan H, Scheler S, Klein I et al (2009) Molecular cytogenetic characterization of the human cerebral microvessel endothelial cell line hCMEC/D3. Cytogenet Genome Res 126(4):313–317
67. Levenberg S, Golub JS, Amit M et al (2002) Endothelial cells derived from human embryonic stem cells. Proc Natl Acad Sci U S A 99(7):4391–4396
68. Sone M, Itoh H, Yamahara K et al (2007) Pathway for differentiation of human embryonic stem cells to vascular cell components and their potential for vascular regeneration. Arterioscler Thromb Vasc Biol 27(10):2127–2134
69. Park SW, Jun Koh Y, Jeon J et al (2010) Efficient differentiation of human pluripotent stem cells into functional CD34+ progenitor cells by combined modulation of the MEK/ERK and BMP4 signaling pathways. Blood 116(25): 5762–5772
70. Haruki H, Sano Y, Shimizu F et al (2013) NMO sera down-regulate AQP4 in human astrocyte and induce cytotoxicity independent of complement. J Neurol Sci 331(1–2): 136–144
71. Sano Y, Shimizu F, Abe M et al (2010) Establishment of a new conditionally immortalized human brain microvascular endothelial cell line retaining an in vivo blood-brain barrier function. J Cell Physiol 225(2):519–528
72. Salman MM, Marsh G, Kusters I et al (2020) Design and validation of a human brain endothelial microvessel-on-a-chip open microfluidic model enabling advanced optical imaging. Front Bioeng Biotechnol 8:573775
73. Takeshita Y, Obermeier B, Cotleur AC et al (2017) Effects of neuromyelitis optica-IgG at the blood-brain barrier in vitro. Neurol Neuroimmunol Neuroinflamm 4(1):e311
74. Spampinato SF, Obermeier B, Cotleur A et al (2015) Sphingosine 1 phosphate at the blood brain barrier: can the modulation of S1P receptor 1 influence the response of endothelial cells and astrocytes to inflammatory stimuli? PLoS One 10(7):e0133392
75. Takeshita Y, Obermeier B, Cotleur A et al (2014) An in vitro blood-brain barrier model combining shear stress and endothelial cell/astrocyte co-culture. J Neurosci Methods 232: 165–172
76. Shimizu F, Schaller KL, Owens GP et al (2017) Glucose-regulated protein 78 autoantibody associates with blood-brain barrier disruption in neuromyelitis optica. Sci Transl Med 9(397): eaai9111
77. Wevers NR, Kasi DG, Gray T et al (2018) A perfused human blood-brain barrier on-a-chip for high-throughput assessment of barrier function and antibody transport. Fluids Barriers CNS 15(1):23

78. Shimizu F, Takeshita Y, Sano Y et al (2019) GRP78 antibodies damage the blood-brain barrier and relate to cerebellar degeneration in Lambert-Eaton myasthenic syndrome. Brain 142(8):2253–2264
79. Delsing L, Herland A, Falk A et al (2020) Models of the blood-brain barrier using iPSC-derived cells. Mol Cell Neurosci 107:103533
80. Katt ME, Mayo LN, Ellis SE et al (2019) The role of mutations associated with familial neurodegenerative disorders on blood-brain barrier function in an iPSC model. Fluids Barriers CNS 16(1):20
81. Appelt-Menzel A, Cubukova A, Gunther K et al (2017) Establishment of a human blood-brain barrier co-culture model mimicking the neurovascular unit using induced pluri- and multipotent stem cells. Stem Cell Rep 8(4): 894–906
82. Canfield SG, Stebbins MJ, Faubion MG et al (2019) Correction to: An isogenic neurovascular unit model comprised of human induced pluripotent stem cell-derived brain microvascular endothelial cells, pericytes, astrocytes, and neurons. Fluids Barriers CNS 16(1):31
83. Rubin LL, Hall DE, Porter S et al (1991) A cell culture model of the blood-brain barrier. J Cell Biol 115(6):1725–1735

Index

A

B

C

Nicole Stone (ed.), *The Blood-Brain Barrier: Methods and Protocols*, Methods in Molecular Biology, vol. 2492, https://doi.org/10.1007/978-1-0716-2289-6,

D

E

F

G

V

W

Z

MIX
Papier aus verantwortungsvollen Quellen
Paper from responsible sources
FSC® C105338

If you have any concerns about our products,
you can contact us on
ProductSafety@springernature.com

In case Publisher is established outside the EU,
the EU authorized representative is:
Springer Nature Customer Service Center GmbH
Europaplatz 3, 69115 Heidelberg, Germany

Printed by Libri Plureos GmbH
in Hamburg, Germany